Contents

Preface

The subject of Impact Engineering is very appealing for engineers, perhaps because it deals with extreme loading of structures, associated with material failure, high noise, explosions and car, train, plane and even ship crashes. It is not surprising then that Impact Engineering is a complex field of mechanical analysis where one requires an increasing specialized degree on visco plasticity, wave propagation, buckling, material behaviour, experimental mechanics, numerical methods, among others.

All these aspects are very difficult to be mastered, especially in a world where one requires more and more a high degree of specialization. It is bearing in mind this context and those alluded difficulties that I set myself the task of writing a book on Impact Engineering that could enable the reader to gain a broader knowledge on the subject from a single volume.

To this end, I had to school myself on various issues, and perhaps this justifies not only the broad coverage of the book but also the some 15 years it took me to write it.

I paid special attention to issues such as formatting, figures quality, writing style, no equation or figure numbering, citations along the text, etc. I also picked up the subjects in a way to cover as much as possible the most important areas of Impact Engineering. The reader will find here theory, experimental mechanics and finite elements. It is so that one can learn such diverse aspects as Fourier Signal Analysis, Vibration, Wave

Propagation, Material Testing, Newton–Raphson procedures, Plasticity, Non-linear finite elements and so on.

This book was written keeping in mind students and professionals who struggle for a better, fair and peaceful society. I did not resist adding comments and problems on ethics. This is very important these days and more so in a field where one can work in the grey area between protecting people and cargo or damaging them.

During writing it became clear to me the responsibility an author has in conveying the contents of a book in an accurate form. I struggled to do so, but I apologise in advance for any mistake or misprint the reader finds in the present volume. At the same time I am thankful if they are brought to my attention for the benefit of future versions.

The present book was followed closely, since the early days, by Prof. Dora Karagiozova, from the Bulgarian Academy of Science, to whom I pay here my homage for being supportive of this idea. Details on buckling and wave propagation benefit from many of her articles and from the discussions we have had. Chapter 6, on buckling of shells, is co-authored with Prof. Karagiozova.

The book was also inspired by the educational attitude towards research and engineering of Prof. Norman Jones; I owe him a lot of what I had learned in the field.

All the beauty and complexity of Computational Mechanics I learned stemmed from the courses taught together with Dr. Larissa Driemeier. Her commitment to an elucidation of all the intricate aspects of non–linear finite elements served as an example to me, and it is reflected in this book, especially in Chapter 9, which she co–authors.

I should mention the late Prof. Carlos Alberto Nunes Dias, who was very sharp in his approach to education on finite elements, numerical methods and vibration.

I take the opportunity to thank all the present and past members of the Group of Solid Mechanics and Structural Impact, GMSIE, in particular Dr. G.B. Micheli, Dr. R.E. Yoshiro, Dr. M.A.C. Gonzales, Dr. R.C. Santiago, Dr. R.T. Yamassaki, R.R.V. Neves, Dr. L. Mazzariol, Dr. A. de Lima, M.H. Shaterzadehyazdi, B. Mussulini, M. Duarte and V. Cruz. I thank also Dr. R.T. Vargas, Dr. R.T. Moura, Prof. M.L. Bitencourt, Prof. R.J. Marczak, Prof. P.A.M. Rojas, Prof. S.P.B. Proença, Prof. E.A. Fancello, Prof. P.T.R. Mendonça, Dr. R.S. Birch, Prof. M. Langseth,

Prof. A.H. Clausen, Prof. M. Brünig, Prof. T.A.H. Coelho and Prof. F.P.R. Martins.

Special thanks go to Dr. P.B. Ataabadi, for his support to the GMSIE activities and for co-authoring Chapter 11.

The support of my Department of Mechatronics and Mechanical Systems Engineering, Polytechnical School and University of São Paulo were fundamental to the completion of this work. Special thanks go to all my co–authors of the various articles we wrote, some of which served as a basis for sections of this book. The Brazilian Research Agencies, FAPESP, CNPq and FINEP are here acknowledged for their financial support to my research.

I am forever greatly indebted to my father David, my mother Luiza, my sister Marilane and my nieces and nephews. They were all fundamental to the completion of this book and here goes my sincere words of thanks to them.

It was the continuous support of my wife, Isabel, my daughter Taís and my son Pedro that enabled this task to be now accomplished.

Special feelings of joy and reverence should be announced due to the arrival of Fábio, my first grandson.

About the author

Marcílio Alves graduated in Mechanical Engineering, Federal University of Santa Catarina, Brazil, in 1983, where he also obtained his Master Degree. His Doctoral Degree is from The University of Liverpool Impact Research Centre in 1996 under the guidance of Prof. N. Jones. Back in Brazil, he established at the University of São Paulo the Group of Solid Mechanics and Structural Impact. He obtained various grants, which supported his research activities in the areas of structural impact, material characterization and non–linear finite elements. He published many articles in scientific journals and conferences, covering theoretical, experimental and numerical aspects of non–linear dynamics. He edited four books on Structural Impact and Solid Mechanics and established a biennial cycles of conferences in Solid Mechanics and in Impact Loading of Lightweight Structures. He is the Editor–in–Chief and Founder of the Latin American *Journal of Solids and Structures* and Associate Editor of the *International Journal of Impact Engineering.* He has been publishing with authors from different universities of Brazil, UK, Germany, Bulgaria, China, South Africa, Belgium, Australia and Norway. He is Professor at University of São Paulo and Executive Director of its Foundation, FUSP.

CHAPTER 1

Introduction to structural impact

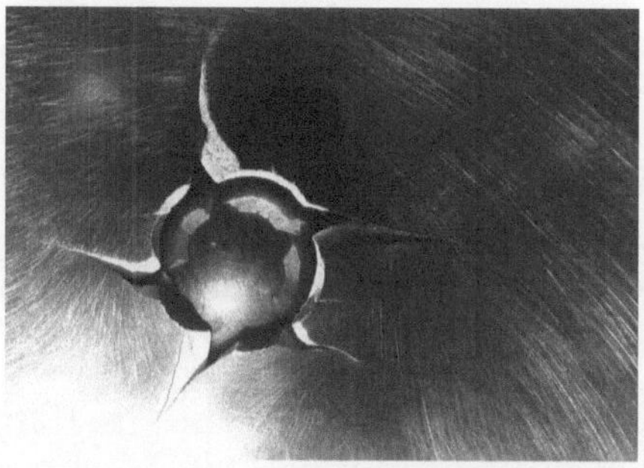

Figure 1.1 A sphere perforating a metal sheet.

Your lively neighbour describes an accident where a young fellow jumped from a trampoline straight into water. This fall, from about 3 meters, caused a bone fracture to the elbow, leading to surgery and a long term recovery. Your neighbour, knowing you are an engineer, asks whether it is possible to estimate the force arising from this impact. As you showed doubt, he goes on asking you about the force a boxer hits his opponent and whether a bird can knock down an air–plane. He says that old cars were stronger, with thicker skins, and therefore crash safer. These questions remain on your mind. You skim through your books on structural analysis and found practically nothing that could assist you

DOI: 10.1201/9781003485292-1

to carry on an intelligent discussion with your neighbour on this topic that one loosely calls impact or shock.

It is not difficult to conclude that some of the questions given above are rather difficult to answer, requiring a specialised knowledge. Also, it is a matter of surprise to become aware of so many such daily events of impact loading. Indeed, the fast loading of structures, objects or live creatures is very common. Examples range from the fall of an egg, pressurised water for metal cutting, a bullet or missile penetrating an armoured tank, explosives in soil for underground mining, the act of running, car, train, air–plane crashes, ship collisions, etc...

Of course that the consequences of an impact event can be a disaster, as shown in the next figure, together with a beautiful image, on the edge of danger, of a flock of birds flying along side an aircraft. It also shows the consequences of a collision between a bird and an aircraft.

(a)

(b)

(c)

Figure 1.2 (a) The Brazilian Space Agency program to design and launch a rocket suffered a major setback in this accident in 2003, when a rocket explosion killed 21 technicians, (b) birds flying together with an aircraft, (c) Leading edge of an air–plane hit by a bird.

Structural impact is a broad and complex subject, requiring sophisticated methods of analysis, computer power and advanced instrumentation.

We can conceive a classification of the various impact events, as in the next figure.

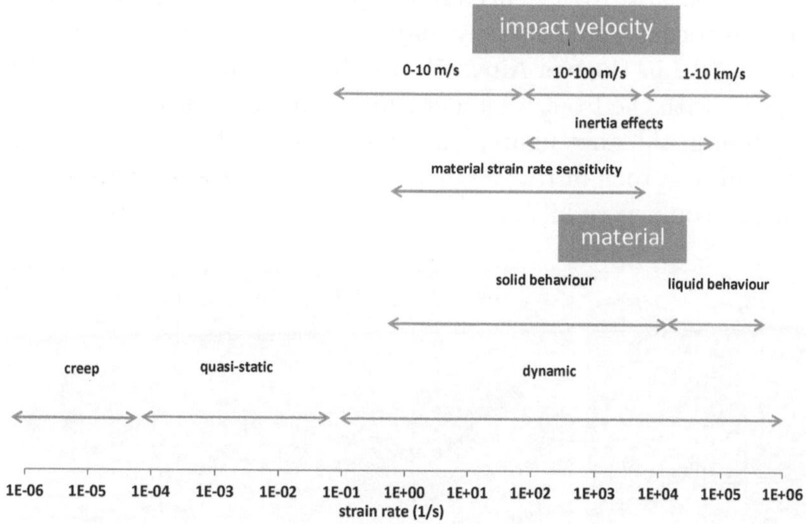

Figure 1.3 A general classification of structural impact based on the strain rate, which is proportional to the impact velocity.

The most obvious approach would be to use the impact velocity as the major variable dictating the type of impact and its consequences. The objects involved in the impact may also serve for classification purposes. It can be seen that impact can take place, for instance, among objects that do no deform much, such as the classic impact between two billiard balls. This rigid impact event is in contrast with the impact of a ball hitting a flexible beam, leading the latter to vibrate. In this elastic impact event, the propagation of waves along the structure is an important feature that needs to be taken into account. Another common type of impact is when the colliding bodies deform permanently, like the impact of two cars. This impact event is usually associated with large permanent deformation of the material.

Scientists at Sandia National Laboratories, in USA, blasted in 1994 a small projectile to a speed of 18 km/s. This is one of the highest velocities ever reached by an object on earth.

In the cases above, the impact velocity is crucial in determining the structure behaviour and we refer to high and hyper velocity impact, the former being, for instance, the impact of a bullet (say 600 m/s) and the latter the impact of a meteorite on a space station (say 20 km/s).

In a general classification some phenomena are manifested in a more pronounced way. For instance, in impact events at low velocity, we shall see that the material strength may increase with the velocity but inertia effects may affect little the overall response of the structure.

An important remark is related to the impact velocity. The velocity of a train some 50 m before stopping in a station is around 25 km/h. This is comparable to the speed of a top marathon runner and could be thought as too slow to cause any major problem in a crash event. Yet, in February 2012 in Buenos Aires, 50 people were killed in an accident at this speed, with the front wagon crushing some 7 m into the second one, as depicted in the next figure. The figure also illustrates the collision of a ship against a rock in Italy, January 2012, at an estimated speed of 25 km/h. 32 people were killed.

(a) (b)

Figure 1.4 (a) A train collision at low velocity. 50 people were killed. (b) A ship collision at low velocity. 32 people were killed.

It is important to distinguish between quasi–static and dynamic analysis. The former is covered in engineering courses but the dynamic analysis is usually restricted to the linear elastic realm. On the contrary, here, we will approach the dynamic response of structures taken into consideration various non–linear features, such as large displacements, material behaviour, strain rate and wave propagation. It will be evident along the chapters that these non–linear dynamic analyses will become too difficult and we will need to resort to numerical methods. This is why we will also learn from this book the finite element method, as specialised to non–linear dynamic events.

It may sound weird to say but reality is a fundamental aspect of structural analysis. By this is meant that the analyst should have a close contact with experiments and with what goes on in the world out of the computer models. So here we equally devote our efforts to the experimental aspects of the linear and non–linear dynamics of structures, including material behaviour. We aim to join the fields of analysis, finite elements and experiments in order to cooperate with a more integrative formation of the engineer, complemented here with some cases related to engineering ethics.

The next figure, which underlies the structure of this book, presents a threefold approach to the activity and formation of an engineer. We have at the core, the system we are dealing with, which is down to analysis, synthesis and/or modelling. From this system a component, a machine or structure, or even a single material is born. The system can be subjected to loads being applied at low to high rates, each one having peculiar characteristics. Broadly speaking, at low rates the problem is of a quasi–static nature. At medium loading rates, the material response to dynamic loading should be considered and at higher rates, one should add inertia effects.

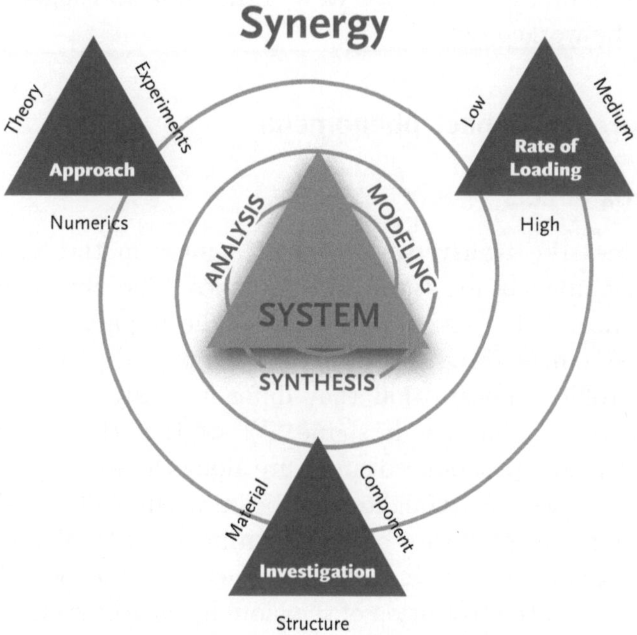

Figure 1.5 An overview of an integrative approach to building knowledge in the area of impact engineering.

The figure also suggests that we can approach a problem using numerical, analytical and/or experimental tools. They are all valid and more and more this integrative approach distinguishes a good engineer. It is noteworthy the spiral underlying this knowledge acquisition strategy. The spiral conveys the fact that the various branches ought to be visited, intermingled, support one another, learn from the errors made in one phase and revisited at a later stage. Hence, the figure represents a summary of the approach of the present book. Of course, it is difficult to balance all this synergy and indeed we will be more concentrated on analysis and modelling, but certainly we will also be experimenting with aspects of synthesis of a system.

As for the sequence of this chapter, we will see some peculiar features of the impact phenomenon next. We will give extra motivation on the study of impact engineering via a section on structural crashworthiness, followed by various comments on experimental techniques. Some comments on analytical and numerical methods are also offered and we argue that analytical methods should underlay the numerical analysis. We briefly present a time line of important contributions to the field of structural impact, entering in the debatable field of engineering ethics. We close the chapter giving an overview of the book and suggesting some problems to be worked out.

1.1 Features of impact phenomena

1.1.1 Inertia effects

Let us consider the transverse impact of a mass in the middle of an aluminium clamped beam at various velocities. The next figure shows some experimental data for this case, with the impact velocity in the range of 70—110 m/s, increasing from the bottom to the top. The various final beam profiles indicate that they differ not only in the maximum central displacement but also in shape. We see that the higher velocity case presents a more pronounced curvature along the span. The lower the impact velocity case, the straighter the beam profile. This inertia effect is also valid for the composite beam shown in the same figure (b), now hit at a velocity of 500 m/s. In this case, we can see the evolution of the beam response and the transition of the more linear to the curved profile.

These aspects can be loosely explained by saying that, in the high impact velocity case, parts of the beam away from the mass contact zone did not have time enough to be accelerated; they were too lazy to move. In this case, the beam density is an important parameter. For the low velocity case, the impact event was relatively long so allowing time for the beam to move like a mechanism, with hinges in the middle and at the supports. We shall see in Chapter 5 that it is possible to solve this problem in an analytical way and to capture these aspects of the impact event.

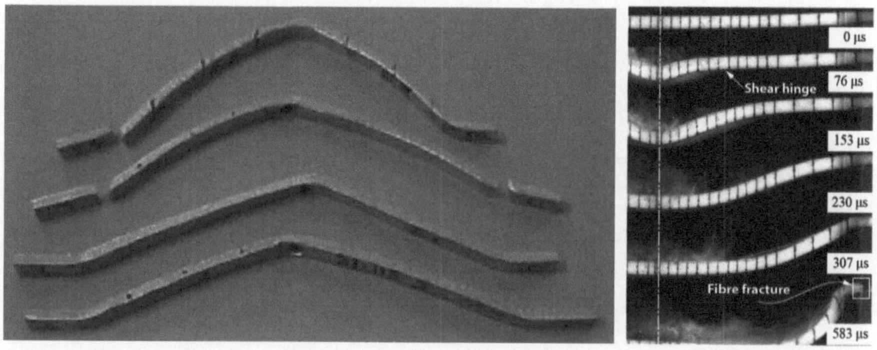

(a) (b)

Figure 1.6 Beam responses to a mass impact at (a) low (arranged in increasing order of impact velocity from bottom to top (73.5, 94.7, 103.0, 113.6 m/s)) and (b) high (500 m/s) velocities. For details see (a) M.N. Mannan, R. Ansari, H. Abbas, Int J Impact Eng 35 (2008) 1201–1212 and (b) K. Karthikeyan, B.P. Russell, N.A. Fleck, M. O'Masta, H.N.G. Wadley, V.S. Deshpande, Int J Impact Eng 60 (2013) 24–36.

In counter distinction with the transverse impact of a beam, let us now consider the axial impact of a mass on a circular tube standing on a platform from its base. The tube is made of aluminium and has a diameter of 23.75 mm, wall thickness of 1.65 mm, height of 106.68 mm. This problem was solved numerically and some results are shown in the next figure. The kinetic energy was kept the same in all the simulations shown, but allowing for different combinations of impact mass and velocity. Quite interesting, for the same initial kinetic energy, the tubes behave very distinctly, as depicted in the figure. The figure also shows the mean crushing force, which becomes smaller as the impact mass increases. Clearly, inertia has an important influence on this phenomenon.

See D. Karagiozova, M. Alves and N. Jones, Inertia effects in axisymmetrically deformed cylindrical shells under axial impact, International Journal of Impact Engineering, p. 1083–1115, 24, 2000

1.1.2 Material response

Of course the material behaviour plays a fundamental part on the overall behaviour of a structure. This is particularly true in the case of impact loadings, where large deformations are expected. Finite strain usually hardens the material, so changing its response to the load.

In the case of an impact event, as a structure deforms, the strains change spatially and along time. We refer to strain rate to convey the time variation of the strain. Consider a beam impacted transversely by a mass at a speed of 7 m/s. This means a height drop of 2.5 m. An analysis of this phenomenon reveals that the strain rate reaches values as high as 120/s=12%/ms, as illustrated in Figure 1.8.

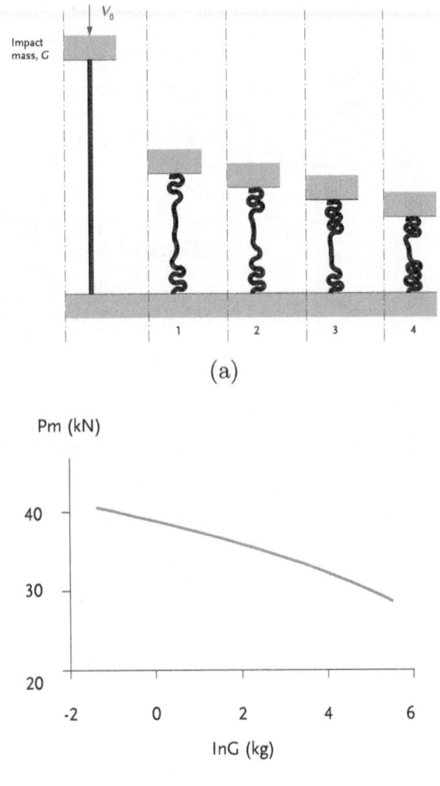

(b)

Figure 1.7 (a) Final buckling shapes of a shell after impacted with different combinations of impact velocity and mass. Only the tube wall is shown. 1– $G = 0.2485$ kg, $V_0 = 130$ m/s; 2– $G = 2.625$ kg, $V_0 = 40$ m/s; 3– $G = 42.0$ kg, $V_0 = 10$ m/s; 4– $G = 262.5$ kg, $V_0 = 4$ m/s. (b) mean crushing force *versus* the striking mass.

Hence, the strain rate in an impact event can be of a quite high magnitude, even when just dropping a mass onto an object. It happens that the strength of a material is affected by the speed one loads it. This effect can be quite significant and for mild steel a strain rate of around 100 s^{-1} can double its yield stress. On the contrary, such a strain rate material sensitivity is not so important for many aluminium alloys, *eg* it doubles its strength for a strain rate as high as 10000 s^{-1}.

In any case, it is clear that it is important to evaluate the strain rate that takes place in a structure subjected to impact loading, so to know its extra strength due to the dynamic loading process. This task can be troublesome due to the fact that the strain rate varies in time and space. Nevertheless, it is possible to obtain some estimation of the strain rate for certain structures and loading conditions. By knowing the strains and strain rates it is possible to gain information on the material strength,

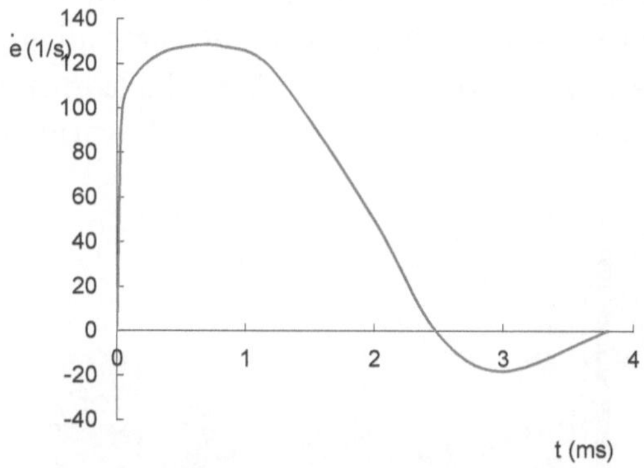

Figure 1.8 Variation of the strain rate along time underneath a beam subjected to the impact of a mass. From J.L. Yu and N. Jones, Numerical simulation of a clamped beam under impact loading, Computer & Structures, p. 281–293, 32, 1989. Remember that a (engineering) strain rate of 100 s^{-1} means that a 1 mm wire would increase its length to 101mm in a second.

ie the stress levels a material can sustain, via the so called constitutive equation.

There are many other aspects of the material behaviour which influences the structure response like strain and kinematic hardening, temperature, to name a few. We defer to Chapter 7 for a more detailed analysis of this topic.

1.1.3 Transient effects

Consider again the drop of a mass on a structure. Immediately after the contact between the drop mass and the structure, local deformations take place. This deformation impinges motion to the neighbouring material, a process that repeats over time thus establishing the propagation of waves, elastic in the beginning, in the structure. These waves interact with the structure boundaries, promoting a complex picture of reflexive, attenuation, change of wave characteristics, etc... Within the next micro–seconds of the impact event, other waves arise, eventually carrying more energy and leading the structure to deform permanently.

Accordingly, a series of important dynamic events take place within the first say 100 μs of an impact. Global long–term response of the structure may lapse 1000 times more, *ie* 100 ms, when all the transient effects, so to say, disappear. Transient effects may define the overall pattern of the final configuration of a structure. This is exemplified in the next figure, where a square cross–section tube is impacted by a moving mass. Depending on the impact velocity and mass, the waves generated travel up and down along the tube, interact with each other in a way that large amplitude strains localise somewhere along the length. This localisation of strains leads to a wrinkle which grows by the further deceleration of the structure, so promoting the various patterns shown in the figure.

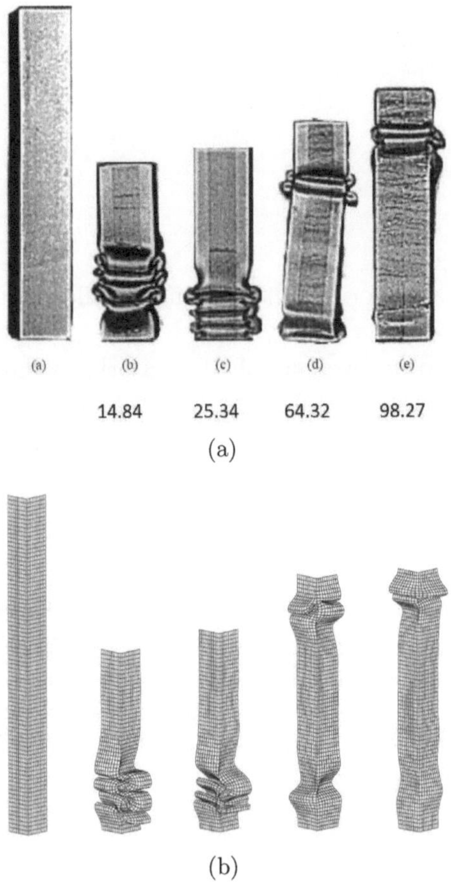

Figure 1.9 (a) Square cross-section aluminium tubes hit by a mass at the velocities indicated below the figure (m/s) and impact energies of approximately 600 J. (b) Quarter of the tubes numerical models subjected to the same velocities as in the experiments. See D. Karagiozova and N. Jones, Dynamic buckling of elastic–plastic square tubes under axial impact—II: structural response, International Journal of Impact Engineering, 30, p. 167–192, 2004.

1.1.4 Stability

In many applications, it is important to obtain a stable response of a component under impact loading. Consider a tube subjected to the drop of a mass. One of the tubes in the next figure buckled without large lateral displacement, so exhibiting a better performance in terms of energy absorption since more of its material is deformed. The other tube, by way of contrast, offers less resistance to the impact mass due to its exaggerated lateral motion, leading to less kinetic energy being absorbed.

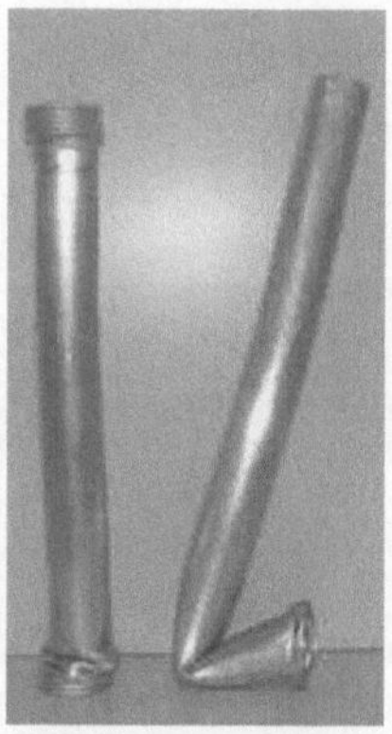

Figure 1.10 Tubes axially impacted by a dropped mass.

It happens, interestingly, that both tubes, made from the same material and having the same dimensions, were subjected to the same impact mass but with velocities only slightly different, $\Delta V = 1$ m/s, *ie* a 5 cm drop height difference. This is to say that the stable/unstable region is very sensitive to the initial conditions, so characterising a bifurcation problem. The study of stability of structures under impact loads is a large and rich subject, with many scientific and practical applications and it will be dealt with in Chapter 6.

1.1.5 Failure

In many situations involving impact loading it is desirable for a structure not to fail, in the sense of local material separation. Hence, the capability of predicting material failure is an important issue to be considered in the design phase of a structure. However, the various failure criteria so far developed do not cover all the cases and a more universal failure criterion does not yet exist.

The situation becomes even more complex in the presence of impact loads, like the impact of a sphere in the circular plate shown in the next figure. The same figure also depicts a finite element simulation of this problem where some resemblance to the experiment is evident. The problem is further complicated by frictional and temperature effects, together with unexpected issues such as the way the plate is held in its supports.

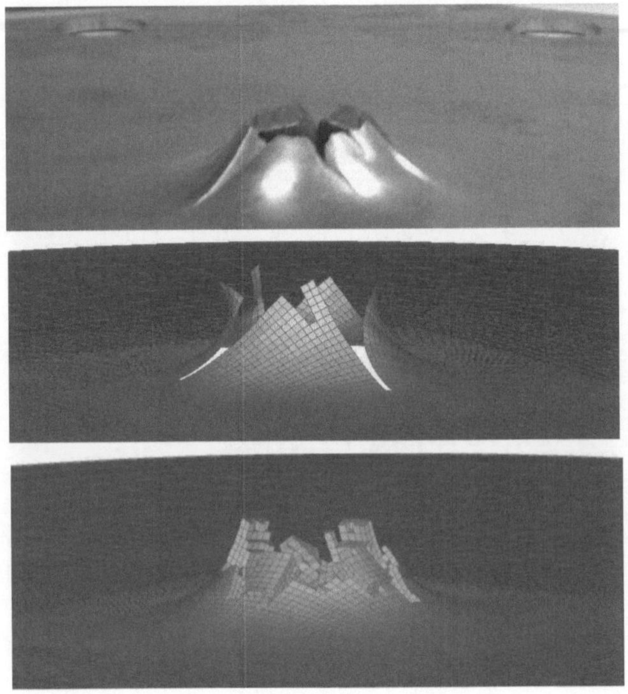

Figure 1.11 Failure of a circular aluminium plate under the central impact at a speed of a hard sphere of over 400 km/h.

1.2 Crashworthiness

Car crash tests are performed worldwide under the coordination of the New Car Access Program, NCAP. For a ship kinetic energy absorption system see L. Wang et al., An impact dynamics analysis on a new crashworthy device against ship–bridge collision, International Journal of Impact Engineering, 35, p. 895–904, 2008.

One useful application drawn from the field of structural impact is the design of structures in crash situations. We use the term crashworthiness not to convey that a structure is worth to be crashed. Rather, the more crashworthy a structure is the better it protects passengers or goods that it transport.

The basic example in this field is related to the performance of cars involved in a collision. The two cars shown in the next figure perform rather distinctly in terms of the safety offered to the occupants. Yet, they are essentially the same model, but one built in 2007 and the other in 2010. It is evident from the photos that the 2010 model was re–engineered so to perform better and to meet government regulations.

The crash energy management is also applied to train and ship collisions. For instance, a device for possible use in preventing train over–ride is based on the useful property of some polyethylene that, under compression, changes from solid to a viscous fluid. The device has good energy absorption and damping properties and when fit in between a

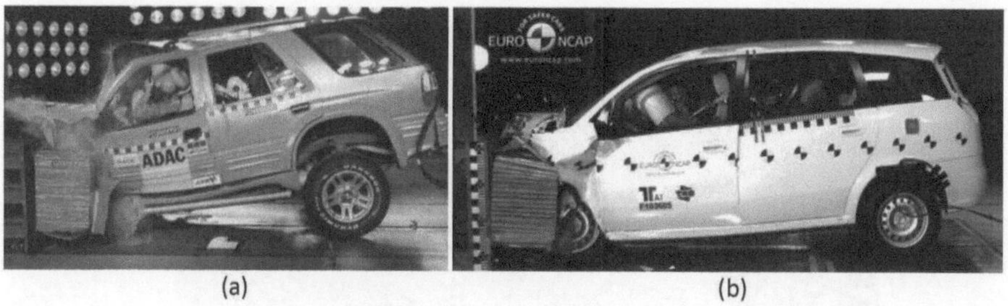

Figure 1.12 Poor (left) and good (right) performances to impact of a car model built in 2007 (left) and 2010 (right).

locomotive and a wagon it avoids overriding, as illustrated in the next figure.

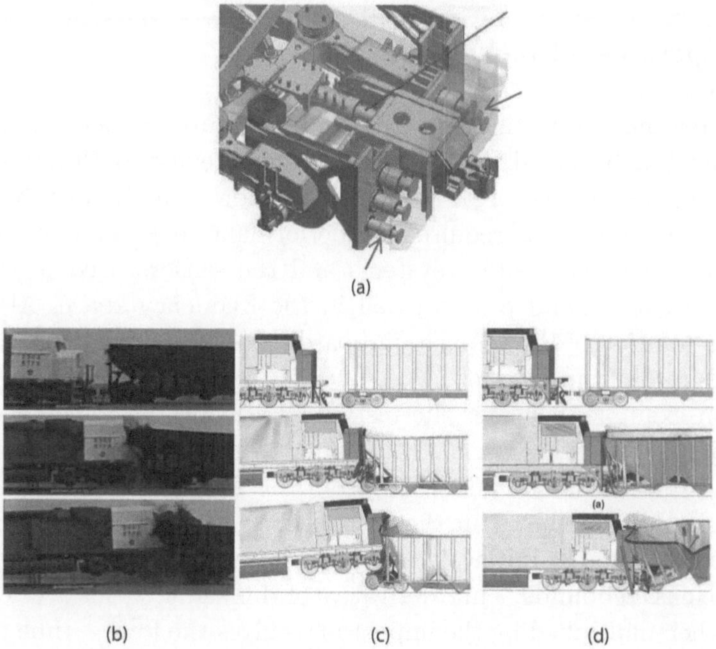

Figure 1.13 The use of an energy absorption device for a train collision. (a) Seven devices were used to absorb some 250 KJ of kinetic energy. (b) Actual train override used to validate the numerical model in (c). (d) Numerical simulation of train collision using the shock absorbers (no overriding). From B.K. Parida, X. Xin, J. Carter, A.K. Zaouk, G. Samavedam, Development of compact shock energy absorber to improve structural crashworthiness, Proceedings of the IMPLAST 2010 Conference, Providence, USA, Society for Experimental Mechanics, 2010.

Out of curiosity, the next figure shows a car that was crushed by a woman who fell from the 23rd. floor of a building, some 80 m high. The woman had to remove one kidney, the spleen and part of one lung, but she did not die. Had she crashed against an undeformable structure, the outcome would have been fatal. This shows the power a structure has, if properly designed, to absorb kinetic energy in a crash event. We will explore this subject in Chapter 6.

Figure 1.14 The fall of a woman from 80 m, the 23rd floor of a building, onto the top of a car.

1.3 Experimental techniques

Various reasons for testing material and structures under a high rate of loading can be listed, like checking a new device, performing tests for comparison with theoretical results, measuring material properties, etc. These dynamic tests require high performance electronic instrumentation. The data acquisition systems and the sensors used to capture dynamic events should perform well in the frequency space. Also, the voltage resolution of the acquisition boards should be very good, at least of 12 bits resolution.

FML stands for fiber—metal—laminate, a sandwiched structure formed by layers of a fabric and metal. It has good fatigue and impact energy absorption properties and it is widely used in aircraft structures.

As an example, consider the task of measuring the force generated by the impact of mass on a FML plate. One smart alternative is to track the velocity of the mass during all the impact event with a laser system, as the one shown in the next figure. This velocity will change continuously from the initial impact value, to zero and to negative values if the tip mass rebounds. Time derivative of this signal yields acceleration which, when multiplied by the impact mass gives the force—time profile. Another advantage is that the time integration of the laser velocity signal gives the displacement, being possible to obtain the force— displacement (and the absorbed energy) of the impact event.

An important issue when processing data from impact events is filtering. Usually, fast signals present some noise which can be eliminated by using hardware or software filters. We should exercise care when filtering a signal since the procedure can eliminate important information of a phenomenon.

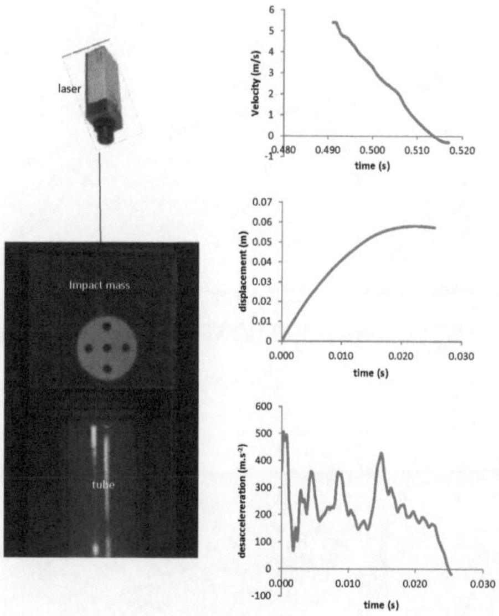

(a)

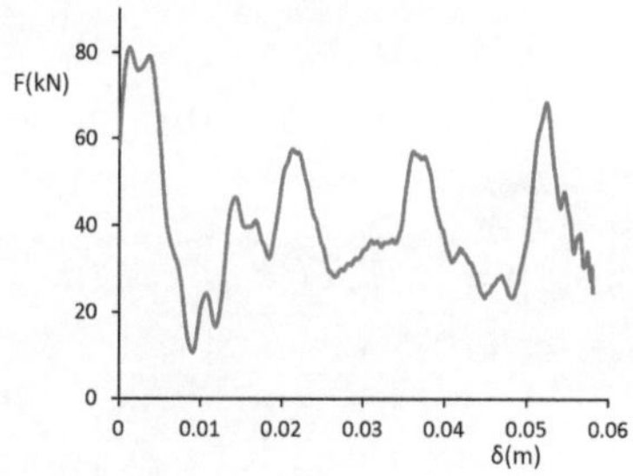

(b)

Figure 1.15 A laser system used to track the velocity of a mass on a tube. (a) tube impact arrangement and the raw velocity signal as measured by the laser, together with its time integration (displacement) and time differentiation (acceleration). (b) Load versus displacement of this test.

This is illustrated in the next figure, where a numerical force—time curve was filtered with a low—pass type of filter. One can see that the peak force changes in time and intensity as different cut–off frequencies are applied.

In structural impact tests, two important devices, among others, are mentioned here. The drop mass tower (also called drop hammer) can be used for impact velocities of up to around 20 m/s, with impact mass as high as 1000 kg. For higher impact masses, a sledge is a good alternative. For higher impact velocities, between 50 m/s and 500 m/s, and small mass, up to 10 kg, the so called gas—gun is used. These systems are shown in Figure 1.17 and are quite useful for testing mid–size structures for a wide range of initial conditions.

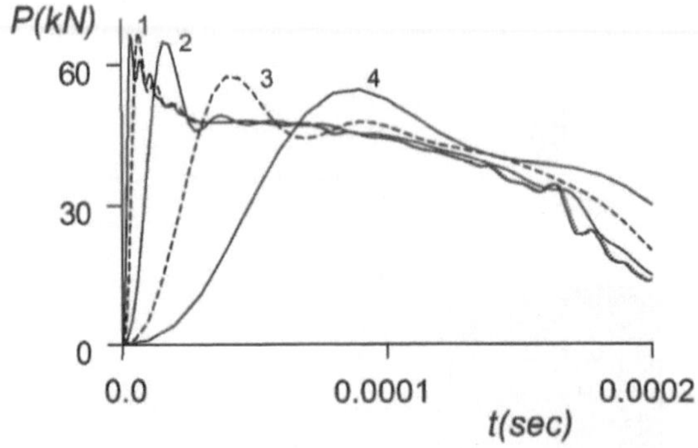

Figure 1.16 Impact force *versus* time for the impact of a 0.4167 kg mass onto a shell at a velocity of 120 m/s. Signal low—pass filtered with cut–off frequencies of 1 – 150 kHz, 2 – 100 kHz, 3 – 50kHz, 4 – 20 kHz.

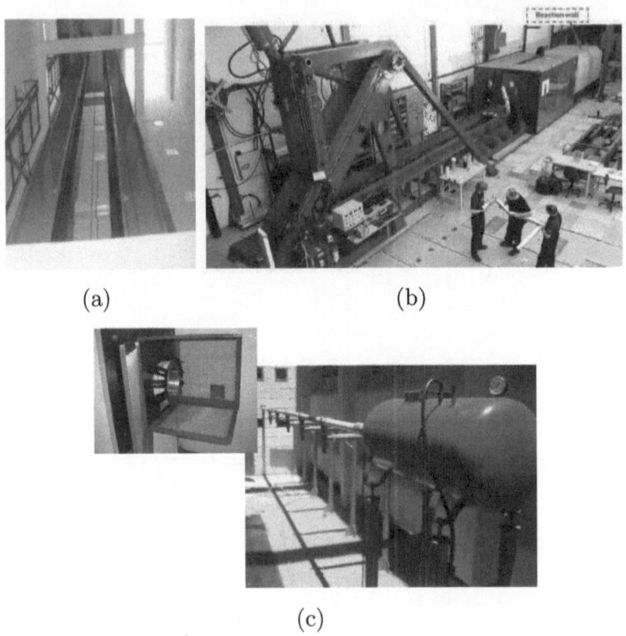

Figure 1.17 (a) Drop tower, (b) high impact mass sledge and (c) gas–gun with details of a safe cage and an anvil holding a circular plate.

As for material tests, a screw–driven tensile test machine is essential for proofing materials at low strain rate. For high strain rate tests, a Hopkinson bar, or Kolsky bar, or elastic wave machine, is the alternative mostly used. They are shown in the next figure. The way these pieces of equipment are used to obtain mechanical material properties at a wide range of strain rate will be detailed in Chapter 7.

Current regulations in our society require that a structure should be tested before putting into operation for the public. This implies that huge structures, such as trains, air–planes, ships, mining equipment, need to

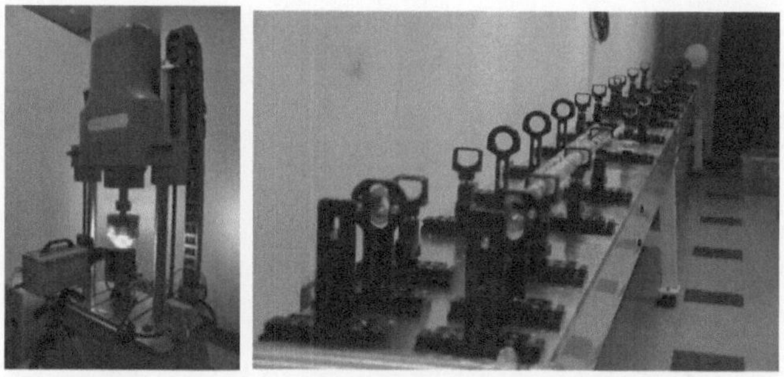

(a) (b)

Figure 1.18 (a) Low–medium speed tensile–torsion test machine and (b) a modern "high velocity" elastic wave machine.

be submitted to destructive crash events. These tests are extremely expensive and in some cases it is more cost effective to test a scaled model, as illustrated in the next figure. These tests in models are used to infer the behaviour of the prototype, but this is only possible if we apply some scaling laws. Common scaling laws fail to perform well when impacted structures are made of strain rate sensitive materials or when one of the dimensions cannot be fully scaled due to manufacturing reasons.

The next figure shows a scaled aircraft and a ship being set up for a test. The collision of the ship with a rigid wall will be detailed in Chapter 10, which will also describe the basic scaling procedures.

Figure 1.19 Models of an airplane and of a ship.

The technique for correcting the impact mass or impact velocity such that the prototype and the model behave the same has some special aspects that we will disclose later. For now, as an illustration, consider the next figure, which shows a tube hit by a dropping mass. It is evident that the buckled configuration of the model, 10 times smaller but graphically enlarged 10 times for easy comparison, does not scale as desired. On the other hand, if a correction procedure is adopted, model and prototype behave nearly the same, as also shown in the figure.

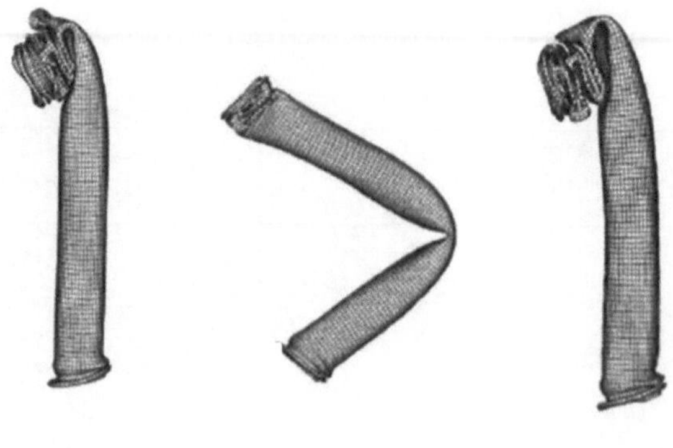

Figure 1.20 Cylindrical tubes axially impacted. The model was scaled down ten times but the figure was enlarged this amount for easy comparison.

1.4 Analytical and computational methods

Analytical methods are used intensively in this book. Wave propagation, impact on beams and shells, material models, plasticity, to name a few, are all studied using mathematical models. These models are translated to differential equations, whose solution is based on standard techniques. Analytical tools are also used for the development of the finite element method; basic algebra, matrices and tensor operations are clearly necessary.

On the other hand, a detailed analysis of the impact loading of a structure is rarely possible without the use of a numerical technique, mainly the finite element method. This is specially true when the structure is complex in shape and in the way its material responds to dynamic loads. We shall see in Chapter 8 how the finite element method can be put to use for these dynamic analyses. In addition, we will also present small finite element codes and the theory underlying them.

As an example of the use of a finite element analysis, consider the complex case of a soft jelly–like body hitting a circular plate at 100 m/s. The plate is made by 24 layers of fibber glass fabric, fixed to each other by epoxy resin. This composite material has some advantages over metals and it has been used in many applications where strength and low weight is required. The next figure shows the impact event just before material failure. Stress and strain magnitudes in each layer of material are difficult to be accessed analytically. Note that in this example, the impact mass changes as the impact velocity increases.

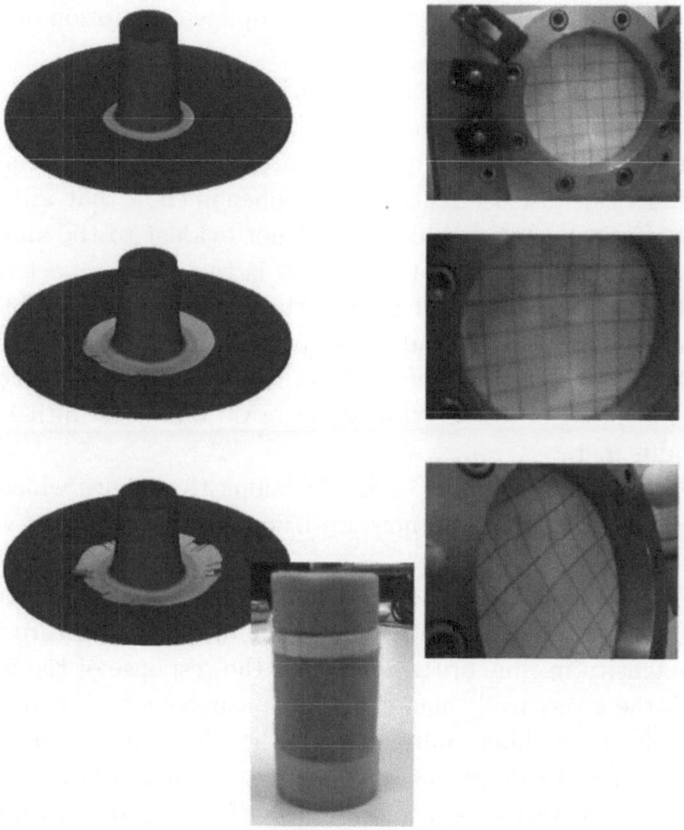

Figure 1.21 Impact of a jelly–like material on a circular composite plate.

1.4.1 Analytical versus numerical solutions

This book will present analytical solutions for somewhat basic structural problems. These solutions can be applied to a large number of real problems and have been most important in the development of structures capable of sustaining impact loads.

On the other hand, it must be recognized that all these methods of analysis have been superseded by the finite element method, FEM. Basically, in the former method, one strives to solve the equilibrium equation of a structure in its differential form while in the FEM the equations are of algebraic nature. This requires simple procedures, *eg* Gauss elimination, to solve a system of equations whose unknowns are the displacements of various points (domain) of a structure. Such a procedure can be carried out along time so one has a complete description of the structure motion in space and time.

The immediate question that arises is why to develop further analytical models. Why not to use exclusively FEM structural analysis?

The answers to these questions are manifold. First of all, analytical solutions can disclose features of a phenomenon that are presented in the numerical solution but that are not evident to the analyst. A good example in the context of this book is the axial impact of a mass on a tube. We shall see in Chapter 6 that it is possible to obtain a simple expression for the peak load, for a given material, geometrical and impact conditions, which are useful in the design of a shock absorber. If one opts for the FEM it is necessary to run various cases until a given design bound is met.

There are many other examples along these lines which will be discussed in this book. For now we bring another point of view into the discussion.

See M.L. Bucalem and K. Bathe, The Mechanics of Solids and Structures — Hierarchical Modelling and the Finite Element Solution, Springer, 2011 and N. Jones, The credibility of predictions for structural designs subjected to large dynamic loadings causing inelastic behaviour, International Journal of Impact Engineering, 53, pp. 106–114, 2013.

The FEM is capable of yielding accurate solutions but at least three aspects should be known in advance: the way the structure is supported, the intensity in time of the loads and the response of the material from which the structure is made. All these aspects are difficult to be known precisely, in striking contrast with the precision of the analyses tool used, *ie* here FEM. Hence, in many cases, given the lack of knowledge on the above variables, a more straightforward analytical solution can give acceptable results.

There is perhaps a more subtle argument for the study of analytical models. Analytical solutions require a dose of imagination. FEM modelling requires operational procedures, greatly simplified by powerful graphical interfaces between the analyst and the computer. Also, if one does not learn the fundamentals of the FEM, via an analytical approach, no further progress is possible in the numerical area.

Many important problems can be solved by elementary equations steamed from basic mechanics laws. To model a problem and to choose the right equations to describe it, hinges on the imaginative process of the analyst. This requires a self–development, an engaging process to acquire knowledge. This is in contrast with the attitude of using software, which allows one to solve real problems with little understanding of the numerical tool at hand. To be provocative, one could say that analytical methods challenge the creativity; numerical methods leave the analyst aside.

This is of course untrue if one really understands what is behind the FEM. In this case, the generality of the method supersedes the analytical methods. That is why we study FEM here in its fundamental aspects.

1.5 A bit of history

History puts knowledge in perspective.

The history of structural impact is mingled with the history of the strength of materials, science and mathematics. A simple and factual account of the subject indicates facts like the following. Newton developed the concept of coefficient of restitution. Also, as early as 1668, the Royal Society of London discussed the laws of impacting bodies, with Wallis discussing the inelastic aspects, while Wren and Huyghens the impact of elastic bodies. Wallis set the concepts of hard, soft and elastic impact, plus the concept of direct impact.

For an account of the history of structures see The History of the Theory of Structures: Searching for Equilibrium, 2nd Edition, Karl-Eugen Kurrer, Wiley, 2018.

Around 1820, Navier investigated the equations of dynamic equilibrium of a solid and with Poisson and Cauchy helped to establish the basics of wave propagation in solids in general. Later in the XIXth. century, Pochamer and, independently, Cree, developed the exact solution of wave propagation in rods. Rayleigh investigated the propagation of surface waves in solids and of waves in plates, which was also studied by Lamb in 1889.

See R. Dugas, A History of Mechanics, Dover Publications, 1988.

A detailed study on the collision of elastic bars was given by St.–Venant and Flamant, and by J. Boussineq, both in the XIX century. S. Timoshenko analysed in 1913 the transverse impact of a sphere on a beam using the Hertz static theory of contact. Donnell in 1930 and, later in 1950, T. Karman and P.E. Duwez, Taylor, Rakhmatulin and others, set the grounds for the propagation of plastic waves, with Davies studying in detail, both theoretically and experimentally, the propagation of waves in bars.

J. Hopkinson (1849–1898) conducted various experiments using a drop mass impacting a collar held by wires. The wire was lead to rupture at its extremes, which was found to depend on the drop height, but not on the drop mass. His son, B. Hopkinson (1874–1918) worked with explosives and devised the use of a bar to measure the pressure produced by explosions. Later, in 1949, H. Kolsky extended this idea to measure dynamic strength material properties. He sandwiched in between two bars the material sample and using explosives at the end of one bar, a pulse was generated, loading the specimen. The use of capacitors, and later, strain gauges, allowed him to measure the amplitude of the various signals generated in this event.

Important books in the area of impact and wave propagation appeared only recently. In 1960, Werner Goldsmith published his book Impact: The Theory and Physical Behaviour of Colliding Solids, which became a standard reference up to the seventies. In 1972, William Johnson published the now classical book Impact Strength of Materials, followed later, in 1982, by the book Impact Dynamics, by J.A. Zukas, T. Nicholas, H.F. Swift and L.B. Greszczuk. In 1989, N. Jones published the book Structural Impact, which became widely used.

As for wave propagation in solids, the books Elastic Waves in Solids (1963), by H. Kolsky, Wave Motion in Elastic Solids (1975), by K.F. Graff, Stress Waves in Non–Elastic Solids (1978), by W.K. Nowacki and Foundations of Stress Waves (2007), by Wang Li–li, all being fundamental and comprehensive references, have secured their place in the history of the structural impact literature.

The student, research and professional can find valuable studies published in two specialized journals, International Journal of Impact Engineering and International Journal of Structural Crashworthiness. The former is the core where advanced studies in impact engineering are presented and the latter has been serving the community by publishing more applied research.

1.6 Engineering ethics

Two basic books on engineering ethics are C.B. Fleddermann, Engineering Ethics, Prentice Hall, 2012, and E.G. Seebauer and R.L. Barry, Fundamentals of Ethics for Scientists and Engineers, Oxford Press, 2001.

The engineer is faced more and more with ethical issues, despite the subject not being, usually, touched upon in technical books. This is unfortunate as much as an engineer working in the field of structural impact faces daily ethical problems, sometimes not readily recognizable as such.

In this book, we will bring every so often questions related to ethics . Ethical questions are not so tamed to be approached in a systematic way. The different moral, religious, political and even economical backgrounds tend to shape one or another position of an individual when facing an ethical problem.

As a conflicting example that involves ethics, consider the case of designing a car. In order to make a popular model, the car body–in–white, *ie* the structure itself, shall be made of a low strength steel, whose crash performance is very low. Numerical simulation and actual crash tests indicate that the car barely meets the regulations of the country where it is to be sold. You are aware that adult passengers sat in the front and children sat in the back seat are likely to die in a car frontal collision at some 60 km/h. You bring this issue to a technical meeting, arguing that the use of a pair of air bags would increase the car final cost by some 100 Euro, which would be justified given the lives that would potentially be saved. Your suggestion is turned down on the basis that this extra cost would imply losses in sales. You walk away from the meeting in conflict. What attitude should you take? Would it be better to work politically to improve the regulations? What about if a family involved in a crash event with this car prosecute you and your company? How to balance your duties to the company and to society?

An even more complex dilemma is military research, where effort is guided towards more efficient weapons. As a reaction to such a trend, one could work on defence, so to counteract the weapons and the cycle is established of growing belligerence. And what about more basic research, say on ballistic, being used for bullet protection or for more efficient bullet perforation?

The examples above pose many questions, whose solutions are many fold. It hinges both in the general progress of society, *eg* regulations and safer roads, and in a personal attitude towards the engineering practice. Ethical dilemmas are explored in this book only by problems. Single answers to them are rarely possible. In this sense, these problems can be seen as provocative, but also yielding some, say, flexibility to the mind, instigating new ways for a balanced solution to the often overlooked ethical duties of the engineer.

In short, we all should be aware that we are responsible for our discoveries and that we give away knowledge to be applied to situations we cannot foresee. Hence, we should be well equipped to discuss ethical issues, specially in the field of structural impact.

1.7 Book organization

We tried to give an overview of some important aspects of impact loading of structures in this chapter. Most of the topics commented on here will be detailed along the book. With this introduction we hope the reader can gather enough motivation to pursue the analyses to come.

We will proceed by studying the impact or rigid bodies in Chapter 2. Basic mechanical laws will be used to investigate some practical problems, which will be exemplified along the text. In Chapter 3, we will study elastic impact of axially colliding bars, exploring the concept of elastic waves. As for the experimental aspect of this chapter, we will introduce various sensors used in the impact field.

In Chapter 4 we will study the dynamics of beams, with emphasis on transverse dynamic loading. We will learn about flexural waves and advanced beam models. Spectral analysis of signals and basic filters are introduced in this chapter.

Chapter 5 will lead the reader to the field of transverse impact on beams and plates that deform permanently. Various types of dynamic loading will be considered and large displacement, transverse shear forces and equivalent pulse concepts will be detailed. We will also present experimental issues when dealing with beam impact.

Plastic waves, buckling of rods and of circular and square shells will be studied in Chapter 6. We will present various aspects of crashworthiness when studying kinetic energy absorption with tubes. We present some experimental results on buckling of shells and their anomalous behaviour.

Chapter 7 explores material response and the various experimental techniques to measure it. Various concepts of plasticity will be learned side by side with constitutive laws and yield criteria. Experimental techniques to measure the material response at low, mid and high strain rates will be presented. The chapter also gives an overview on ductile material failure criteria.

Chapter 8 presents the Finite Element Method, FEM, for linear analysis of structures. It introduces explicit and implicit time integration algorithms for rods and it expands it to linear beam finite elements. Concepts of connectivity, transformation matrix, isoparametric finite elements and numerical integration are also given.

A nonlinear truss and beam finite elements are detailed in Chapter 9, side by side with the Newton–Raphson procedure. Plasticity is studied in the context of Computational Mechanics, leading the way to the concept of deformation gradient and others from this field. Chapter 10 gives an account of scaling structures, a useful technique for inferring the behaviour of actual size structures from tests performed in scaled models. Modern concepts will be learned and examples will assist us in understanding them. The last chapter of the book, Chapter 11, gives a series of application, case studies, of impact problems that will serve to show how the various concepts learned along the book can be used.

I hope you enjoy!

1.8 Problems

1. A bar of length L hanging from the ceiling is suddenly loaded when a collar hits its free end with a velocity V_0. Evaluate the average engineering strain rate, $\dot{e}$, in the bar.

2. A constitutive equation describing the influence of the strain rate, $\dot{e}$, on the dynamic flow stress, σ_d, of a mild steel is given by $\sigma_d = 300\dot{e}$. What should the height of a bar of length L be vertically released so that, when arrested by the collision with the floor, its strength doubles?

3. The component of a car designed using a commercial finite element code failed. Further enquires revealed that a fault on the code was overlooked by the engineer, who then arrived at the wrong part strength. Who should be blamed for the part failure: the engineer or the software company?

4. You are working on the design of a car to be sold worldwide. Your team arrived at a model whose main column, when made of a high strength steel gives a good overall crash performance. For low income countries your car can be fitted with a low strength steel and even so meeting the local regulations, but leading to a very bad crash performance. Do you see any conflict in this design? If positive, how could it be solved? If negative, would you support a change in the local regulations so as to improve the car crashworthiness?

CHAPTER 2

Rigid body impact

Figure 2.1 A deformable ball, hitting the face of a player, can be treated as a rigid body in some basic analysis.

Impact, collision and shock are words, in the context of mechanics, that carry the idea of a sudden change in the motion of interacting bodies. We refer to impact when attention is given to one body only, like the case of a car hitting a rigid barrier. The term collision is applied when the involved bodies have somewhat similar masses and their motion after the event is sought. Shock can refer both to impacting objects or to phenomena connected to explosions; it is distinct from shock wave. In this book, we sometimes intermingle these concepts, but we use mostly the word impact to convey this very notion of a sudden change in motion of objects, components, structures, etc...

Impact, press closely into something, from Latin *impactus*.

DOI: 10.1201/9781003485292-2

In this chapter we deal with the impact of particles and rigid bodies. The concepts to be seen here are helpful in many practical cases, *eg* in forensic studies of car collisions. Broadly speaking, the impact between two bodies involves a change in motion caused by forces acting in a very short interval, from the time the bodies come into contact until they separate. We use the word rigid to convey the notion that changes in the bodies shape are not relevant to the analysis. What is usually sought are their velocities and directions after collision. In a rigid impact analysis, phenomena like wave propagation and temperature changes are thoroughly disregarded. To solve this category of problem we will use the principle of impulse and linear momentum conservation.

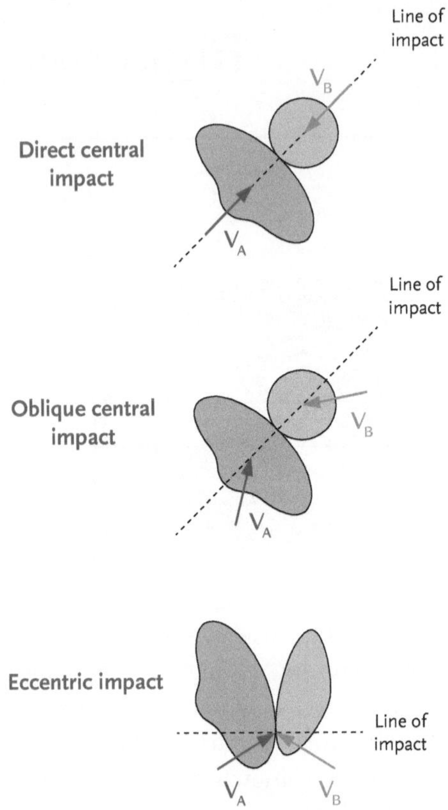

Figure 2.2 Basic types of in–plane rigid body impact.

The figure above depicts a broad classification of rigid body impact. The line of impact is the common normal to the point of contact. When this line passes through the center of mass of the particles or bodies involved in the collision, we refer to it as central impact. Clearly, for the impact of particles, it is only possible to have central impact. If the velocities of the particles are collinear with the line of impact, we refer to direct central impact. Otherwise, we refer to oblique central impact. For bodies which cannot be modelled as a particle, upon impact, they may rebound and turn in the case of no central impact. This is the case of eccentric impact.

A rigid body may experience a soft or hard collision, according to the shape and properties of the bodies involved in the impact. The next figure illustrates these concepts and shows at once that the change in velocity results in the raise of a force and transfer of energy.

According to Leonardo da Vinci, force is the dying of motion. See Max Jammer, Concepts of Force, Dover, 1962.

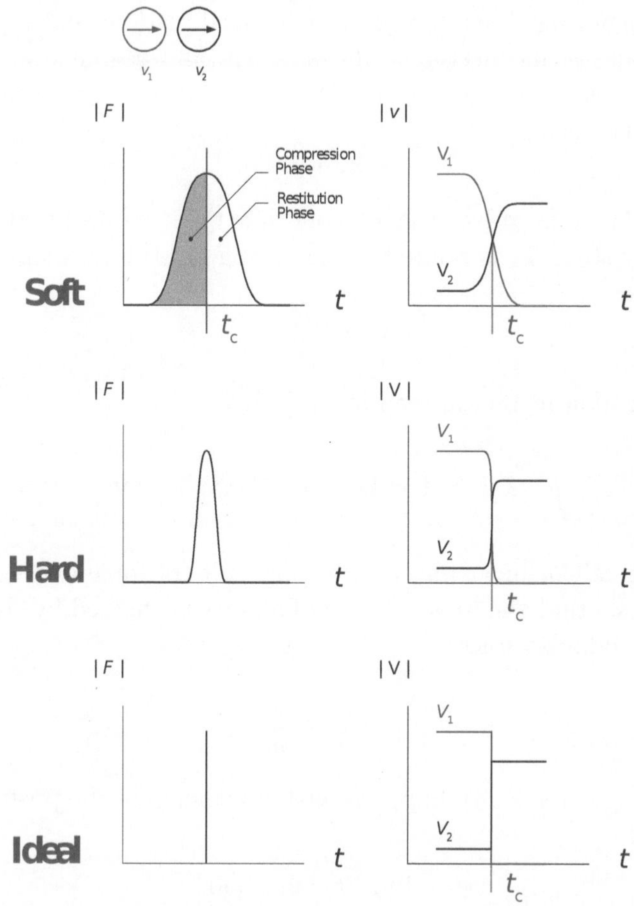

Figure 2.3 Idealisation of force and velocity behaviour during collision of two bodies.

During the impact of two objects, the local pressure in the contact area increases with time, followed by a decrease, *ie* energy is stored in the compressed body and is then released. The bodies start to separate, moving away from each other and possibly rotating. Also, part of the stored energy in the compression phase can be dissipated, mainly in the form of plastic energy. To take these local phenomena into account, we use a single parameter called coefficient of restitution.

2.1 Impulse and momentum of a single particle

The collision of two masses is one of the first impact problems we learn in mechanics and also one of the first impact problems studied in history.

The tendency an object has to move in one direction can be quantified by the product of its inertia, *ie* its mass, m, and velocity, $\mathbf{v}$, with the bold face notation used for vectors. We define linear momentum, $\mathbf{L}$, of a mass m as the vector

$$\mathbf{L} = m\mathbf{v}.$$

We also know from Newton's second law that the resultant of forces acting in a system, $\mathbf{F}$, is related to the change in the linear momentum according to

$$\mathbf{F} = \frac{\mathrm{d}\mathbf{L}}{\mathrm{d}t},$$

whose integration in the interval $[t_i : t_f]$ gives

$$\int_{t_i}^{t_f} \mathbf{F}\mathrm{d}t = \bar{\mathbf{I}} = \mathbf{L}_f - \mathbf{L}_i = m\mathbf{v}_f - m\mathbf{v}_i$$

with $\bar{\mathbf{I}}$ being called linear impulse, having units of force and time. This equation shows that the final velocity of a body is affected by the change in the linear impulse since

$$\mathbf{v}_f = \mathbf{v}_i + \frac{\bar{\mathbf{I}}}{m},$$

but this change in velocity happens with no change in the position, $\mathbf{s}$, of a body since

$$\mathbf{s} = \lim_{t_0 \to t_1} \int_{t_0}^{t_1} \mathbf{v}\mathrm{d}t = 0.$$

It is interesting to note that the impulse is the variation of the kinetic energy, K, with the velocity. Also, the impulse, *ie* the variation of the linear momentum, is the area under the force–time curve. An average value of the impact force can be estimated by dividing the impulse by the time interval when the change in velocity occurred. As a matter of fact, the only way to quantify the force in a rigid body impact is to have information on the impact duration, a variable difficult to measure.

■ KNOCKING A WALL

Some children playing outdoors are trying to knock down an old wall by throwing clay balls at it, as in the next figure. None of them succeeded in demolishing the wall so they asked you to help. You notice that the balls the children were throwing were too wet so you search for a dry one and ask a child to throw it against the wall. The wall smashes. How can this be explained?

Figure 2.4 A dough ball hitting a wall.

Suppose the initial momentum of the dough ball is 1. Because the dough is soft, upon hitting the wall, it nearly sticks to it so the final momentum is zero and the total change in momentum is close to -1. For the hard ball, the initial momentum is again 1. But since the ball is harder, it bounces back when hitting the wall, having a residual momentum of, say -0.5, with a total change in momentum of -1.5. This extra amount ends to be decisive in knocking down the wall.

■

Load cells will be dealt with in Chapter 3.

As a complement to the problem above and at the same time introducing some experiment, the next figure shows pictures of a steel sphere and a soft polymeric one, which were let to fall from the same height on the top of a load cell. Load *versus* time profile is also shown, being evident that the peak load is higher for the hard ball, *ie* higher impulse. We should also pay attention here to the shown force and time scale, being evident that the hard impact occurs in a much shorter time interval than the soft impact case. The load intensity is clearly also very different.

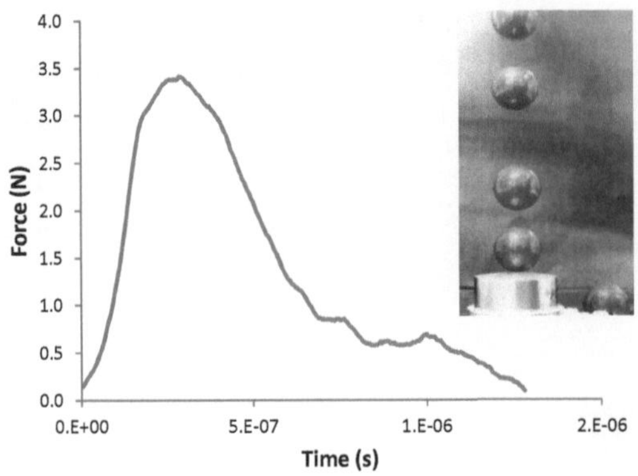

(a)

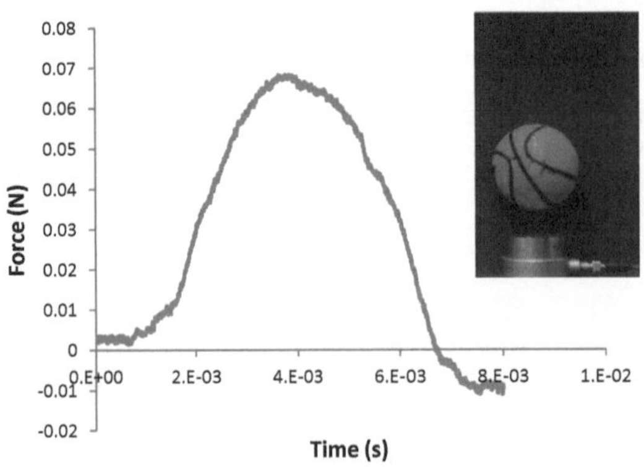

(b)

Figure 2.5 Force–time curve of a (a) hard and (b) soft ball hitting a load cell.

There are situations where all external impulsive forces in the system are zero or can be neglected. In this case, linear momentum is conserved and we can simply write

$$\mathbf{L}_i = \mathbf{L}_f.$$

This occurs, for instance, in the central impact of two bodies of mass m_A and m_B, possessing velocities v_A and $v_B < v_A$, as in the next figure, with vector bold face notation dropped for unidimensional problems. If these two bodies adhere to each other after the impact event, they will possess the same final velocity, v'_{AB}, obtained from the conservation of the linear momentum as

$$v'_{AB} = \frac{m_A v_A + m_B v_B}{m_A + m_B}$$

and we refer to perfectly plastic impact.

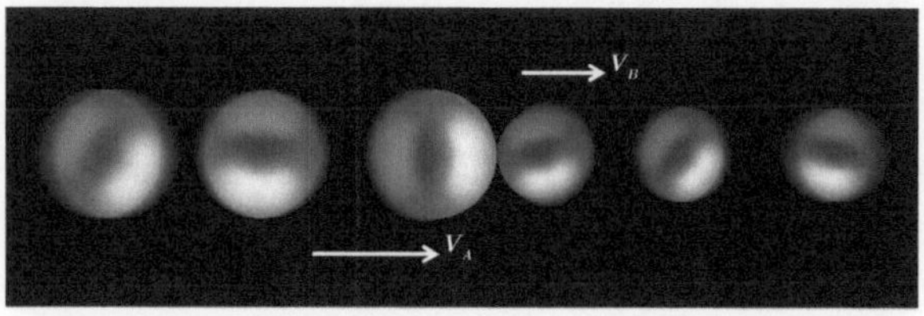

Figure 2.6 Central impact of two bodies.

On the other extreme case, when no kinetic energy is lost in the impact event, we refer to perfectly elastic impact and combination of the conservation of kinetic energy, *ie* $m_A v_A^2 + m_B v_B^2 = m_A v_A'^2 + m_B v_B'^2$ plus conservation of the linear momentum, *ie* $m_A v_A + m_B v_B = m_A v'_A + m_B v'_B$ leads to

$$v_A + v'_A = v_B + v'_B,$$

so allowing the determination of the final velocities as

$$v'_A = \frac{(m_r - 1)v_A + 2v_B}{1 + m_r} \qquad v'_B = \frac{2m_r v_A + (1 - m_r)v_B}{1 + m_r},$$

without any further consideration of the contact phenomenon, *eg* friction, with $m_r = m_A/m_B$.

For a more comprehensive study on the collision of spheres, see C. Wu, L. Li and C. Thornton, Energy dissipation during normal impact of elastic and elastic–plastic spheres, International Journal of Impact Engineering, 28, 929–946, 2003 and H. Minamoto and S. Kawamura, Moderately high speed impact of two identical spheres, International Journal of Impact Engineering, 38, 123–129, 2011.

■ COLLISION BETWEEN BUS AND CAR

A bus of mass m_b collides against a stopped car of mass m_c, with the wreckage displacing a distance d on a road with a kinetic friction coefficient μ. Estimate the speed of the bus before the collision and the average contact force as a function of the collision time, Δt.

Energy equilibrium gives the common speed of the wreckaged, v'_{bc},

$$\mu(m_b + m_c)gd = \frac{1}{2}(m_b + m_c)v_{bc}'^2 \qquad \rightarrow \qquad v'_{bc} = \sqrt{2\mu gd}$$

and linear momentum conservation gives the bus speed,

$$m_b v_b = (m_b + m_c) v'_{bc} \qquad \rightarrow \qquad v_b = \frac{m_b + m_c}{m_b} \sqrt{2\mu g d},$$

with g being the gravity acceleration.

The impulse experimented by the bus is

$$\bar{I}_b = (v'_{bc} - v_b) m_b,$$

giving an average force of

$$F_{av} = -\frac{\sqrt{2\mu g d}}{\Delta t} m_c.$$

Note that, to evaluate the impact force, it is necessary to know the collision time, which is not always easy to measure or estimate.

■

As argumented by N.W. Murray in When it Comes to the Crunch: The Mechanics of Car Collisions, World Scientific, 1994.

The fact that the linear momentum is conserved even when kinetic energy is dissipated, like in the example above, can be explained considering that, during the impact event, the bus is applying a pushing force to the car, which then increases its velocity. At the same time, the car slows down the bus by applying the same force but in the opposite direction. The friction force acting in the short time interval the colliding masses are joined together can be disregarded. Since the time interval is common to both masses, so are the action and reaction forces. The impulse should also be the same as well as the momentum of the masses before and after impact.

2.2 Coefficient of restitution

The final velocities of the bodies in those perfectly plastic and elastic impact events seen above are particular cases. In general, the bodies neither stick one to each other nor rebound without loss of kinetic energy. In this more general case, we again have conservation of the linear momentum if no external impulsive force is applied, but not necessarily of energy. It is evident that the final velocities of the colliding bodies depend not only on their mass and initial velocities but also on their shape and flexibility.

Consider the next figure, which shows a playing ball at various instants of its collision against a hard surface. It is possible to see that there is an initial stage where the ball is compressed further and further, followed by a restitution of this stored energy, lasting until the ball starts to leave the surface. Evidently, this restitution phase depends on the material characteristics of ball and surface. To take this into account in an average and simple sense, we define a scalar, called coefficient of restitution, e, as the ratio of the impulse generated in the restitution, r, and compression, c, phases ,

The coefficient of restitution was first suggested by I. Newton in his Principia. For more details on rigid body dynamics see R.W. Soutas–Little, D.J. Inman and D.S. Balint, Engineering Mechanics, Dynamics, Cengage Learning, 2009.

$$e = \frac{\int \mathbf{F_r} dt}{\int \mathbf{F_c} dt}.$$

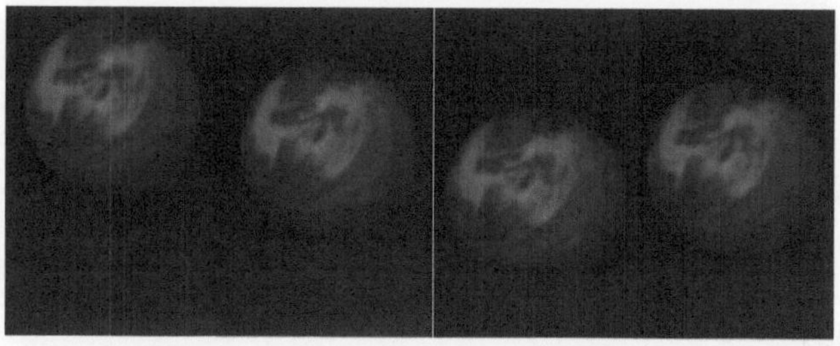

Figure 2.7 A playing ball at various instants hitting a hard surface.

In the case of two colliding particles on a horizontal frictionless plane, the impulse—momentum equation for particles A and B, in the compression and restitution phases are (positive to the right)

$$-\int_{t_1}^{t_u} \mathbf{F}_c \mathrm{d}t = m_A u - m_A v_A, \quad -\int_{t_u}^{t_2} \mathbf{F}_r \mathrm{d}t = m_A v'_A - m_A u \quad \text{for } A$$

and

$$\int_{t_1}^{t_u} \mathbf{F}_c \mathrm{d}t = m_B u - m_B v_B, \qquad \int_{t_u}^{t_2} \mathbf{F}_r \mathrm{d}t = m_B v'_B - m_B u \qquad \text{for } B,$$

with u being the common particles velocity at the maximum deformation and $'$ associated with the respective post impact velocity. Applying the definition of the restitution coefficient it follows, by eliminating u, that

$$e = \frac{v'_B - v'_A}{v_A - v_B},$$

which, together with the equation of the linear momentum conservation gives the final body velocities

$$v'_A = \frac{(m_r - e)\, v_A + (1 + e)\, v_B}{1 + m_r}$$

and

$$v'_B = \frac{(1 + e)\, m_r v_A + (1 - e m_r)\, v_B}{1 + m_r},$$

with $m_r = m_A/m_B$.

We can check that, for $e = 0$, the final velocities of bodies A and B are the same, meaning that after collision they stick to each other. This is the case of perfectly plastic impact. In opposition to the perfectly plastic impact, we have the elastic impact, $e = 1$, with the bodies travelling apart in opposite directions with the same velocity difference they approached each other.

■ A BALL HITTING A WALL

What is the final velocity, v_b, of a football hitting a wall with a velocity V_0?

Considering that the wall has an infinity mass, $m_r = \infty$, the last two equations give

$$v_b = -eV_0.$$

We see here that the coefficient of restitution is an empirical parameter, function in this case of the ball internal pressure, among other variables.

■

Let us examine the dropping of the ball shown earlier, but now in connection with the compression and restitution phases. Suppose that the ball is released from a height H (see next figure). The ball will bounce back and forth many times, always returning to a height lower than the previous one. After elapsing enough time, the ball will stop jumping.

The time for the first fall is $t_0 = \sqrt{2H/g} = V_0/g$. The time for the ball to go from the bottom to the top and hit the floor again is $t_A = 2v_A/g = 2eV_0/g$, such that $t_n = 2v_n/g = 2ev_{n-1}/g$. Hence, the total time event lasts

$$t = \sqrt{\frac{2H}{g}}\left(1 + 2e + 2e^2 + 2e^3 + \ldots\right) = \sqrt{\frac{2H}{g}}\frac{1 + e}{1 - e},$$

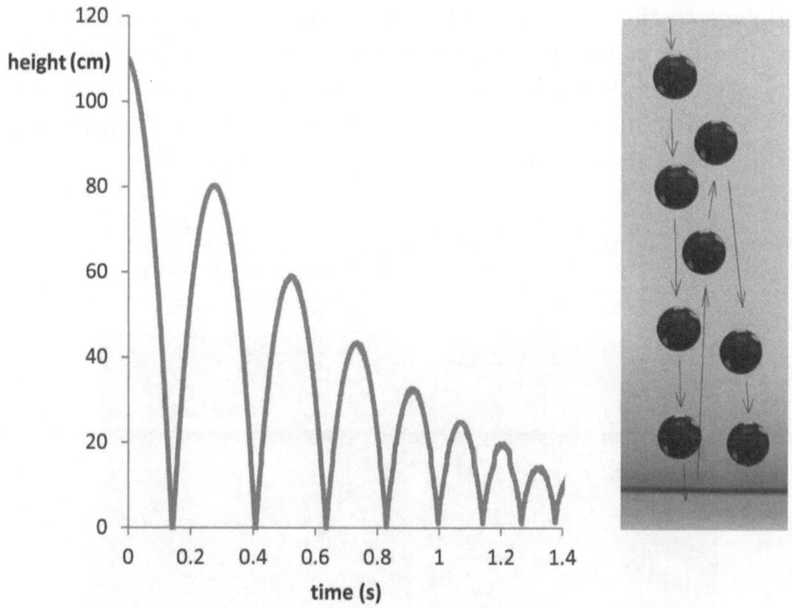

Figure 2.8 The motion of a bouncing ball as registered by a high speed camera.

and the coefficient of restitution is

$$e = \frac{t - \sqrt{2H/g}}{t + \sqrt{2H/g}}.$$

For the particular ball filmed above, the total time was measured to $t = 1.55$s, when releasing the ball from $H = 1.1$ m and this gives $e = 0.53$

Since in a collision event the final velocities of the bodies differ from the initial ones, it is interesting to enquire about the loss of kinetic energy during the collision. The variation of the kinetic energy of two impacted bodies is given by

$$\Delta K = \frac{1}{2}\left(m_A v_A^2 + m_B v_B^2 - m_A v_A'^2 - m_B v_B'^2\right),$$

or

$$\Delta K = \frac{1-e^2}{2}\frac{m_r}{1+m_r}\left(v_A - v_B\right)^2,$$

when substituting the expressions for v'_A and v'_B and after some algebra.

Here we notice that $\Delta K = 0$ only if $e = 1$, *ie* elastic impact. In all other cases, there will be some loss of kinetic energy. Note again that linear momentum is conserved when no external impulse is applied.

To explore this equation, let us investigate two problems. The first one is depicted in the next figure and it deals with a drop hammer rig. Obviously, it is desired the drop mass, G, to transfer all its kinetic energy, K_G, to the specimen under test, *ie* we do not want to set in motion the anvil, m_a. We can assume inelastic impact, $e = 0$, so the loss of kinetic energy is

$$\Delta K = \frac{1}{2}\frac{G m_a}{G + m_a} V_0^2 = \frac{m_a}{G + m_a} K = \frac{K}{1 + m_r}.$$

Figure 2.9 Some 4 kg dropping mass with its tip hitting at 4 m/s the mid span of a beam clamped at both extremes. The whole rig stands on a anvil of over 40 tons so that $\Delta K \approx 0.3\%$.

Accordingly, the energy of the dropping mass is more effectively dissipated into the specimen when the anvil mass is very large. That is why the drop hammer in the above figure has some 40 ton anvil mass. This gives, according to this model, a kinetic energy loss of 0.28% for an impact mass of 111 kg released from 2 m.

In the case of a pile driver we have the opposite situation. We want to transfer all the kinetic energy of the dropping mass, K, to the anvil mass, m_p, here in the role of the still pile. The loss of kinetic energy in the impact event amounts to

$$\Delta K = \left(1 - e^2\right) \frac{m_p}{G + m_p} K.$$

Hence, the higher the hammer kinetic energy the better, which require a high impact mass G, or a high initial velocity, V_0. Clearly, it is also interesting to have a high value for the coefficient of restitution, leading to a smaller loss of kinetic energy.

■

THREE BARS IMPACT

Two equal long bars are in end contact and resting on a frictionless surface when they are end impacted by a third equal bar travelling with a velocity V_0. What are the final velocities of the three bars assuming $e = 0$.

$$mV_0 = (m + m + m)v_f \qquad \rightarrow \qquad v_f = \frac{V_0}{3}.$$

This problem is far from being that simple and we will investigate it further in the next chapter in the context of wave propagation.

■

The force levels in a collision event can be obtained if we measure the force itself, or the collision time and the total impulse of the event. These are somewhat difficult tasks, but possible to be performed. Take for instance the case of the forces involved when hitting a tennis ball. A force platform, basically a plate supported by load sensors, can be used to measure the force during rebounding. Together with a device to measure the ball impact speed, it is possible to measure the contact time, the coefficient of restitution and the force—time profile, as indicated in the next figure for a 57.7 g tennis ball.

2.3 Oblique central impact

We refer to oblique central impact when the direction of the velocities of two colliding bodies does not coincide with the line of impact. If friction can be disregarded, then the tangential, v_t, velocity before and after impact is the same, *ie*

$$v_{A_t} = v'_{A_t} \qquad \text{and} \qquad v_{B_t} = v'_{B_t}.$$

We also know that conservation of the linear momentum holds, allowing us to write in the normal, n, direction

$$m_A v_{A_n} + m_B v_{B_n} = m_A v'_{A_n} + m_B v'_{B_n}.$$

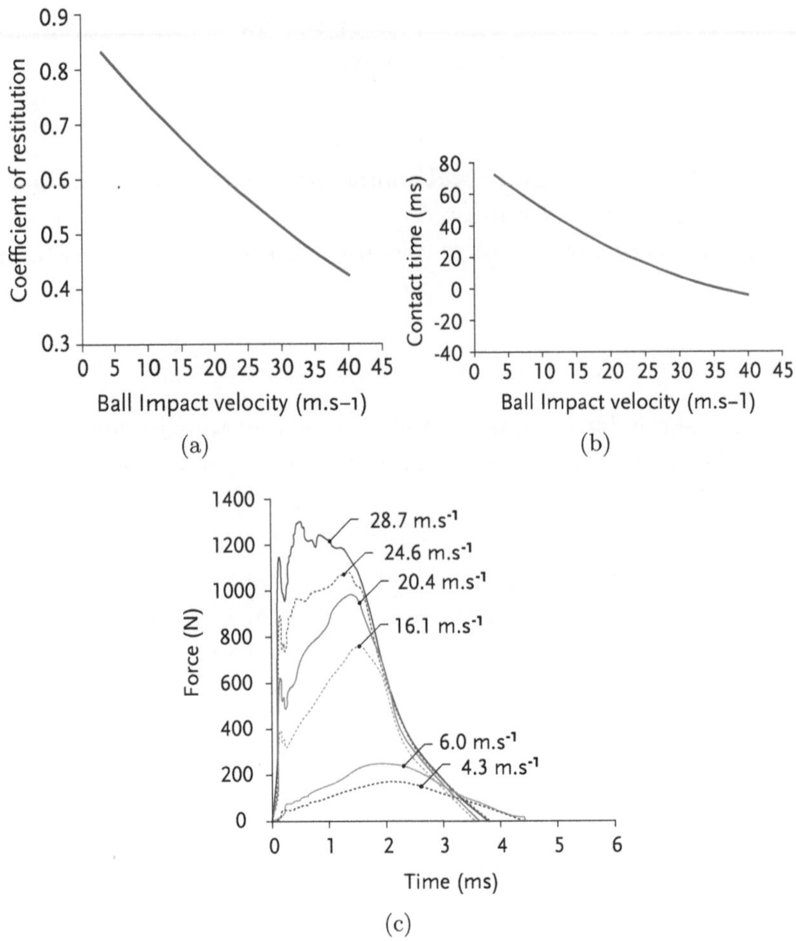

Figure 2.10 (a) Coefficient of restitution, (b) contact time and (c) force imparted by a 57.7 g tennis ball when hitting a still racket *versus* the impact speed or time. Trend lines of experimental data plotted for (a) and (b). Adapted from S.J. Haake, M.J. Carre and S.R. Goodwill, The dynamic impact characteristics of tennis balls with tennis rackets, Journal of Sports Sciences, 21:10, 839–885, 2003.

As in the direct impact case, we need a further equation, which is obtained by multiplying the pre–collision relative velocity by the coefficient of restitution e and equating to the final relative velocity, both in the normal direction,

$$v'_{B_n} - v'_{A_n} = e(v_{A_n} - v_{B_n}).$$

■ IMPACT OF TWO DISKS

Consider the impact of two flat disks moving on a frictionless surface as in the next figure and evaluate their final velocities. Assume a coefficient of restitution e.

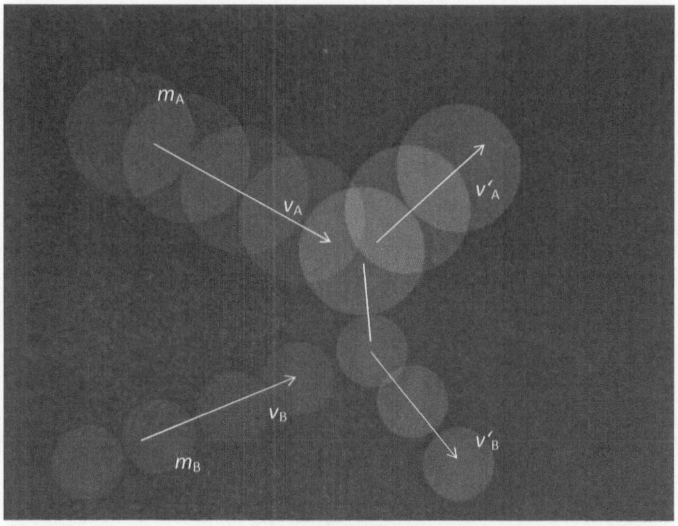

Figure 2.11 Oblique impact of two disks moving on a frictionless surface.

Since $v'_{B_n} - v'_{A_n} = e(v_{A_n} - v_{B_n})$, we write for the normal direction, n,

$$m_A v_{A_n} + m_B v_{B_n} = m_A v'_{A_n} + m_B[v'_{A_n} + e(v_{A_n} - v_{B_n})],$$

from which it follows, with $m_r = m_A/m_B$, that,

$$v'_{A_n} = \frac{1}{1+m_r}[(m_r - e)v_{A_n} + (1+e)v_{B_n}]$$

and

$$v'_{B_n} = (1+e)m_r v_{A_n} + (1 - em_r)v_{B_n},$$

with $v_{A_n} = v_A \sin\alpha$ and $v_{B_n} = v_B \sin\alpha$, being α the angle between v_A and v_B. Of course that $v_{A_t} = v'_{A_t}$ and $v_{B_t} = v'_{B_t}$

■

There are some practical problems where the colliding objects have their motion somehow restrained, as illustrated in the next figure. Again as for the frictionless case, the impulsive forces are of internal nature in the contact region of bodies A and B along the line of impact and are in equilibrium. There is also an impulsive vertical force between the horizontal surface and body A. To determine the post–impact velocities, we know that, in the case of the figure, the velocity of block A is horizontal. The direction and magnitude of the post–impact velocity of body

B comes from the conservation of linear momentum,

$$m_A v_A + m_B v_{B_x} = m_A v'_A + m_B v'_{B_x}$$

and from the use of the coefficient of restitution,

$$v'_{B_n} - v'_{A_n} = e(v_{A_n} - v_{B_n}).$$

This is illustrated in the next example.

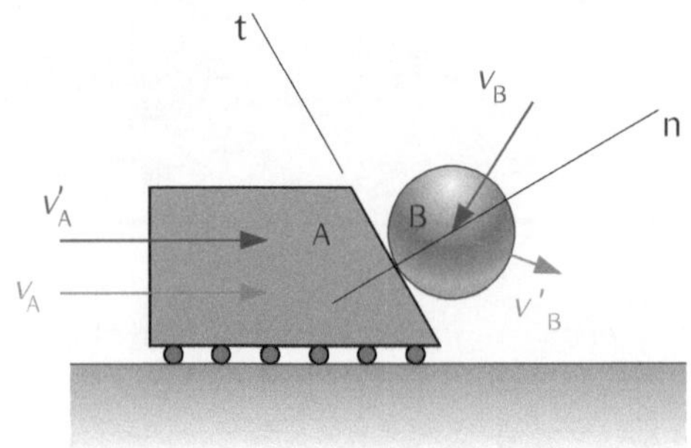

Figure 2.12 Impact of an object on a constrained block.

REACTIVE IMPULSE AT THE SUPPORT

Many interesting problems, similar to this one, are discussed in F.P. Beer, E.R. Johnston, D.F. Mazurek, P.J. Cornwell, E.R. Eisenberg, Vector Mechanics for Engineers, McGrawHill, 2009.

The collision of two spheres of radius R_A and R_B held by light rods are used in an experiment to determine the coefficient of restitution between the spheres by measuring the reactive impulse at the support of rod A, see next figure. Obtain an expression for e as a function of the measured impulse. What are the final velocities of the spheres?

Since the total momentum is conserved, we can write

$$mv_A + \bar{\mathbf{I}} = mv'_A,$$

which becomes

$$m_A v_A - F\Delta t \cos\theta = m_A v'_A \quad \text{and} \quad F\Delta t \cos\theta = m_B v'_B$$

for masses A and B, or

$$v'_A = v_A - \frac{\bar{I}_m}{m_A}\frac{1}{\tan\theta} \quad \text{and} \quad v'_B = \frac{\bar{I}_m}{m_B}\frac{1}{\tan\theta}$$

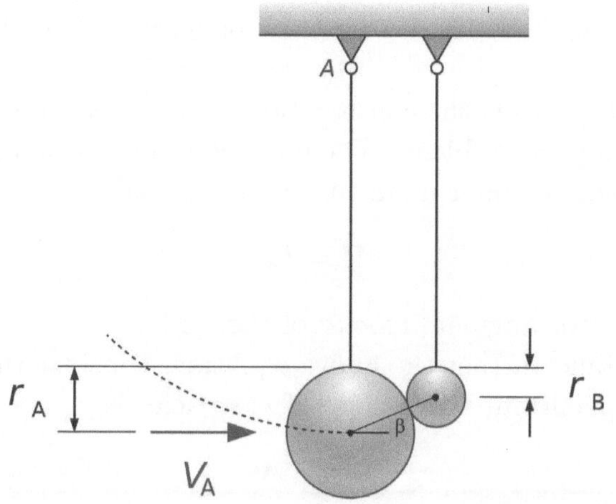

Figure 2.13 Collision of two spheres held by light rods.

with

$$\bar{I}_m = F\Delta t \sin\theta = R_A \Delta t,$$

being the vertical measured impulse at the rod support A and R_A the reaction.

Now, applying the definition of the coefficient of restitution, it follows that

$$e = \frac{v_B' - v_A'}{v_A} = \frac{\bar{I}_m}{v_A} \frac{2\sqrt{r_A r_B}}{r_A - r_B} \left(\frac{1}{m_B} + \frac{1}{m_A} \right) - 1,$$

where use was made of the fact that $\tan\theta = (r_A - r_B)/(2\sqrt{r_A r_B})$.

■

2.4 In-plane rigid body dynamics: eccentric impact

So far, for the problems we have dealt with, the shape of the colliding objects was not relevant in the analysis. This is not the case if we consider a bullet hitting a hanged rod and a hanged plate. The maximum angles for these two shapes can be rather different. Consider also the case of a rotating body, a wheel, impacting a corner. The consideration of its spin is fundamental for a correct description of the event.

Hence, it is necessary to take into account not only the mass of the colliding bodies but also the way this mass is distributed. This is to say that we need to take the moment of inertia, I, of the body into consideration.

We have so far used the conservation of the linear momentum, $\mathbf{L}$, to solve various impact problems. The body counterpart of a particle linear momentum is the angular momentum, $\mathbf{H}$, defined by

$$\mathbf{H} = I\omega,$$

ω being the vector angular velocity of the body.

To understand further the above equation, consider the next figure. The linear momentum, $\mathbf{L}$, for a set of i particles is

$$\mathbf{L} = m_i \mathbf{v}_i$$

$\times$ is the vector product symbol. The double indice i means summation.

and their angular momentum about an origin, $\mathbf{H}_o$, reads .

$$\mathbf{H}_o = \mathbf{r}_i \times m_i \mathbf{v}_i.$$

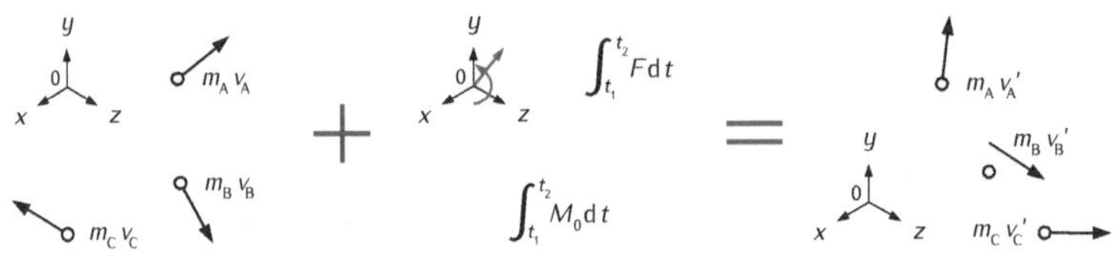

Figure 2.14 A system of particles connected by internal forces.

The same results hold for the reference frame being at the center of mass of the body so we omit the index.

Time derivative of the above equation gives the counterpart of $\dot{\mathbf{L}} = m_i \mathbf{a}_i$ ($\mathbf{a}_i$ is acceleration) for a rigid body, *ie*

$$\dot{\mathbf{H}}_o = \mathbf{M}_o,$$

with $\mathbf{M}_o$ being the moment of all the particles around the origin o. This equals zero if there is no external force acting on the system, leading to the conclusion that, in this case, both linear and angular momentum are constants, *ie* they are conserved.

We previously found for a particle that its momentum at a time t_1 plus the impulse of external forces between t_1 and t_2 equals the momentum at t_2. This holds also for a set of particles and for a rigid body. Integrating the previous equation, we have

$$\mathbf{H}_1 + \int_{t_1}^{t_2} \mathbf{M} dt = \mathbf{H}_2,$$

being implied that **M** contains the contribution of all the moments acting in the body. Equipped with this equation we can solve various interesting problems.

■ A ROTATING BALL

Analyse the motion of a rotating ball approaching a flat surface, as in the next figure.

Adapted from Advanced Dynamic, S. Timoshenko and D.H. Young, McGraw-Hill, NY, 1948.

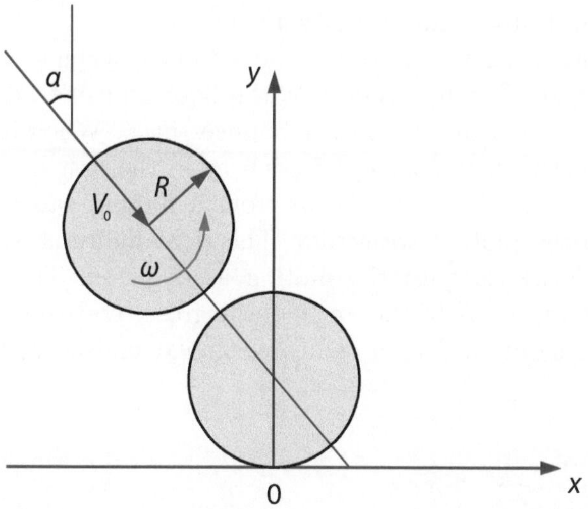

Figure 2.15 A rotating ball hitting a flat surface.

The impulse at the contact point in the x and y directions are

$$\bar{I}_x = m\left(v'_x - V_0 \sin\alpha\right) \qquad \text{and} \qquad \bar{I}_y = m\left(v'_y + V_0 \cos\alpha\right)$$

and the angular momentum with respect to the centre of gravity of the ball is

$$I_z\left(\omega' - \omega\right) = \bar{I}_x R,$$

with $I_z = 2mR^2/5$ being the mass inertia of the ball (sphere) of mass m and radius R.

Considering that the impact is not completely inelastic, the final vertical velocity is $v'_y = eV_0 \cos\alpha$. If slipping at the point of contact is prevented by friction then $\omega' R + v_x = 0$. These equations give

$$\omega' = \frac{2}{7}\omega - \frac{5}{7}\frac{V_0}{R}\sin\alpha$$

and

$$v'_x = -\omega' R = \frac{5}{7}V_0 \sin\alpha - \frac{2}{7}R\omega,$$

such that

$$\frac{v'_x}{v'_y} = \frac{1}{e}\left(\frac{5}{7}\tan\alpha - \frac{2}{7}\frac{R\omega}{V_0\cos\alpha}\right).$$

For the expression within the brackets to be negative it is necessary that

$$\omega > \frac{5V_0\sin\alpha}{2R},$$

which gives the angular velocity above which the ball rebounds in the direction opposite to its initial motion.

The next figure shows two tests. The left one is when a ball is thrown counter clockwise. The ball then hits the floor and instead of jumping further from the launching point, it bounces back. When hitting again the floor, it moves in the same direction as in the first launch, to only bounce back again. A set of photos from a filming camera is shown, together with the plotted trajectory. The right figure is when the ball is launched clockwise, with the motion as depicted. The figure gives quantitative indications of the angle (in degrees) and time (in s). This information was obtained by a dedicated image analysis software.

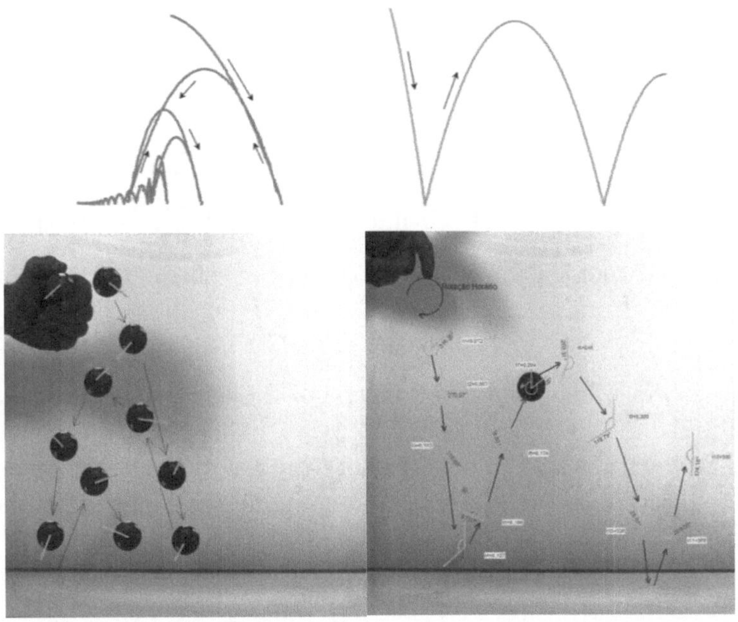

Figure 2.16 A rotating ball launched counter clockwise (left) and clockwise (right).

■

■ A BULLET HITTING A ROD

We use this example to introduce he concept of center of percussion. The target is to calculate the angular velocity and the support reactions of a rod of length L and mass m_r subjected to the impact of a mass m_b at l and travelling at a speed v_b. Let us calculate also the bullet velocity just after the impact, v_b', assuming a coefficient of restitution e.

The angular momentum of the horizontal component of the bullet velocity immediately before the collision equals the angular momentum of the rod plus of the bullet just after impact, see next figure, from which it follows that

$$m_b v_b l \cos\alpha = m_b l v_{b_x}' + m_r \bar{v}_r' L/2 + \bar{I}\omega'.$$

The use of the coefficient of restitution yields

$$v_r' - v_{b_x}' = e v_b \cos\alpha,$$

which, together with the previous equation gives

$$\omega' = \frac{3 l m_b v_b \cos\alpha(1+e)}{3 m_b l^2 + m_R L^2} \quad \text{and} \quad v_{b_x}' = v_b \cos\alpha \left(\frac{3 m_b l^2 - m_R L^2 e}{3 m_b l^2 + m_R L^2}\right).$$

The final velocity is the vector addition of v_{b_x}' and $v_{b_y}' = v_b \sin\alpha$.

The reactive impulse in the x direction comes from $m_b v_b \cos\alpha + F_x \Delta t = m_S v_{b_x}' + m_R \bar{v}_R'$, giving

$$\bar{I}_x = F_x \Delta t = (1+e) m_b m_R v_b \cos\alpha L \left(\frac{3l/2 - L}{3 m_b l^2 + m_R L^2}\right),$$

making it clear that, for elastic impact, the impulsive force is the double as for plastic impact, in line with what we had concluded in the opening example of this chapter, *ie* the children knocking down a wall.

It is interesting to enquire for the above example in what position, l_{cp}, should the bullet hit the rod such that all its impact energy moves instantaneously and horizontally on the rod, *ie* with no rotation around the support. We just impose $F\Delta t = 0$, giving

$$l_{cp} = \frac{2L}{3}.$$

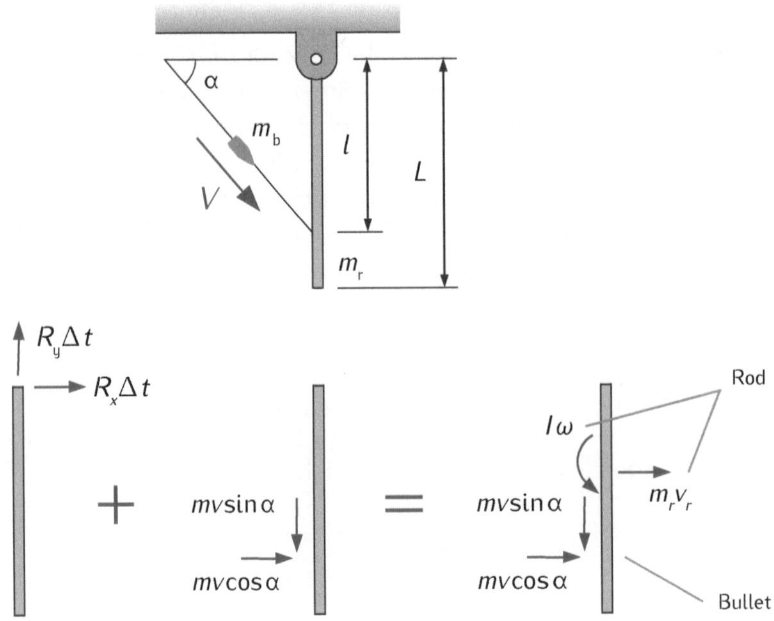

Figure 2.17 A bullet striking a rod.

See Emiliano Mucchi, On the sweet spot estimation in beach tennis rackets, Measurement, Vol. 46, p. 1399–1410, 2013.

The fact that the rod does not rotate when struck at this point l_{cp} means that, if a person is holding the rod (at the support), it would not feel much of the impact. Such a point in the rod is called center of percussion and it is the best position one should hit a ball or hold a hammer in order to deliver most efficiently the impact energy. The next figure helps to understand this concept further.

■

■ A WHEEL CLIMBING A STEP

What is the minimum angular velocity the wheel shown in Figure 2.19 should have so that it climbs the step? Assume there is no slipping between the cylinder and the horizontal plane.

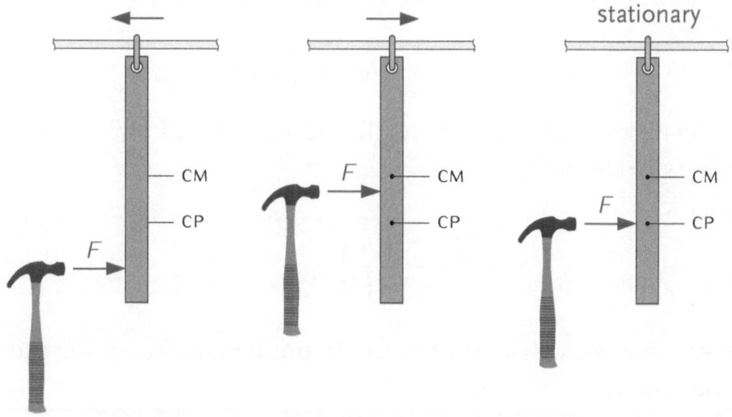

Figure 2.18 Center of percussion.

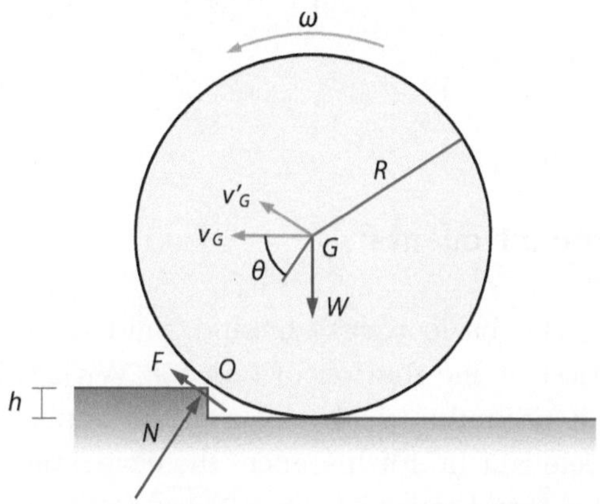

Figure 2.19 A wheel going over a curb.

The angular momentum around the contact point O just before and after impact should be the same. Hence, we write

$$H_o = I_G\omega + mRv_G \sin\theta$$

and

$$H_o' = I_o\omega' = I_o\frac{v_G'}{R} = \left(\frac{I_G}{R} + mR\right) v_G',$$

so it follows that

$$v_G' = \frac{I_G + mR(R-h)}{I_G + mR^2} v_G,$$

which is the velocity of the disk just after hitting the kurb.

The disk starts to climb the step, increasing its potential energy so that
$$\frac{1}{2}I_o\omega'^2 = mgh + \frac{1}{2}I_o\omega''^2.$$
Hence, the residual velocity of the disk after climbing, when defining the radius of gyration as $k_o = \sqrt{I_o/m}$, is
$$v''_G = \sqrt{\left(1 - \frac{Rh}{k_o^2}\right)^2 v_G^2 - 2gh\frac{R^2}{k_o^2}}.$$
If v''_G is set to zero, we obtain the minimum limit velocity for the disk to climb the step as
$$v_{G_{min}} = \frac{2gh\frac{R^2}{k_o^2}}{\left(1 - \frac{Rh}{k_o^2}\right)^2}.$$

■

2.5 Filming impact events

As we have seen, the basic thrust behind rigid body impact analysis is the determination of the motion of objects. With this approach it is possible to have a preliminary idea of a given impact problem, which may prove to be relevant in, for instance, the early stage of a component design or in forensic studies.

In terms of experiments, the determination of the trajectories of bodies can be made is various ways. A laser Doppler velocity meter or accelerometers can be used to obtain relevant information of the velocity and acceleration possessed by objects. These devices will be studied in the next chapter. For now, we present what may become in the near future the most used general purpose measurement system, motion analysis using recorded images from a video camera.

Image processing software requires good resolution images, obtained by filming the event. Current hand held amateur cameras can easily film at a speed of 60 frames per second (fps) with a high resolution. This filming speed can capture details where some features of an impact event are at play. However, in the majority of impact cases, it is necessary to film at a much faster speed, typically in the range of 1000 fps to 50000 fps.

Most of the current filming cameras are based on an image (light) sensor referred to as CMOS sensor, recording the images to a digital memory. In order to uncover details of a filming sequence, the resolution of the camera must be good and image sizes as high as 4 Mpixels are common today. If one films at 10000 fps, files of 40 Gb are generated at each second, which need to be transferred to the PC memory. Hence, we need to exercise care in not recording what is not relevant.

This poses the question of when we should start filming, *ie* when we trigger the camera. In most cases, it is not feasible to start the camera manually and we should rely on various types of trigger systems. For instance, we can set the camera to start recording when a sound is captured by a microphone. In this case, the microphone signal is fed to the input trigger channel of the camera. Other types of trigger include signals coming from light emission diodes, LED, which, when disturbed by the passage of an object, *eg* a projectile, generates a voltage peak.

It is also possible to set a high speed camera in a continuous filming mode and choose it to record a certain number of frames before the trigger event. This pre–trigger technique allows the use of the output of an instrument used to monitor the impact event itself, like a load cell. Once a certain load level is reached, the camera receives the signal and dumps a number of frames recorded prior to and after that instant into its memory.

The data transfer from the camera to the PC can be made in various ways, like using an optical fibre cable. The concern here is the necessity of transferring a large amount of data in a short time.

It is important to note that filming at high speed requires very good illumination of the scene. If common lights are used, they not only overheat the objects being filmed but also flick at the mains frequency. So, other types of illumination are required,*eg* a light–emitting diode (LED) system. Currently, such systems can provide 3.0E6 lux, equivalent to some four times the bright sunlight at a summer mid–day.

2.5.1 Motion analysis

Motion analysis of a film provides time–dependent data of the planar and spatial movement of objects. The set–up is simple, all that is needed is to film the event, or even taking sequential pictures of it. This is all possible nowadays using devices as simple as mobile phone cameras or, at the other extreme, high speed filming.

The use of sophisticated algorithms based on filtering enable us to track an object without attaching markers of any kind to it. All is detected by image analysis. In terms of impact engineering, the idea is to film the event and post–processing it by software. Computer programs are available that, by detecting the position and knowing the time lapsed, can calculate the velocity and acceleration of the targets. These tracking techniques can also be used to measure the deformation of a structure or material sample. This requires a more specific technique called Digital Image Correlation, DIC, that we will have the chance to use in Chapter 7.

We can illustrate this motion track technique here with the impact of two identical spheres. It has been found for the studied case that the coefficient of restitution and the contact time change with the impact velocity, V, according to

H. Minamoto and S. Kawamura, Moderately high speed impact of two identical spheres, International Journal of Impact Engineering, 38, 123–129, 2011. See also A. Aryaei, K. Hashemnia and K. Jafarpur, Experimental and numerical study of ball size effect on restitution coefficient in low velocity impacts, International Journal of Impact Engineering, 37, 1037–1044, 2010.

$$e = 0.62V^{-0.15} \qquad \text{and} \qquad t_c = 80V^{-0.18}.$$

Note then that the coefficient of restitution in not a constant, as assumed in the first approximation models we solved in this chapter. Indeed, it should be clear by now that e is just a scalar used in an oversimplified analysis of some impact problems.

The next figure shows some experimental data adapted from this reference.

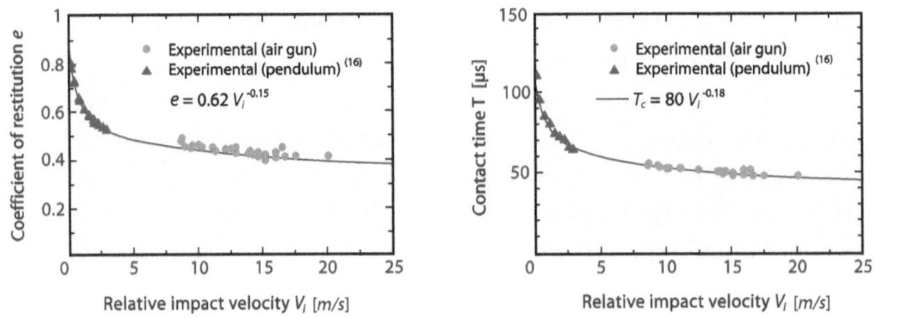

Figure 2.20 Coefficient of restitution and contact time for the impact of two spheres.

2.6 Closure

The aim of this chapter was to give a brief overview of the basic mechanical laws applied to the impact of rigid bodies of simple shape. These laws are tamed to be programmed and from the rigid body shapes it is possible to readily calculate the basic inertia properties. Indeed, there is some software that performs a rigid impact analysis. Some of which take advantage of their finite element structure so that they incorporate rigid body impact analysis features.

But the main interest of this book is to enable one to gain ground on the study of the impact of bodies which deform with the load, which vibrate, dissipate energy, etc... So we leave this area of fundamental mechanics now to occupy ourselves with the area of impact engineering where not only geometry is implied but also material behaviour.

2.7 Problems

1. Two cars are involved in an accident, as shown in the next figure. Considering that the wreckage was moved a distance d at an angle α, calculate the velocities of the cars prior to the collision.

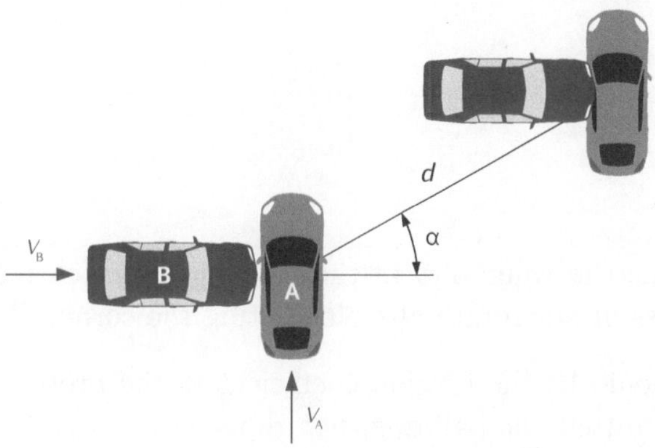

Figure 2.21 The collision of two cars.

2. A bullet of mass m_b is fired against a wood block of mass G, which then slides a distance d on a surface with a friction coefficent μ, see Figure 2.22. What is the bullet velocity?

3. Show that the impulse is the variation of the kinetic energy with the velocity.

4. Show that the total distance travelled by the bouncing ball analysed on page 37 is $s = h\left(1+e^2\right)/\left(1-e^2\right)$.

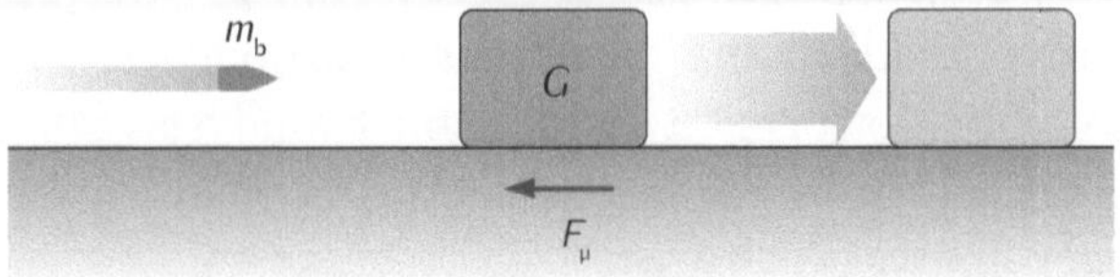

Figure 2.22 A block as a device to measure the bullet velocity.

5. Suppose a projectile with mass m_1 travelling at a velocity v_1, see next figure. The projectile hits the stationary pendulum of mass CM and sticks to it, making the pendulum to raise a distance h from its original position, rotating an angle α. Obtain an expression for the bullet impact velocity as a function of α.

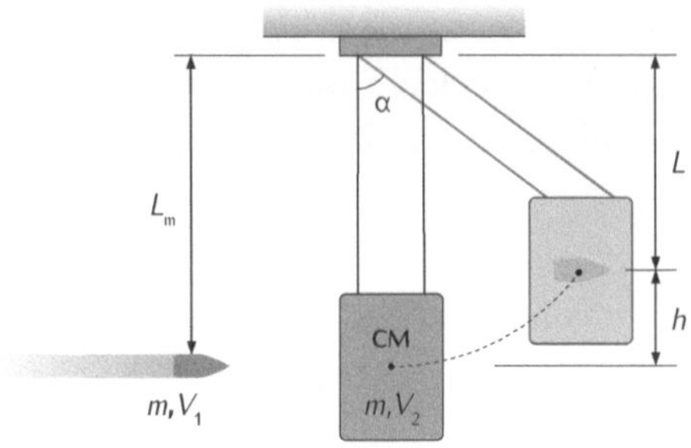

Figure 2.23 A bullet speed measuring device.

6. Determine the value of b in the next figure which makes the rod with mass m not rotate after impacting the corner P.

7. What should be the friction coefficient in the problem studied on page 45 so that the ball does not slip.

8. Considering a row of n spheres held by strings, as in Figure 2.25, obtain the velocity of the last sphere assuming an initial impact velocity V_0 and a coefficient of restitution e between any pair of spheres.

9. A block moving on a frictionless surface with velocity v collides against stationary identical block with an angle α between the common normal direction and the path of the travelling block. Evaluate the deflected angle of the two blocks as a function of the restitution coefficient e.

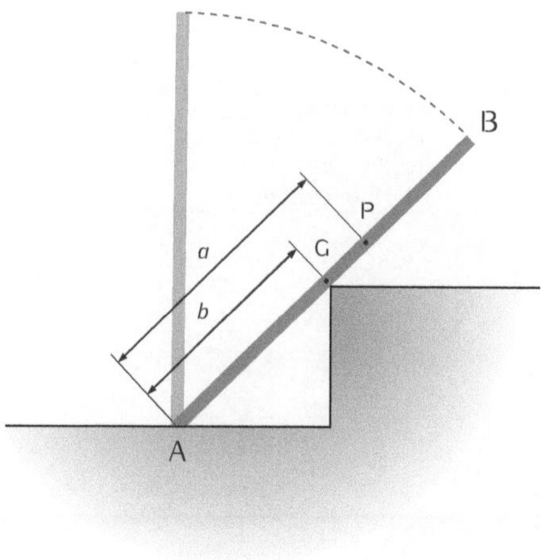

Figure 2.24 A rod hitting a corner.

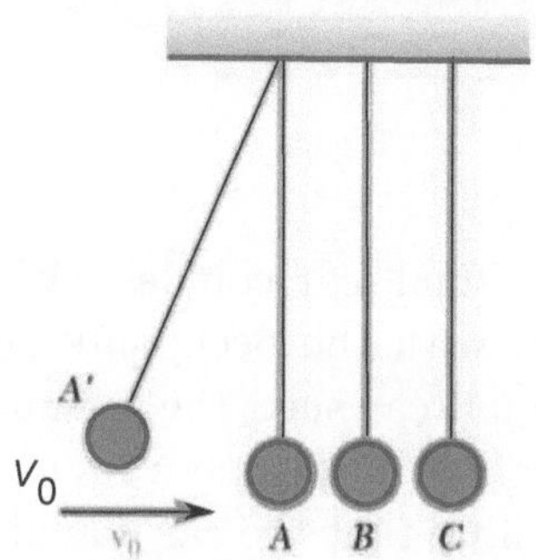

Figure 2.25 A row of spheres being hit by the first one.

10. Based on Figure 2.26, determine the final velocity of block A and sphere B as a function of the angles shown.

11. The block shown in Figure 2.27 of mass m_b and travelling at a velocity V_0 hits the sphere of mass m_s, hanging by string of length L. Evaluate the angle θ and the distance travelled by the block after collision, knowing that the friction coefficient between the block and the surface is μ and that the coefficient of restitution between the sphere and the block is e.

12. The manufacturer of a small car claims that its car is crash worthy, meeting the legislation regarding crash tests. Indeed, by colliding the car against a rigid barrier, the dummies in it did not suffer high deceleration and the car was rated as a five star car. In an

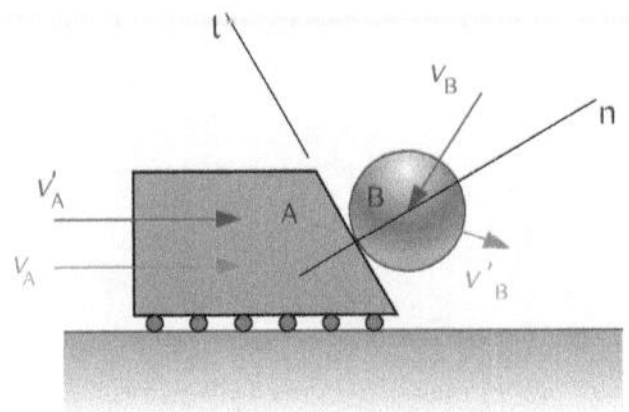

Figure 2.26 A block hit by a sphere.

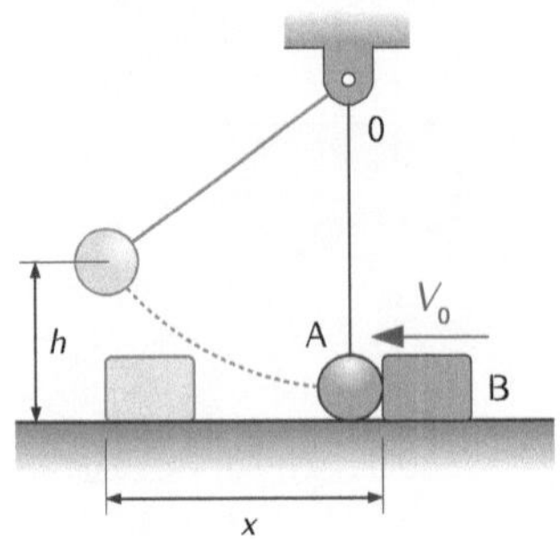

Figure 2.27 A block hitting a sphere.

actual accident, this small car collides with a large family car of mass 3 times larger, with the occupants being seriously injured. The owner of the small car sued the manufacturer, claiming that a five star vehicle cannot impinge on its occupants such a degree of injuries. Do you think that the car manufacturer is well covered by the results of the crash test? Was the consumer misled by the star rated system or his/her technological ignorance was to blame as the cause of this situation? Is it right to blame the large car manufacturer?

CHAPTER 3

One–dimensional elastic waves and impact of bars

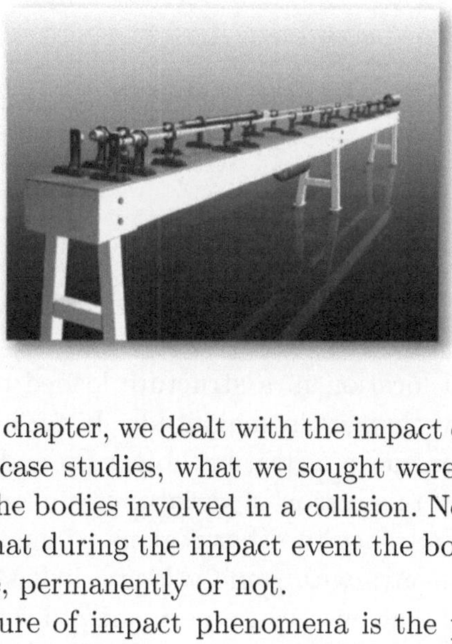

Figure 3.1 This *elastic wave machine* is used for testing materials and operates based on bars impact and uni–dimensional wave propagation.

In the previous chapter, we dealt with the impact of undeformed bodies. In the various case studies, what we sought were the final velocities and trajectory of the bodies involved in a collision. No consideration was given to the fact that during the impact event the bodies and structures change their shape, permanently or not.

A distinct feature of impact phenomena is the propagation of disturbances in a structure when it is loaded dynamically. In many cases,

DOI: 10.1201/9781003485292-3

this transient aspect of the loading process can be neglected at once. For instance, the impact of a heavy mass on a light cantilever can be analysed with no consideration of the wave propagation. On the other hand, many important practical analyses are only meaningful if these disturbances are thoroughly considered. This is the case, for example, of the axial impact of a mass on a tube, when the peak load is sought, as we shall see in Chapter 6. Also, it is the knowledge of the wave propagation theory that allows one to perform material tests at a high loading rate, a topic to be seen in Chapter 7.

We introduce in this chapter many important features of impact phenomena where the propagation of waves is at play. We start by introducing the concept of waves, the dynamic equilibrium equation for a rod and strain–stress waves. We use these concepts to analyse the impact of bars together with the boundary effects on reflection of waves. We also introduce free and forced vibration of bars and the important phenomenon of dispersion of waves, both of material and geometric nature. In the experimental part of the chapter, we describe various motion transducers used in the impact field.

3.1 Strain–stress waves

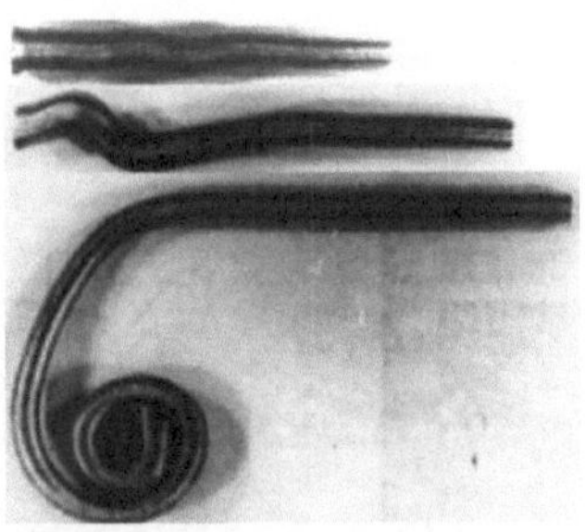

Figure 3.2 The effect of plastic waves on rods impacting a rigid wall. From M. B. Valentine and G. D. Whitehouse, The buckling failure modes of rods with large length–to–diameter ratios impacting rigid plates. In G. C. Sih, editor, Dynamic Crack Propagation, p. 709–721, 1973.

At a given fixed location in a structure loaded dynamically, deformation continuously changes in magnitude. In an analogy with waves in the sea, which can move a floating object up and down, we use the word wave to convey this notion of variable strain–stress in a structure. These disturbances advance to other regions of the structure. A strain or stress wave propagates with some velocity, it has a magnitude, it is a power carrier, and may even cause permanent deformation in a structure, as shown in the previous figure. Waves of high amplitude, capable of straining the material such that it changes permanently its original shape, are called plastic waves and will be studied in Chapter 6. For now, we consider a varying disturbance in time with amplitude small enough not to cause permanent deformation of the material. We refer to it as an elastic wave.

Let us start by analysing the simplest case of a wave propagating in a straight rod, see next figure. A small compressive disturbance applied in the rod will, so to say, push the material ahead, which in turn will move the next material in front, and so on. We have then a compressive propagating pulse, causing a compressive stress, $-\sigma$, at, say, x. As the pulse travels, the stress level at $x = x + \mathrm{d}x$ will change from $-\sigma$ to $\sigma + (\partial\sigma/\partial x)\mathrm{d}x$, such that force equilibrium gives

$$-\sigma A_0 + \left(\sigma + \frac{\partial\sigma}{\partial x}\mathrm{d}x\right) A_0 = \rho A_0 \mathrm{d}x \frac{\partial^2 u}{\partial t^2}$$

or

$$\frac{\partial\sigma}{\partial x} = \rho\frac{\partial^2 u}{\partial t^2},$$

where ρ is the rod density.

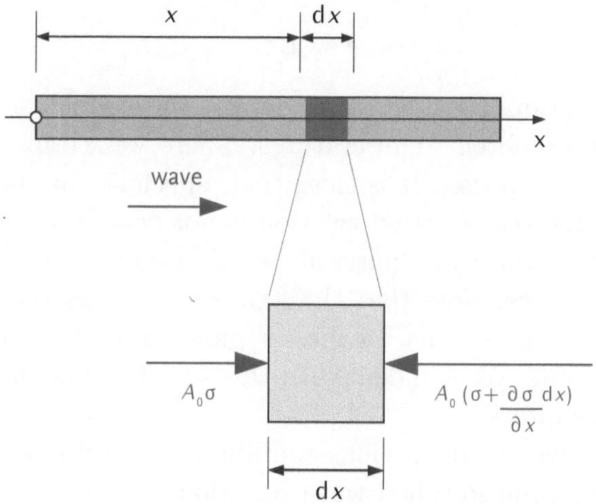

Figure 3.3 Equilibrium of a rod.

Observe that the above equation does not give any information regarding the deformation the wave causes when propagating in the rod. We can evaluate such a variable by invoking the so called *constitutive law*. Differently from the previous chapter, where no attention was given to the material the bodies were made of, we now consider the case when the material is elastic.

A body is said to be elastic when it returns to its original shape after any external load is removed. This means that a load F_1 stretches a body of some displacement u_1 and a load $2F_1$ stretches the body of u_2. This is equal to $2u_1$ for a *linear* elastic material. In any case, after load removal, the body returns to its initial shape.

It is more convenient to work with stress and strain, rather than with load and displacement, given that the former set (latter) does not depend on the area (length) of a rod axially loaded. To our purposes in this chapter, we will work with a uniaxial stress definition simply given by load, F, over initial area, A_0,

No distinction is made in this chapter between initial and current area.

$$\sigma = \frac{F}{A_0},$$

with the strain measure

$$\varepsilon = \frac{\mathrm{d}u}{\mathrm{d}x}.$$

We then define the material constant E, the elastic modulus, so as to relate linearly stress and strain according to

$$\sigma = E\varepsilon.$$

As an illustration, the next figure depicts the elastic region of a stainless steel at two different temperatures. There we see in the detail the elastic modulus definition. It is clear that, specially for the higher temperature case, the stress–strain relation is not exactly linear.

The elastic modulus is a useful concept but detailed measurement reveals that it is a strain dependent parameter for many metals. See Z. Chen, HJ Bong, D. Li, RH Wagoner, The elastic–plastic transition of metals, International Journal of Plasticity, 83, pp. 178-–201, 2016.

Of course that there are materials which are more or less linear than the one in the figure. Note that the figure also shows the behaviour of the material at large strains. In these region the load removal does not lead to zero strains. We will deal with this so called plastic behaviour in the following chapters.

Returning now to our previous equilibrium equation and changing σ by $E\varepsilon$, we obtain the so called wave equation,

$$\frac{E}{\rho}\frac{\partial^2 u}{\partial x^2} = \frac{\partial^2 u}{\partial t^2}.$$

Note that we started with an equilibrium equation and by applying the material constitutive law to it we arrived at the above governing

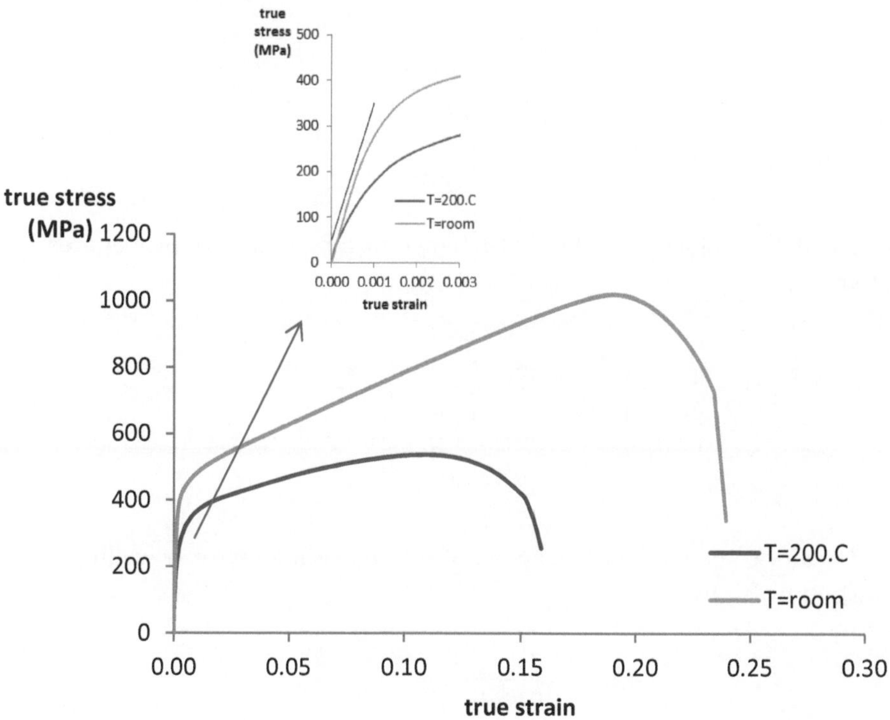

Figure 3.4 Stainless steel behaviour at two different temperatures. The detail shows the elastic region.

equation of a propagating disturbance in a rod made of a linear elastic material. Of course we aim to solve this partial differential equation to find u at some point, x, in the rod at some instant, t.

We can find the solution to this equation by inspection. Considering the function

$$u = f\left(x - \sqrt{E/\rho}\, t\right)$$

and differentiating it twice with respect to time and space, we can see that such a solution satisfies the wave equation. Another function that is also the solution to the wave equation is

$$u = g\left(x + \sqrt{E/\rho}\, t\right).$$

Examining closely one of these solutions, say $u = f$, at $t = t_1$ and $x = x_1$, the resulting magnitude is u_1. Such a magnitude is also attained at $t = t_2$ when $x = x_2$, as depicted in the next figure, so that

$$u_1 = f\left(x_1 - \sqrt{E/\rho}\, t_1\right) = f\left(x_2 - \sqrt{E/\rho}\, t_2\right),$$

or

$$\sqrt{\frac{E}{\rho}} = \frac{x_2 - x_1}{t_2 - t_1},$$

implying that the ratio,

$$c = \sqrt{E/\rho}$$

is in fact the velocity, c, the disturbance of amplitude u_1 propagates in a bar.

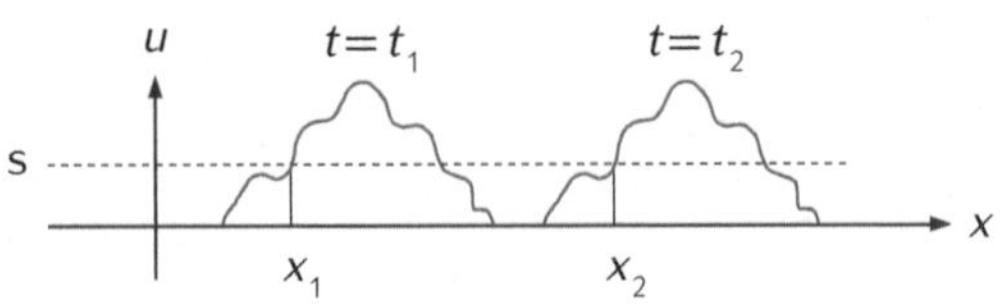

Figure 3.5 A propagating disturbance.

It can be shown that the speed of a longitudinal wave travelling in a continuum is given by

$$c_c = \sqrt{\frac{1-\nu}{(1+\nu)(1-2\nu)}\frac{E}{\rho}},$$

which is 16% larger than c for elastic rods of Poisson ratio, ν, equal to 0.3.

The next table lists typical data for the speed of the elastic wave, c, in some materials. It also lists basic mechanical properties like density, elastic modulus, Poisson ratio, yielding and ultimate stress .

The Poisson ratio is the ratio between transverse and longitudinal contraction of a rod. The yielding stress is the stress limit beyond which the material present permanent deformation after load is removed.

The lateral motion of the bar due to the passage of the wave will cause a non–uniform stress distribution in the cross–section, which then does not remain plane. In practice, such an effect becomes important only for short wave lengths, *ie* high frequency waves. Here no consideration is given to the interaction of a wave with the boundaries of a rod.

For instance, when a bar is laterally constrained such that no lateral deformation is allowed, one has a limiting case of a plane stress problem. In this case, it can be shown that the ratio between the bounded velocity wave, c_b, and the one in a free media, c_c, is

See W. Johnson, Impact Strength of Materials, Edward Arnold, 1972

$$\frac{c_b}{c_c} = \sqrt{\frac{1-\nu}{(1+\nu)\,(1-2\nu)}}.$$

Note the high speeds a wave can reach for high Poisson ratios.

TABLE 3.1 Uni–dimensional wave speed an other mechanical properties of various materials. These constants should be read as average magnitudes since they can vary widely.

Material	E GPa	ρ kg/m^3	c m/s	ν	σ_y MPa	σ_u MPa
Acrylic	3	1180	1594	0.35	124	70
Aluminium	70	2700	5092	0.22	95	110
Beryllium	303	1844	12819	0.03	240	370
Brass	97	8670	3345	0.34	112	37
Cast iron	130	7810	4080	0.21	100	1650
Copper	130	8930	3815	0.34	33	210
Diamond	1200	3520	18464	0.20	1600	2800
Epoxy resin	2.41	2600	963	0.40	85	97
Glass	62	2400	5083	0.25	-	33
Gold	77	19300	1997	0.42	205	220
Ice	9	897	3168	0.33	-	1
Iron	211	7850	5184	0.29	100	350
Lead	14	11400	1108	0.42	19	32
Magnesium	44	1740	5029	0.35	105	205
Molybdenum	330	1022	17969	0.38	415	515
Nickel	207	8880	4828	0.31	59	317
Platinum	171	21500	2820	0.39	140	180
Plexiglass	3	1190	1588	0.37	80	100
Porcelain	104	2400	6583	0.17	-	130
PVC	3	2500	1095	0.40	52	59
Quartz	76	2650	5355	0.17	-	48
Steel	200	7870	5041	0.29	300-1000	500-2000
Titanium	116	4500	5077	0.34	880	950
Tungsten	400	19250	4558	0.28	750	980
Uranium	208	18950	3313	0.30	200	500
Zinc	85	7100	3460	0.33	300	400

It is most important to realise that the wave speed is different from the speed experienced by a particle disturbed by a wave. To gain some insight on this particular, consider the next figure. The wave, by passing in the row of masses and springs, displaces one mass, which displaces the next, and so on. The masses move a small distance with a velocity v, returning to their equilibrium position. The row of masses move one after the other as the wave sweeps through with a velocity c.

Since the material particles change their rest position, there is some straining in the material and, hence, stress. The stress level caused by a wave propagating in a linear elastic material can be quantified considering that

$$u = f\left(x - ct\right), \qquad \frac{\partial u}{\partial x} = \frac{\partial f}{\partial x}, \qquad \frac{\partial u}{\partial t} = v = -c\frac{\partial f}{\partial x},$$

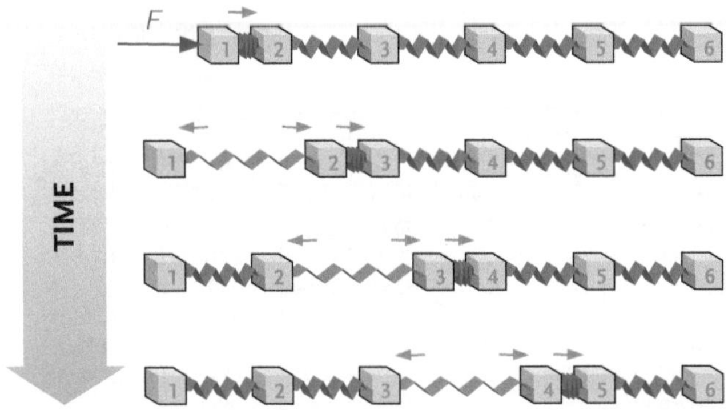

Figure 3.6 A row of masses and springs being disturbed by the sweeping of a wave.

from which it follows

$$\sigma = E\frac{\partial u}{\partial x} = -E\frac{v}{c} = -\rho c v,$$

with ρc being called mechanical impedance.

The above result is an important one: a wave causes stress whose magnitude depends on the velocity of the particle. It is also evident from the previous figure that a longitudinal disturbance moving the particles in the same (opposite) direction it travels is a compressive (tensile) wave. Transverse movement of particles causes material shearing and the associate waves are called shear or torsional waves and are not considered here.

Observe that a free bar travelling in the space with a velocity v_b will have all its particles moving with that velocity and yet the stress in the bar is, of course, zero. The stresses only rise in the bar when there is a gradient of velocity, when a wave has travelled through it, so causing its particles to gain a velocity v_p, such that the stress in this case is $\sigma = \rho c(v_p - v_b)$. In what follows, we refer to the particle velocity as the magnitude of the velocity discontinuity across a wave front.

■ STRESS IN A WIRE

The wire of mass m and density ρ in the next figure is loaded by a falling mass, G, hitting the collar with a velocity V_0. Calculate the stress level in the wire by assuming that all kinetic energy is absorbed uniformly by the rod. What is the stress level using the wave theory?

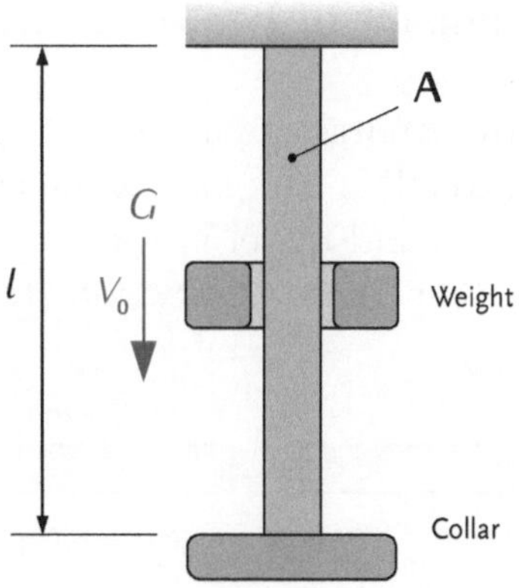

Figure 3.7 A mass hitting a bar.

The strain energy in the wire when equated to the kinetic energy gives

$$Al\int \varepsilon \mathrm{d}\sigma = \frac{Al\sigma^2}{2E} = \frac{GV_0^2}{2} \quad \rightarrow \quad \sigma = \sqrt{\frac{GE\rho}{m}}V_0.$$

The stress level according to the wave theory is

$$\sigma_d = \sqrt{E\rho}V_0,$$

so that

$$\frac{\sigma_d}{\sigma} = \sqrt{\frac{m}{G}}.$$

For a small striking mass, G, the actual stress can be many times larger than the one obtained by the strength of materials approach. Note that the stress level depends largely on the impact velocity.

■

The problem above is not completely solved because wave reflection at the extremes of the bar has not been taken into account. We will come back to this problem in the sequence and will be able to understand what the Hopkinsons, father and son, at the turn of the nineteenth–twentieth century, concluded: that the wire breaks at the suspension point and not at the impact point. They found that the wire rupture does not depend on the falling mass but rather on its velocity.

3.2 Axial collinear impact of two equal semi–infinite bars

Let us now examine the frontal impact of two semi–infinite bars, as depicted in the next figure. The bars have same cross–section areas, are made of different materials and travel in the same direction at different velocities, with $v_2 > v_1$. The final velocity and stress levels are sought.

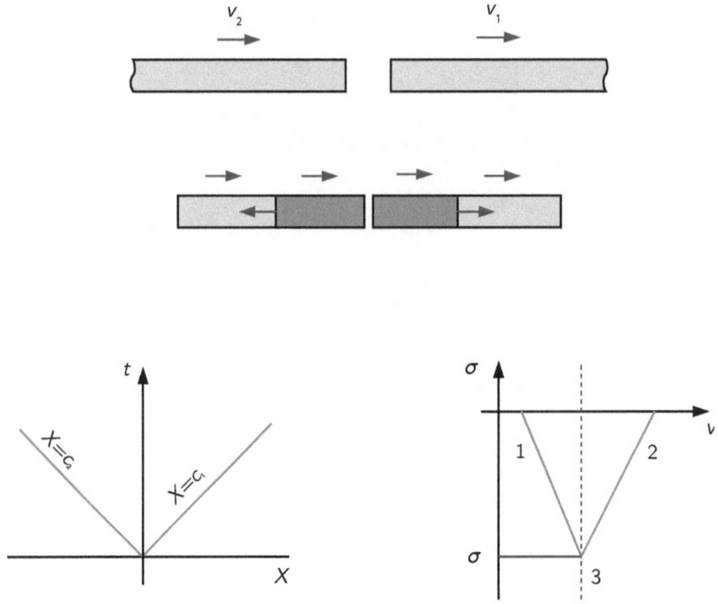

Figure 3.8 Impact of two semi–infinite bars and their characteristic plane.

Upon impact, two elastic compressive waves are generated and travel in the bars with velocities c_1 and c_2 in the directions shown in the previous figure. Since momentum is transferred in the impact event from one bar to the other, the particles speed in the material engulfed by the waves is the same, v_f, and so are the stress levels in this region, a result of force equilibrium. We can then write,

$v_f < v_2$ and the waves cause compressive (negative) stresses. These equations can be deduced using stress–velocity plots.

$$\sigma_f = \rho_1 c_1 (v_1 - v_f) = \rho_2 c_2 (v_f - v_2),$$

from which the final velocity is

$$v_f = \frac{\rho_1 c_1 v_1 + \rho_2 c_2 v_2}{\rho_1 c_1 + \rho_2 c_2}.$$

Substituting this back in the stress equation yields

$$\sigma_f = \frac{v_2 - v_1}{\frac{1}{\rho_1 c_1} + \frac{1}{\rho_2 c_2}}.$$

Observe that the problem does not depend on the length of the bar (or the mass) and the focus is on the behaviour at the contact region. Also, the approach is somewhat distinct from the analysis of the impact of rigid bodies. It is interesting to note that in this impact event the bars do not separate after the initial impact. This is due to the fact that the compressive waves do not return to the impact interface since the bars are semi–infinite in length.

We can plot the stress state in the bars as a function of the velocity, as also depicted in the previous figure. We observe that bar 1 (2) increases (decreases) the velocity to the final value of v_f, with the stresses growing suddenly (discontinuously) from 0 to σ_f. We also plot in the same figure the space–time diagram of this impact event (also called Lagrange diagram or characteristic plane), which has as vertical axis, time, and as horizontal, the distance, *ie* a given position of the wave front, with the slope indicating the wave speed in the rod.

It is worth noting the limit case when the bars are identical and are heading one to each other at the same velocity, $v_1 = -v$ and $v_2 = v$, so that $v_f = 0$ and $\sigma_f = -\rho c v$. This is the same case as when one bar hits a rigid wall.

■ STRESS IN A DROPPED ROD

Calculate the stress level reached in an aluminium rod which was dropped vertically from a height of 1.8 m and impacted a steel anvil.

Using the above formula one obtains at once

$$\sigma = \frac{\sqrt{2g1.8}}{\frac{1}{2660\times5094} + \frac{1}{7757\times5155}} = 60\text{MPa}.$$

This stress magnitude is such that the bar will not be dented in the impact end since plastic strains in aluminium occur only at higher stress levels, say 200 MPa.

For a steel bullet impinging normally upon a sheet of water at a velocity of 500 m/s, the stress level in the bullet is

$$\sigma = \frac{500}{\frac{1}{1000\times1500} + \frac{1}{7757\times5155}} = 723\text{MPa}.$$

Such an intensity easily deforms the cartridge permanently.

■

3.3 Reflection of waves

Planar waves reaching a material interface in a solid body may generated shear and dilatational waves.

We explore in this section what happens when a uniaxial wave reaches the end of a bar. We analyse the two most simple cases, namely when the wave reaches a rod with a free and a fixed end. Consider the situation when a uniaxial compressive wave reaches the end of a free rod. In this case, we argue that there should be no stress there since the end is free. We need then to inquire into which situation the stress level can be set to zero even when elastic waves are arriving at the end of the bar. This can be accomplished by adding a hypothetical tensile pulse of the same magnitude as the compressive pulse but travelling in the opposite direction, as illustrated in the next figure.

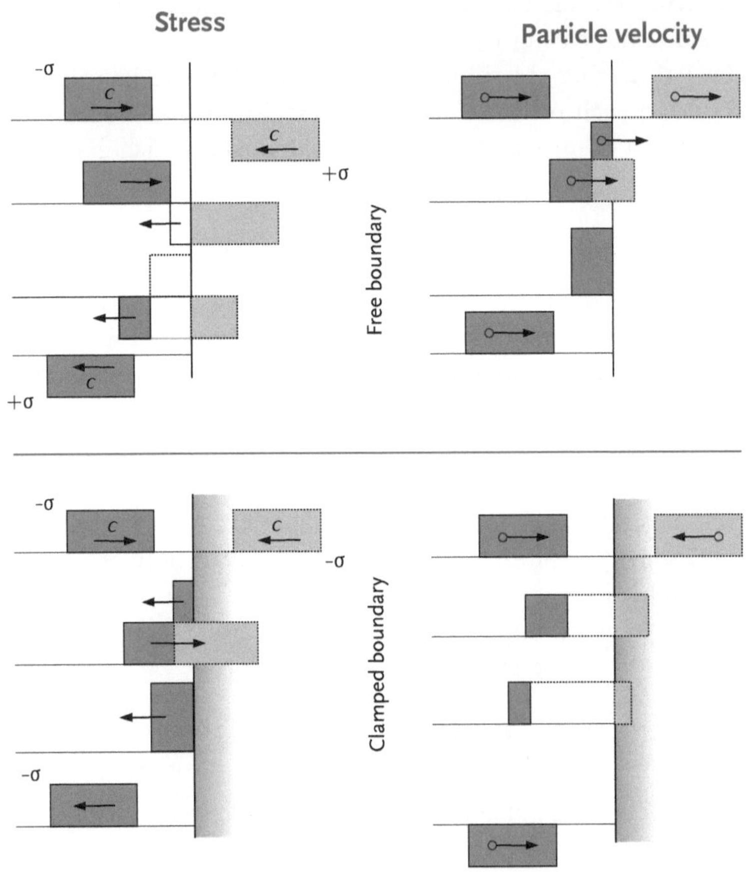

Figure 3.9 Reflection of waves in free (top) and fixed (bottom) boundaries. Adapted from M.A. Meyers, Dynamic Material Behavior, Wiley, 1994.

When the two pulses overlap each other at the free end, the net stress is zero. Remember now that the compressive pulse moves the particles in the same direction it propagates, while tensile waves move the particles opposite to their propagation direction. Applied to the present case, the net velocity of the particles at the free end will double, a consequence of the stress level to be zero at this very same free end. It should now be clear that a compressive pulse is reflected as a tensile pulse in a free end and that the speed of the particles engulfed by the wave returning on itself at the boundary doubles.

Consider now the case of an elastic wave reaching a fixed end of a bar. If the bar end is fixed, the particles speed there must be zero. This is only possible, when applying the previous reasoning of wave superposition, if the reflected wave moves the particles in the opposite direction to the movement caused by the incident wave, as depicted in the previous figure. This leads to the conclusion that a tensile pulse is reflected as a tensile pulse by a fixed end. Therefore, the stress–strain amplitude of the reflected pulse doubles.

A more mathematical way of obtaining the stress amplitudes and particles velocities in a boundary is by considering that there the incident, i, reflected, r and transmitted, t, velocities and stresses should satisfy

$$v_i = v_t - v_r, \qquad \sigma_i = \sigma_t - \sigma_r,$$

which gives

$$\sigma_t = \sigma_i \left(\frac{2\rho_2 c_2}{\rho_1 c_1 + \rho_2 c_2} \right) \qquad \sigma_r = \sigma_i \left(\frac{\rho_2 c_2 - \rho_1 c_1}{\rho_1 c_1 + \rho_2 c_2} \right),$$

and

$$v_t = v_i \left(\frac{2\rho_1 c_1}{\rho_1 c_1 + \rho_2 c_2} \right) \qquad v_r = v_i \left(\frac{\rho_1 c_1 - \rho_2 c_2}{\rho_1 c_1 + \rho_2 c_2} \right),$$

when using $v_i = \sigma_i/\rho_1 c_1$, $v_t = \sigma_t/\rho_2 c_2$ and $v_r = -\sigma_r/\rho_1 c_1$, with the negative indicating that the tension wave displaces the particles in the opposite direction of its propagation direction.

In the case an incident wave encounters a free surface, we set $\rho_2 = 0$ giving

$$\sigma_t = 0,\ \sigma_r = -\sigma_i \qquad \text{and} \qquad v_t = 2v_i,\ v_r = v_i.$$

For the clamped rod end case, it is as if the incoming wave reaches a rigid surface, *ie* $\rho_2 \to \infty$, so that

$$\sigma_t = 2\sigma_i,\ \sigma_r = \sigma_i \qquad \text{and} \qquad v_t = 0,\ v_r = -v_i.$$

■ WHERE A WIRE BREAKS

Considering again the problem of a falling mass hitting from a height H the end of a wire, is the string more likely to break at its support or at the impact point?

Upon impact of the mass with the stopper, a (assumed elastic) wave of intensity

$$\sigma = \rho c\sqrt{2gH}$$

will be generated, propagating upwards. When it meets the fixed end, this wave will double in amplitude.

This stress level may be high enough to cause the wire to break at the support.

■

3.4 Impact of two finite bars

Consider now the impact of two finite bars of length L. Firstly, let us take two identical bars heading one to each other at the same speed, V_0. When the two bars impact, the contact interface is immediately brought to rest and two identical compressive pressure pulses are generated, as illustrated in the next figure. As this loading wave travels in the bar, it brings its particle velocity to zero.

The two compressive waves propagate each one in one bar up to their ends. At time $t = L/c$, the bars are stationary, with all their particles still and fully compressed. This means that all the kinetic energy, $AL\rho V_0^2/2$, from the one bar motion has been changed to elastic strain energy, $AL\sigma^2/2E$.

The compressive waves are then reflected by the free ends as tensile waves of same absolute amplitude. The tensile waves, in their way to the impacted interface, unload the now compressed bar (so the name unloading waves).

As the waves sweep through the bars, the direction of the particle speed reverses. At $t = 2L/c$, the bars are completely unloaded and start to move away from each other at a velocity V_0 so that the coefficient of restitution is 1, as it could be inferred from a rigid body impact analysis.

Let us investigate the more complex case of two bars, made from the same material, same cross-section but different lengths, colliding heads on with the same velocity, V_0, as in Figure 3.11.

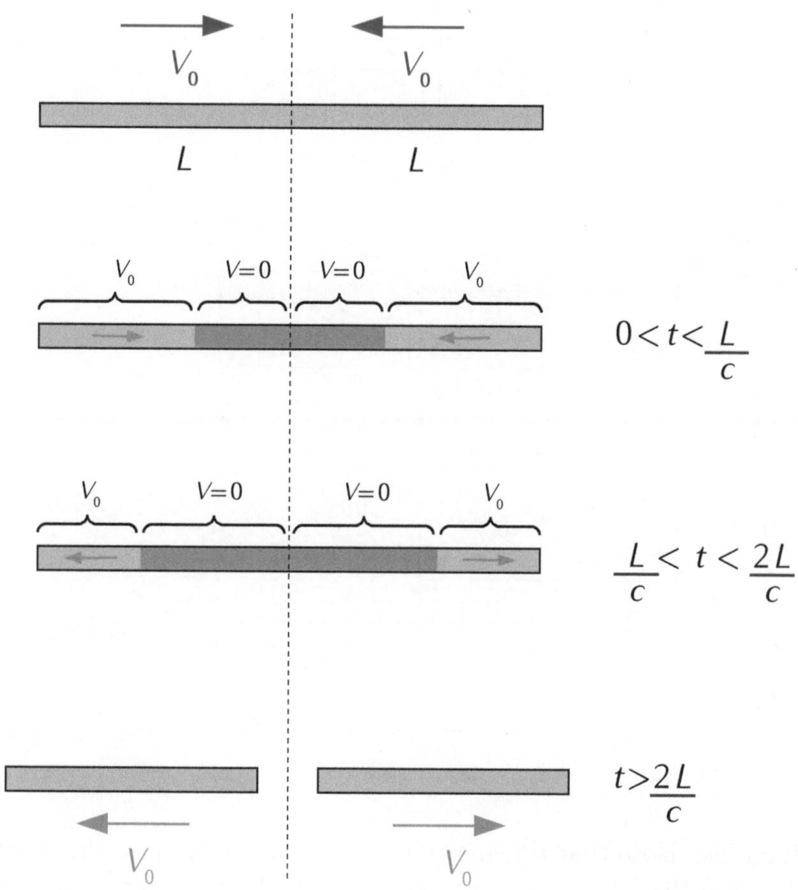

Figure 3.10 Two equal bars approaching and impacting each other with velocity V_0.

As before, equal lengths of the bars are brought to rest. However, the pulse in the shortest bar arrives first at the interface, after being reflected as a tensile pulse in the short end. At this moment, the bars are still in tight contact. The arriving pulse of the shortest bar comes unloading it, *ie* displacing the particles to the right.

Because the particles in the shortest bar at the interface are moved to the right, the particles of the left longer bar at the interface, being compressed, are free to follow such a motion, *ie* move to the right. This means that contact between the two bars does not cease at $t = 2L_2/c$. It does at $t = 2L_1/c$, when the unloading wave in the longest bar arrives at

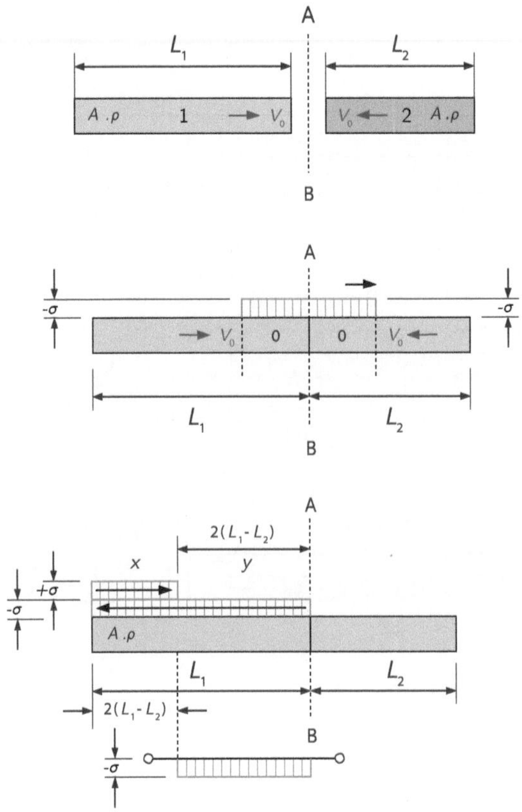

Figure 3.11 Head on impact of two different finite bars.

the interface. Note that this incoming wave cannot jump to the shortest bar because the latter is stress free and with the particles velocities all reversed in relation to the initial configuration, *ie* the short bar rebounds at a velocity V_0. If the long bar were infinite, the shorter bar would never bounce back, maintaining contact indefinitely.

The longer bar has now two waves going to one extreme and bouncing back to the other, all this while it is moving as a rigid body in the same direction as the shorter bar or opposite to it, depending on its length. In this particular case, the longer bar vibrates as it travels, while the short bar moves as a rigid body.

If the two approaching bars were of the same length, they would be stress free at $t = 2L/c$. In the present case though, the shorter bar is completely unloaded at $t = 2L_2/c$. At this instant, the wave in the long bar may have had time to return from its left extreme, unloading part of it, such that the net compressive stress encompasses a length of $2L_2 - L_1$, as shown in the previous figure. This stressed part of the bar did not yet acquire a rigid body velocity and its kinetic energy is $A\rho(2L_2 - L_1)V_0^2/2$. If this is added to the current kinetic energy of the short bar, $A\rho L_2(-V_0)^2/2$, we will have a total value that does not match the kinetic energy before impact. This difference, of $A\rho(2L_1-2L_2)v^2/2 = A2(L_1 - L_2)\sigma^2/2E$, is just the elastic strain energy accumulated in the stressed longer bar.

For the particular and peculiar case when $L_1 = 2L_2$, depicted in the next figure, the velocity of the bars according to the rigid body mechanics in Chapter 2, with $e = 1$, resulting in

See W. Johnson, Impact Strength of Materials, Edward Arnold Press, 1972.

$$v_{1_f} = \frac{V_0}{3} \quad \text{and} \quad v_{2_f} = -\frac{5V_0}{3},$$

which seem reasonable values if it were not for the fact that they are wrong. We now know that at $t = 2L_2/c$, the short bar is stress free and all its particle speed is reversed. This unloading wave jumps to the longer bar such that, at $t = 2L_2/c + L_1/2c = 1.5L_1/c$, the half left hand of the long bar is stress free and the particles have velocity pointing to the left.

On the other hand, the other half right hand of this same long bar is also stress free but with the particles velocity pointing to the right. At $t = 2L_1/c$, the long bar will be all in tension. The half left hand of the long bar will be moving to the right and the other right half moving to the left. At $t = 2.5L_1/c$ the bar is again stress free and at $t = 3L_1/c$ fully compressed. Of course that, at $t = 2L_1/c$, the short bar starts to move to the right, rebounding from this impact event at a velocity $-V_0$. In summary, this particular case of collision has such a geometry that the long bar simply vibrate in its natural frequency, c/L, without moving, which is somewhat unexpected and certainly different from the results the rigid body analysis can give. This is to say that, if one hits a bar with another one half of its length, the shorter will bounce back while the longer one will remain still. Note that the coefficient of restitution for this problem is

$$e = -\frac{\text{vel. sep.}}{\text{vel. app.}} = -\frac{0 - V_0}{V_0 - (-V_0)} = \frac{1}{2}.$$

Let us now briefly study the collinear impact of two bars of different lengths and materials, moving to each other at different velocities, see Figure 3.13.

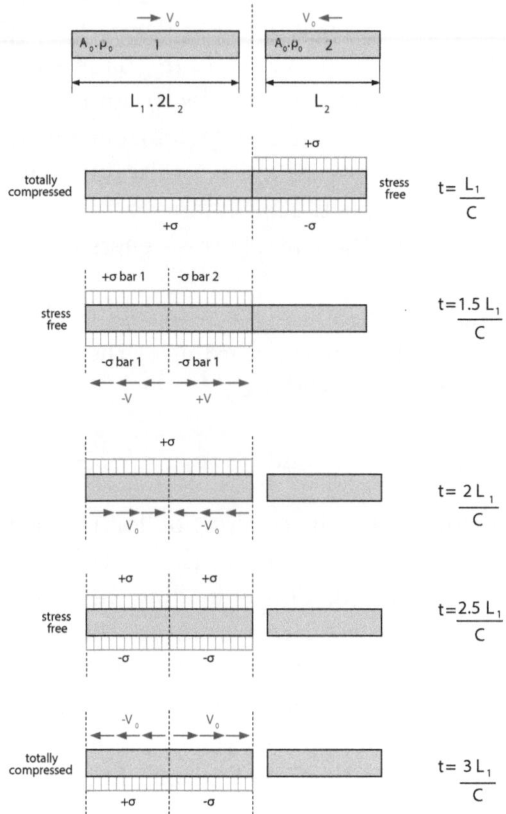

Figure 3.12 Impact of two bars with $L_1 = 2L_2$.

Force equilibrium at the impact interface requires that

$$A_1\sigma_1 = A_2\sigma_2 \quad \rightarrow \quad A_1\rho_1 c_1 (v_1 - v) = A_2\rho_2 c_2 (v - v_2),$$

v being the common velocity of the bars region engulfed by the compressed elastic waves given by

$$v = \frac{v_2 + \frac{A_1\rho_1 c_1}{A_2\rho_2 c_2} v_1}{1 + \frac{A_1\rho_1 c_1}{A_2\rho_2 c_2}}.$$

It follows that the stresses are

$$\sigma_1 = \frac{\rho_1 c_1 v_1}{1 + \frac{A_1\rho_1 c_1}{A_2\rho_2 c_2}} \left(1 - \frac{v_2}{v_1}\right), \quad \sigma_2 = \frac{\rho_2 c_2 v_2}{1 + \frac{A_1\rho_1 c_1}{A_2\rho_2 c_2}} \left(\frac{v_1}{v_2} - 1\right) \frac{A_1\rho_1 c_1}{A_2\rho_2 c_2}.$$

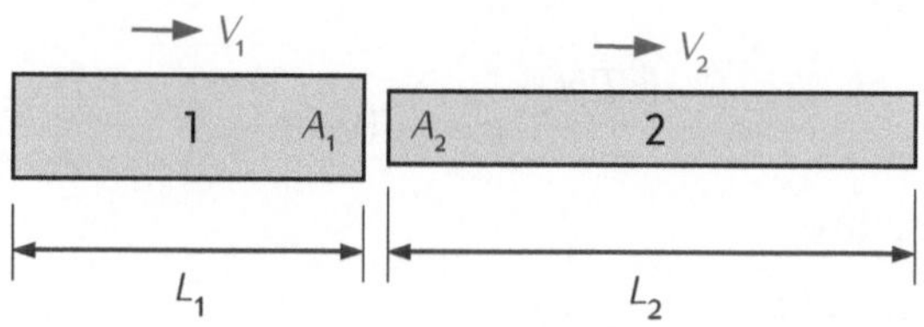

Figure 3.13 Colinear impact of two different bars.

In the special case that $\rho_1 c_1 = \rho_2 c_2$, the velocity around the impact region is

$$v = \frac{v_2 + v_1 A_1/A_2}{1 + A_1/A_2}.$$

3.5 Free vibration of bars

The wave equation and its wave solution seen before are rather useful to study the propagation of waves. We now know that the waves propagating back and forth in a finite bar will cause it to vibrate.

There are many practical examples where, instead of colliding, a given bar may be loaded by some pulse or other initial condition and its dynamics is of interest to us. We can expand our analysis of waves in bars by exploring the so called free vibration solutions to the governing equation

$$\frac{E}{\rho}\frac{\partial^2 u}{\partial x^2} = \frac{\partial^2 u}{\partial t^2}.$$

Obviously, the displacement depends on time and position and it can be expressed as a combination of two functions, $U(x)$ and $T(t)$, such that,

$$u(x,t) = U(x)T(t).$$

Accordingly, the wave equation now reads

$$\frac{c^2}{U(x)}\frac{\mathrm{d}^2 U(x)}{\mathrm{d}x^2} = \frac{1}{T(t)}\frac{\mathrm{d}^2 T(t)}{\mathrm{d}t^2}.$$

Note that its left side depends only on x, while the right one on t. This is only possible if these sides equal to a constant term, that we set as $-\omega^2$ for convenience. This leads to the equations

$$\frac{\mathrm{d}^2 U(x)}{\mathrm{d}x^2} + \frac{\omega^2}{c^2}U(x) = 0$$

and

$$\frac{d^2T(t)}{dt^2} + \omega^2 T(t) = 0,$$

The solution to these equations can be found by inspection. For a more rigorous proof, see any basic book on calculus.

whose solutions are

$$U(x) = a_1 \cos\frac{\omega x}{c} + a_2 \sin\frac{\omega x}{c} \quad \text{and} \quad T(t) = a_3 \cos\omega t + a_4 \sin\omega t,$$

giving

$$u(x,t) = \left(a_1 \cos\frac{\omega x}{c} + a_2 \sin\frac{\omega x}{c}\right)(a_3 \cos\omega t + a_4 \sin\omega t)$$

with the constants a_1, a_2, a_3, a_4 being obtained from the boundary conditions, *ie* the way the rod is held in space, and from the initial conditions, *ie* the way the rod is set in motion at $t = 0$, respectively.

Consider, as an example of application of the above solution, a horizontal bar of length L supported by strings and lightly hit at one of its end. This will set it to vibrate and the aim is to calculate the natural frequencies and vibration modes of the bar in its free condition.

The boundary conditions for this case is that of a free–free bar, meaning that at the extremes there is no stress. Since $\sigma = E\varepsilon$ and $\varepsilon = du/dx$, we can write

$$\frac{dU(0,t)}{dx} = 0 \quad \text{and} \quad \frac{dU(L,t)}{dx} = 0.$$

The first equation at $x = 0$ implies that

$$\frac{dU(0)}{dx} = \frac{-a_1\omega}{c}\sin\frac{\omega x}{c} + \frac{a_2\omega}{c}\cos\frac{\omega x}{c} = 0 \to a_2 = 0$$

and the second boundary condition yields

$$\sin\frac{\omega L}{c} = 0,$$

since $a_1 = 0$ implies a trivial solution. The above equation has the solution

$$\frac{\omega_n L}{c} = n\pi, \qquad n = 1, 2, 3, \ldots$$

ie the bar can vibrate at various natural frequencies, each one associated to the respective natural mode

$$U_n(x) = a_1 \cos\frac{n\pi x}{L}.$$

The complete motion of the bar is given by

$$u(x,t) = \sum_{n=1}^{\infty} a_1 \cos \frac{n\pi x}{L} T(t),$$

with $T(t)$ being the harmonic function seen before. Observe that this solution depends on a_1, obtained once a known excitation is applied to the bar

The above example indicates that, for each natural frequency, also called eigenvalues, there is an associate natural mode of vibration.

The natural frequency is a function of the material and geometry. Therefore, it is possible to have structures with very high frequency, *ie* stiff, but made of more flexible materials.

Note the use of

$$u(x,t) = \sum_{n=1}^{\infty} U_n(x) T_n(t),$$

ie the bar will vibrate in a way that entails a combination of the basic vibration modes. This advanced idea is known as the expansion theorem. A rigorous proof of this theorem is not given here but we indicate that the eigenvectors are orthogonal and form a basis in the n-dimensional space. The solution of a linear dynamic problem is a combination of these eigenvectors, the weight of each vector being an unknown determined by the boundary and initial conditions of a given problem.

It is very important to note that this vibration solution for the wave equation leads to the setting in the bar, or in other more complex structures, of a standing wave. This is a wave that travels forth and back in the rod, establishing its vibration pattern.

We enquire now what it would be the speed of this standing wave, c_s, in a clamped rod of length L

A standing wave does not change its shape on time, only its magnitude. See also G. Szuladziński, Formulas for Mechanical and Structural Shock and Impact, CRC Press, 2010.

The natural mode of vibration of this both end clamped bar is

$$U_n(x) = a_1 \sin \frac{n\pi x}{L},$$

and its frequency equation reads

$$\sin \frac{\omega_n L}{c} = 0,$$

from which

$$\omega_n = \frac{n\pi c}{L}.$$

Now, the mode of vibration establishes a standing wave whose period is

$$\frac{nT}{2}$$

which is the time, t, a disturbance takes to travel the whole bar length at a speed c_s, such that

$$t = \frac{L}{c_s} = \frac{nT}{2},$$

or

$$c_s = \frac{2L}{n}\frac{1}{T} = \frac{2L}{n}\frac{\omega_n}{2\pi} = \frac{2L}{n}\frac{1}{2\pi}\frac{n\pi c}{L} = c.$$

This is valid for a clamped bar and similar reasoning for other configurations will show the same, so that the speed of the standing wave is the same as the speed of the elastic wave.

■ NATURAL FREQUENCIES OF A ROD–MASS

A mass G is attached at the free end of a vertical rod fixed on a ceiling, see next figure. Evaluate the influence of the mass on the natural frequencies of the rod.

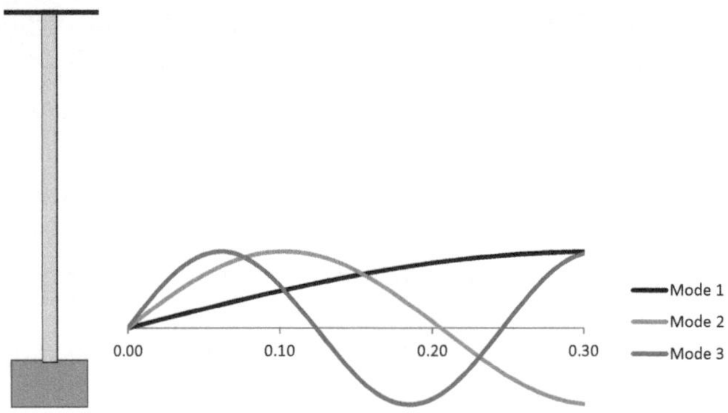

Figure 3.14 The first three natural modes of axial vibration of a 30 cm long, 10 mm diameter, steel bar with an attached mass of 500 g. The associated natural frequencies are 4108.3 Hz, 12324.8 Hz and 20557.3 Hz.

The figure indicates that the boundary conditions at the bar support is

$$u(0,t) = 0$$

and at the mass extreme is

$$F(L,t) = A\sigma(L,t) = AE\frac{\partial u(L,t)}{\partial x},$$

which equals to

$$= -G\frac{\partial^2 u(L,t)}{\partial t^2}.$$

By using these conditions in the equilibrium equations we obtain the transcendental frequency equation

$$\beta_n \tan \beta_n = m_r, \quad n = 1, 2, \ldots$$

with $\beta_n = \omega_n L/c$ and $m_r = m/G$, the ratio between the rod mass, m, and the suspended mass, G. As before, each n represents one of the infinity natural frequencies, an eigenvalue.

For a 30 cm long, 10 mm diameter, steel bar with an attached mass of 500 g, the previous figure shows the first three natural frequencies. Note that the amplitudes shown in the figure should be interpreted as a measure of the no absolute displacements of the various portions of the bar as it is elastically stretched and compressed.

■

■ A MASS RELEASED FROM A ROD

A bar is vertically fixed on a ceiling and it has elastic modulus E, density ρ and cross section A. At its opposite extreme it is loaded by a mass G via a massless string, as depicted in the next figure. Obtain the dynamic displacement field of the bar when the string is cut at $t = 0$.

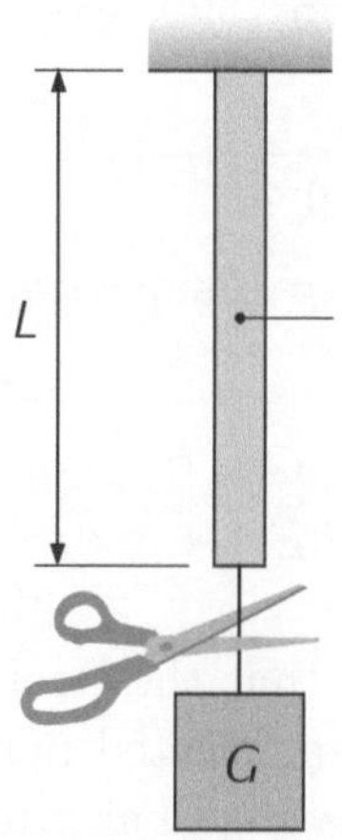

Figure 3.15 A vertical bar loaded by a mass.

A known boundary condition for this problem is that, at the support,

$$u(x = 0, t) = 0,$$

which implies that, in our general solution $u(x, t)$ seen before, $a_1 = 0$.

We also know that, once the string is cut, the bar is set in motion. The bar end at $x = L$ is now free to vibrate and there is no more stress there, so that

$$\frac{\mathrm{d}u(L,t)}{\mathrm{d}x} = \sum_{n=1}^{\infty} \frac{\omega_n}{c} a_2 \cos \frac{\omega_n L}{c} T_n(t) = 0,$$

giving the natural frequencies of the bar as

$$\omega_n = \frac{n\pi c}{2L}, \qquad n = 1, 3, 5, \ldots$$

The bar motion reads now

$$u(x,t) = \sum_{n=1}^{\infty} \sin \frac{n\pi x}{2L} \left(a_3 \cos \omega_n t + a_4 \sin \omega_n t\right),$$

with the constant a_2 being incorporated by a_3 and a_4.

We also have that the strain ε caused by the hanged mass, the bar weight not being considered, affects the axial displacement such that, at $t = 0$,

$$\varepsilon x = \sin \frac{n\pi x}{2L} a_3,$$

which can be multiplied by $\sin \frac{n\pi x}{2L}$ and integrated from 0 to L to give

$$a_3 = \frac{8\varepsilon L}{n^2\pi^2} (-1)^{\frac{n-1}{2}}.$$

Finally, considering that $a_4 = 0$ since $\mathrm{d}u(x,0)/\mathrm{d}t = 0$, we have the final motion of the bar as

$$u(x,t) = \sum_{n=1,3,5,\ldots}^{\infty} \frac{8L}{n^2\pi^2} \frac{Gg}{EA} (-1)^{\frac{n-1}{2}} \sin \frac{n\pi x}{2L} \cos \frac{n\pi ct}{2L},$$

where use was made of the fact that the strain is $\varepsilon = \sigma/E = Gg/EA$.

The next figure plots for some elected parameters ($L = 1$ m, $A = 1$ mm^2, $G = 1$ kg) the motion of a steel bar after the string is cut.

We see from this solution that, after the string is cut, the bar will not vibrate at a definite frequency but rather with a pool of natural frequencies, each one cooperating with a given amplitude to the overall motion of the bar. However, we note that the cooperation of high natural modes is small so that with only the first term the series above gives

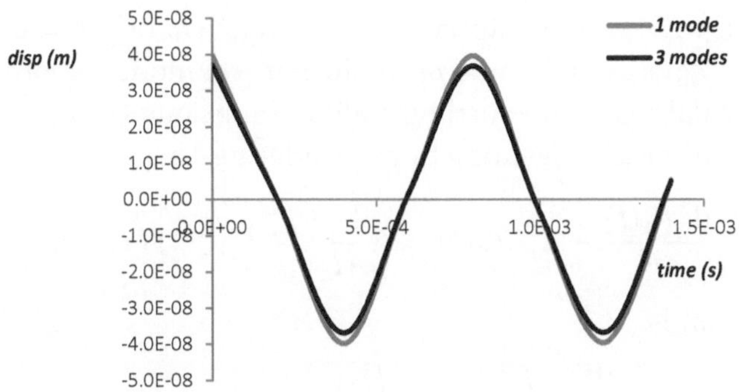

Figure 3.16 Motion of the extremity of a bar after a mass connected there is cut.

already a very good estimation of the bar motion, as indicated in the figure.

■

3.6 Forced vibration of bars

In the examples of the previous section, loading was imposed to the bar via an initial condition. It was not necessary to have in the wave equation the force term, $f(x,t)$. For a forced vibration however, $f(x,t)$ comes into play and we need to deal with the partial differential equation

$$\frac{E}{\rho}\frac{\partial^2 u(x,t)}{\partial x^2} + \frac{1}{\rho A} f(x,t) = \frac{\partial^2 u(x,t)}{\partial t^2}.$$

To solve it generically, let us assume, as we did before, that the solution is of the type

$$u(x,t) = \sum_{n=1}^{\infty} U_n(x) T_n(t),$$

which can be substituted in the governing equation, multiplied by $U_m(x)$ and integrated along the bar length, resulting in

$$\sum_{n=1}^{\infty}\left[-\frac{E}{\rho}T_n(t)\int_0^L \frac{\mathrm{d}^2 U_n(x)}{\mathrm{d}x^2}U_m(x)\mathrm{d}x + \frac{\mathrm{d}^2 T_n(t)}{\mathrm{d}t^2}\int_0^L U_n(x)U_m(x)\mathrm{d}x\right]$$
$$= \frac{1}{\rho A}\int_0^L U_n(x) f(x,t)\mathrm{d}x.$$

Now, we will see in Chapter 4 that $\int U_n(x)U_m(x)\mathrm{d}x = 0$ for $n \neq m$ and $\int U_n(x)U_m(x) = 1$ for $n = m$, *ie* the natural modes of vibration form an orthonormal basis, this orthogonality conditions being also valid for the derivatives of the eigenfunctions, rendering the above equation as

$$\frac{\mathrm{d}^2 T_n(t)}{\mathrm{d}t^2} - \omega_n^2\, T_n(t) = \frac{1}{\rho A L}\int_0^L U_n(x) f(x,t)\mathrm{d}x.$$

This is known as the Duhamel integral.

whose solution is

$$T_n(t) = \frac{1}{\rho A \omega_n}\int_0^L U_n(x)\int_0^t f(x,\tau)\sin\omega_n(t-\tau)\mathrm{d}\tau\mathrm{d}x,$$

with τ in the present context being an auxiliary variable. The final sought forced solution becomes then

$$u(x,t) = \sum_{n=1}^{\infty}\frac{U_n(x)}{\rho A\omega_n}\int_0^L U_n(x)\int_0^t f(x,\tau)\sin\omega_n(t-\tau)\mathrm{d}\tau\mathrm{d}x.$$

■ A BAR LOADED BY A FORCE

Obtain the ensued motion of the free–free bar suddenly pulled in its extreme $x = L$ by a force F.

We note first that the vibration modes for this free–free bar is $U_n(x) = a_1\cos n\pi x/L$, with the natural frequencies $\omega_n = n\pi c/L$. Normalizing $U_n(x)$ according to

$$\int_0^L U_n(x)\mathrm{d}x = 1,$$

we have that $a_1 = \sqrt{2/L}$ and we obtain, from the general expression for the forced solution,

$$u(x,t) = \sum_{n=1}^{\infty}\frac{\sqrt{\frac{2}{L}}\cos\frac{n\pi x}{L}}{\frac{n\pi c}{L}}\int_0^L\sqrt{\frac{2}{L}}\cos\frac{n\pi x}{L}\left[\int_0^t\frac{F}{\rho A}\sin\frac{n\pi c}{L}(t-\tau)\mathrm{d}\tau\right]\mathrm{d}x.$$

The time integral above amounts to $\frac{F}{\rho A}\frac{L}{n\pi c}(1-\cos\frac{n\pi c}{L}t)$ and the integral along the length is simply $U_n(x = L) = \sqrt{\frac{2}{L}}(-1)^n$ since the force is concentrated. Hence, the vibrational motion of this beam is given by

$$\bar{u}(x,t) = \frac{2LF}{\pi^2 c^2\rho A}\sum_{n=1}^{\infty}\frac{(-1)^n}{n^2}\cos\frac{n\pi x}{L}\left(1-\cos\frac{n\pi c t}{L}\right).$$

Observe that any force applied to the beam will also move it as a rigid body and this can be quantified by solving the equilibrium equation for a rigid rod

$$\rho AL\frac{\mathrm{d}^2\bar{\bar{u}}(t)}{\mathrm{d}t^2} = F,$$

whose solution

$$\bar{\bar{u}}(t) = \frac{F}{2\rho AL}t^2$$

needs to be added to $\bar{u}(x,t)$ to give the total motion of the bar as

$$u(x,t) = \frac{F}{2\rho AL}t^2 + \frac{2LF}{\pi^2c^2\rho A}\sum_{n=1}^{\infty}\frac{(-1)^n}{n^2}\cos\frac{n\pi x}{L}\left(1-\cos\frac{n\pi ct}{L}\right).$$

The figure plots this solution for some specific parameters, a 10 mm diameter, 500 mm long aluminium bar under a force of 10 N. An inset is provided that shows the delay for the wave to reach the support.

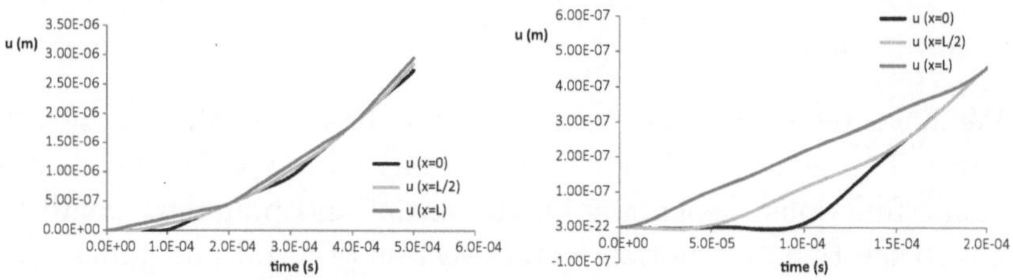

(a) (b)

Figure 3.17 The motion of a free–free bar under an end force applied at $x = L$. (a) Overall behaviour, (b) detail of the wave delay in reaching $x = 0$.

■

This last example suggests that the hammering of the bar of length L can be modelled similarly, the application of a force, F, on a short interval, Δt, such that the bar of mass m (more precisely, its center of gravity) will move with a constant velocity $v = F\Delta t/m$. This impulse will compress a slice of the bar of length $c\Delta t$, which will propagate forward until the other extreme.

Hence, in the interval $0 < t < \Delta t$, the left face of the bar will be moving to the right at the velocity v above. The motion of this face stops though at $t = \Delta t$ because the wave has passed. The latter will progress, moving all the particles to the right as it sweeps through the bar. At a time $t = L/c$ the wave reaches the right extreme of the bar, which then is moved during Δt, after which it stops.

Similar comments are given by K.F. Graff, Wave Motion in Elastic Solids, Dover, 1975 and by H. Kolsky, Stress Waves in Solids, Dover, 1963, from which the figure here was redrawn.

The wave returns and reaches the other end and such a motion continues the bar to move in jumps. The next figure helps to understand further this impact event.

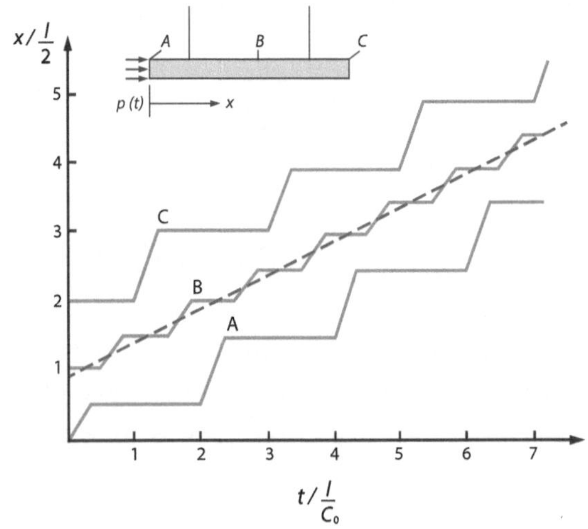

Figure 3.18 The motion of a free–free bar under a step pressure pulse.

We have seen that the wave equation has the solution $u(x,t) = f(x \pm ct)$. This general solution is particularized to spatial and temporal harmonic functions, depending on the initial and boundary conditions. This is to say that the motion a cross section of a rod undergoes is such that the particles oscillate around an equilibrium position, with the limiting case that some points will not oscillate at all due to geometrical restrictions, such as at a clamped boundary. Hence, the various points in the rod domain will oscillate at different amplitudes, leading to a spatial representation of a wave that does not move along the length. These standing waves are of various frequencies, so that the final motion of a linear structure will be the superposition of these various standing waves, each with one frequency. We refer to vibration instead of wave propagation for this harmonic response. Of course that, if a given harmonic excitation has a frequency that coincides with one of the natural frequencies of a structure, then the latter will vibrate only at this frequency. We say that the structure response is tuned to the excitation frequency. We will further discuss in Chapter 4 the case of non–harmonic excitation in beams.

3.7 Visco–elastic waves: dispersion

Hookean materials, *ie* the ones which obey, in the uniaxial case, the constitutive law $\sigma = E\varepsilon$, are not able to represent many materials we daily use in devices and structures well. Take for instance a polymeric string. If a mass is hanged by it, it may displace further and further as time passes by and this is indicative of the so called visco–elastic material response, not contemplated by the Hook law for the lack of the parameter time. Indeed, there are a myriad of materials; science and technology strive to develop constitutive laws so structures made with them can be correctly analysed. The next figure gives an account of some common material models.

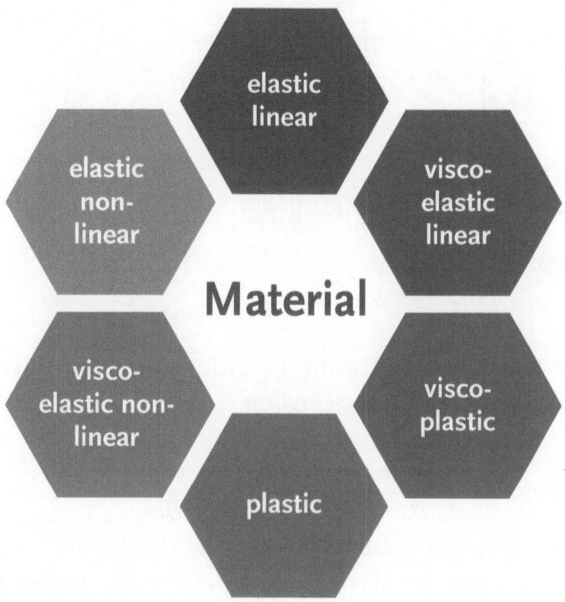

Figure 3.19 Some material models.

To model the behaviour of materials that are time dependent, a common practice is to use a set of springs and dashpots, assembled in a way thought to represent some measured response. For instance, the use of one spring and one dashpot working in parallel, known as Kelvin–Voigt model, can represent the phenomenon of creep, defined as the solid deformation under the action of a, usually constant, stress whose magnitude is lower than the flow stress.

This section is based on the book Foundation of Stress Waves, W. Li–li, Elsevier, 2007.

According to the next figure, the strain experienced by the spring and dashpot are the same and the total stress is added, *ie*

$$\varepsilon = \varepsilon_E = \varepsilon_\eta, \quad \sigma = \sigma_E + \sigma_\eta \rightarrow \sigma = E\varepsilon + \eta\dot{\varepsilon},$$

use being made of the linear Hooke's law, $\sigma_E = E\varepsilon$, and the linear viscous law, $\sigma_\eta = \eta\dot{\varepsilon}$. This is a linear differential ordinary equation, whose solution is

$$\varepsilon = \frac{\sigma}{E}\left(1 - e^{-t/\tau}\right), \quad \tau = \frac{\eta}{E},$$

indicated in the next figure.

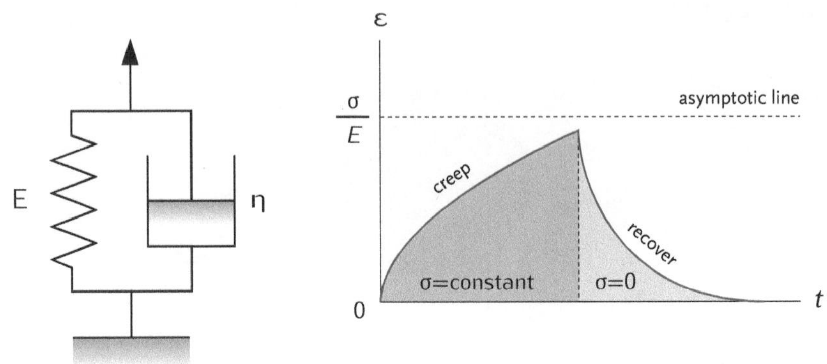

Figure 3.20 The Kelvin–Voigt material model and its response on time.

Recall now that the equilibrium equation for a bar with no external load is $\partial\sigma/\partial x = \rho\partial^2 u/\partial t^2$ which, using $\sigma = E\varepsilon + \eta\dot{\varepsilon}$, with $\varepsilon = \partial u/\partial x$, yields

$$c^2\frac{\partial^2 u}{\partial x^2} + c^2\tau\frac{\partial^3 u}{\partial x^2 \partial t} = \frac{\partial^2 u}{\partial t^2}.$$

Observe that the wave solution used for the linear elastic material, $u = f(x \pm ct)$ does not satisfy the above equation. Instead of applying the method of separation of variables, we reason that the solution should have an oscillatory character of the type

This strategy will be also used on other occasions.

$$u(x,t) = \bar{A}e^{i(\omega t - k_1 x)},$$

with $\bar{A}$ being the wave amplitude, ω the radial frequency and k_1 the wave number, a concept to be detailed in the next chapter. Substituting this expression in the equilibrium equation it follows that

$$\rho\omega^2 = k_1^2(E + i\eta\omega) = (k + i\alpha)^2(E + i\eta\omega),$$

where use was made of the fact that k_1 must be a complex number, otherwise the equality above would be possible only if $\eta = 0$.

If we now equate the real and imaginary part of both sides of this equation we arrive at

$$k = \sqrt{\frac{\omega^2}{2c^2(1+\omega^2\tau^2)}\left(\sqrt{\omega^2\tau^2+1}+1\right)}$$

and

$$\alpha = \sqrt{\frac{\omega^2}{2c^2(1+\omega^2\tau^2)}\left(\sqrt{\omega^2\tau^2+1}-1\right)},$$

which can be substitute in the assumed solution

$$u(x,t) = \bar{A}e^{-\alpha x}e^{i(\omega t - kx)}.$$

Note that α depends on the frequency and causes the wave to decrease its amplitude. Accordingly, α is called attenuation factor.

This phenomenon of attenuation indicates that different frequency components present in a pulse travel each one with a distinct velocity, with its amplitude decreasing with the distance.

Now, knowing that $k = \omega/c$, we obtain the wave velocity in this material as

$$c_{KV} = c\sqrt{\frac{2\,(1+\omega^2\tau^2)}{1+\sqrt{1+\omega^2\tau^2}}},$$

which reduces to c when $\tau = 0$, *ie* when the material is elastic linear.

We arrive then at the important conclusion that the various frequency components (harmonics) of a pulse will travel each one with a certain velocity. This causes a change in the shape of the original pulse set in this kind of material.

We refer to wave dispersion to convey this phenomenon. Another source of dispersion is geometrical and it will be presented in the next section.

3.7.1 Geometrical dispersion

In deriving the equilibrium equation for bars, we did not take notice of its radial expansion, which occurs in most real materials, as quantified by the Poisson ratio, ν. This lateral expansion requires energy, expected to be greater for high frequency components, since the particles are forced to move radially very fast.

To quantify this kind of dispersion, let us note that the radial velocity, v, and acceleration, a, of the particles in a xyz Cartesian reference frame, the origin being in the center of the rod cross–section with x along the rod main axis, are given by

These equations indicate that the rod cross–section is no longer planar.

$$v_y = \frac{\partial u_y}{\partial t} = -\nu y \frac{\partial \varepsilon_x}{\partial t} = -\nu y \frac{\partial v_x}{\partial x}$$

and

$$a_y = \frac{\partial v_y}{\partial t} = -\nu y \frac{\partial a_x}{\partial x},$$

with similar expression for v_z and a_z.

The transverse kinetic energy is

$$\frac{1}{A\mathrm{d}x} \int \frac{1}{2} \rho (v_y^2 + v_z^2) \mathrm{d}x\mathrm{d}y\mathrm{d}z = \frac{1}{2} \rho \nu^2 r_g^2 \dot{\varepsilon}_x^2,$$

with r_g being the rotational radius of the generic cross–section with respect to the bar main axis x. This energy is to be added to the axial kinetic energy, so that energy equilibrium gives

$$\sigma_x \dot{\varepsilon_x} = \frac{\mathrm{d}}{\mathrm{d}t} \left(\frac{1}{2} E \varepsilon_x^2 \right) + \frac{\mathrm{d}}{\mathrm{d}t} \left(\frac{1}{2} \rho \nu^2 r_g^2 \dot{\varepsilon}_x^2 \right),$$

from which it follows that

$$\sigma = E\varepsilon + \rho \nu^2 r_g^2 \ddot{\varepsilon},$$

with the subscript $_x$ being dropped.

If we now substitute this equation in the equilibrium one we obtain

$$c^2 \frac{\partial^2 u}{\partial x^2} = \frac{\partial u^2}{\partial t^2} - \nu^2 r_g^2 \frac{\partial^4 u}{\partial x^2 \partial t^2},$$

which reduces to the unidimensional equation if we set $\nu = 0$, *ie* when no transverse effect is considered.

As before, we introduce in this equation the harmonic wave

$$u(x,t) = Ae^{i(\omega t - kx)},$$

with $k = 2\pi/\lambda$ being the wave number, to obtain the wave velocity in a rod when considering transverse inertia, c_ν, as

$$c_\nu = \sqrt{c^2 - \nu^2 r_g^2 \omega^2} = \frac{c}{\sqrt{1 + \nu^2 r_g^2 k^2}},$$

which reduces to

$$c_\nu \approx c\left[1 - \nu^2 \pi^2 \left(\frac{a}{\lambda}\right)^2\right],$$

for a cylindrical rod of radius r, $r_g = r/\sqrt{2}$, and assuming $\nu^2 r_g^2 k^2 \ll 1$. This equation is referred to as the Rayleigh approximation. It shows that short waves, *ie* high frequency waves, propagate slower than low frequency waves. Hence, each harmonic of a given pulse will travel at its own velocity, causing the wave to change its shape as it moves in a rod. Note that this effect is of geometrical nature, in contradistinction with the material dispersion seen above for the Kelvin–Voigt model. Note also that the geometrical and material dispersion work one contrary to the other, since for the visco–elastic model, high–frequency harmonics travel faster than low–frequency ones.

We could go further in our analysis by exploring the propagation of elastic waves in linear elastic cylindrical bars starting from the general equilibrium equations for an elastic body, instead of the simple axial equilibrium used here. Such an exact solution is somewhat more involved and it is not presented in this book.

3.8 Measurement of motion

In the theoretical results shown before concerning the application of a sudden load to a bar, we showed the existence of an overall rigid body motion superposed to the vibration.

Concentrating on measuring small displacements, let us assume that the bar is 20 cm long and that the maximum allowed strain is 0.1%, so to assure that the material behaves elastically. In this case, the maximum end face displacement would be $\Delta L = 0.1/100 \times 20 = 0.2$ mm. We could wish to have a transducer that records this signal discretizing its amplitude in, say, 10 points, which gives increments of $\Delta L = 200$ μm. We need then to be able to measure amplitudes as small as that. On the other hand, we want to track the motion as time passes by and for a stress level of say 100 MPa, the particle speed would be some 7 m/s, giving a reading time interval of $\Delta t = \Delta L/V \approx 30\mu s$. This require a digital electronic board capable of acquiring at least 2×30 thousand points per second, *ie* a sample rate of 60 kHz.

We have then two aspects to take into consideration when designing a measurement system for this task. One aspect is the sensor, which should be accurate and responsive. The second aspect is the data recording system. As for the sensors, three types will be discussed in the next subsections. We will discuss later in this chapter on the data recording system and the concept of sample rate. This will allow us to understand the factor 2 in the sample rate calculation above.

3.8.1 Foil strain gauges

In terms of sensors applied to the impact of bars, we consider first the strain gauge. This is a coiled resistance embed in a polymeric film that can be glued in the place where strains are to be measured. Since the waves generate strains, this device can be used to measure the wave amplitudes.

There are various arrangements for strain gauges, as seen in the next figure, depending on the expected load configuration and on the sought compensation, *ie* to remove from the readings undesirable influence, like bending when one wants to measure axial strains in a rod.

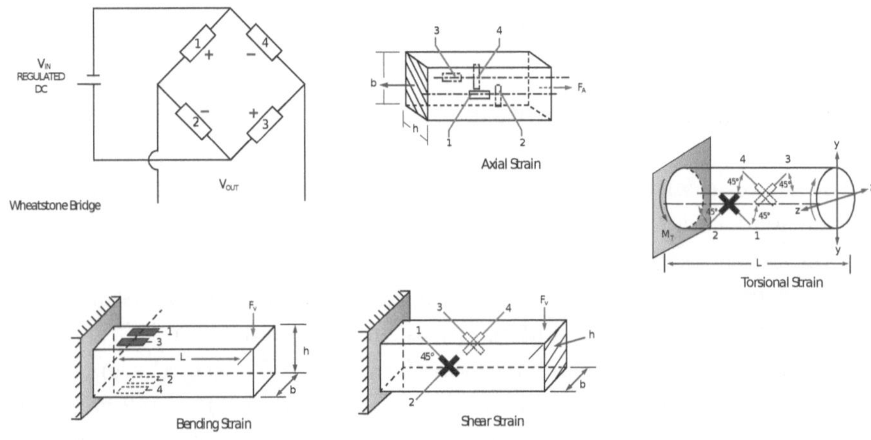

Figure 3.21 Strain gauge principle and various arrangements.

In any of these cases, what we really measure with these gauges is the change in voltage caused by a change in the electrical resistance, ultimately due to a change in length of the region where the strain gauge is glued. The electrical resistance is a function of the gauge length, L, the wire cross section area, A, and the resistivity, $\bar{\rho}$, according to

$$R = \bar{\rho}\frac{L}{A}.$$

Taking the derivative of the resistance with respect to the length and assuming volume conservation of the wire, *ie* volume before and after straining is the same, $AL = (A + \Delta A)(L + \Delta L)$, we obtain, disregarding high order terms,

$$\mathrm{d}R = (2\bar{\rho}/A)\mathrm{d}L \quad \rightarrow \quad \frac{\mathrm{d}R}{R} = G_f\frac{\mathrm{d}L}{L} \quad \rightarrow \quad \varepsilon = \frac{\Delta R/R}{G_f},$$

with G_f being the gauge factor. The typical value of $G_f = 2$ is an approximation and each manufacturer will indicate the actual value for a given type of strain gauge. There are also semiconductor strain gauges, based on the piezoresistive effect. Their gauge factor can be as high as 150 but they are more sensitive to temperature variations and their resistance/strain relationship is nonlinear. Typical values for the foil strain gauge resistance are 120 Ω and 350 Ω and they can be made of alloys like Constantan (copper–nickel) and Nichrome V (nickel–chrome). A strain gauge has a very high frequency response, reaching values as high as 100 kHz. This is to say that they can capture signals that change fast in time, which is the case for the end rod motion here.

The stress level in the bar described in the beginning of this section, of magnitude of 100 MPa, gives a strain of around 1.4E-3=1400 μ strain. This means some 0.34 Ω resistance variation in the strain gauge. Adopting a maximum current of 50 mA, the voltage change would be only 17 mV. To "quantitise" the amplitude of this signal in, say, 100 points means that we need to have an instrumentation of some 170 μV of sensitivity, which is not cost effective. An easier way around this is to use a strain gauge placed in one or various arms of the so called Wheatstone bridge. This doubles or quadruples the output signal, which can then be amplified. Amplification of over 1000 times can be readily obtained with integrated circuits.

Balance of the Wheatstone bridge shown in the next figure, *ie* $V_o = 0$, means that $i_o = 0$. This implies that $i_1 = i_2$, or $V_{BA} = V_{DA}$, and $i_3 = i_4$, hence $V_{BC} = V_{DC}$, which gives $i_1R_1 = i_3R_3$, $i_2R_2 = i_4R_4$

and $R_1R_4 = R_2R_3$. When this condition for bridge balance is met, the voltages in the arms are $V_{BC} = R_2i_1 = V_iR_2/(R_1 + R_2)$ and $V_{DC} = V_iR_4/(R_3 + R_4)$, so the output voltage is $V_o = V_{BC} - V_{DC} = [R_2R_3 - R_1R_4]/[(R_1R_2)(R_3R_4)]V_i$, with V_i being the input voltage. Allow now each resistance to vary, caused by strain, and we will have

$$V_o = V_i\frac{R_3R_4}{(R_3 + R_4)^2}\left(\frac{\Delta R_2}{R_2} - \frac{\Delta R_1}{R_1} + \frac{\Delta R_3}{R_3} - \frac{\Delta R_4}{R_4}\right)(1 - \eta),$$

with η a constant of cross–products of resistances and close to zero for small ΔR. It follows that

$$\frac{V_o}{V_i} \approx \frac{G_f}{4}\left(\varepsilon_2 - \varepsilon_1 + \varepsilon_3 - \varepsilon_4\right).$$

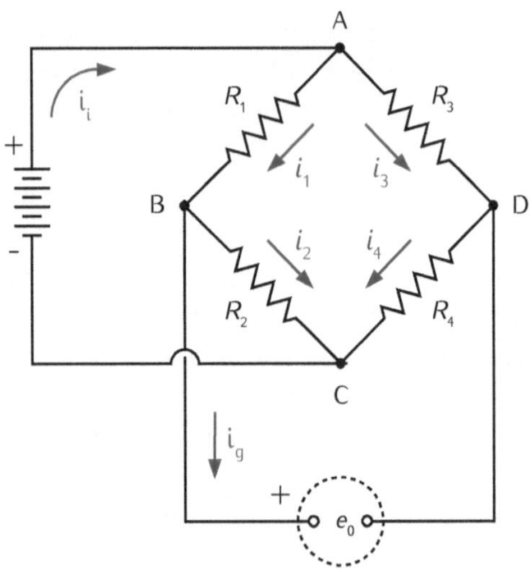

Figure 3.22 The Wheatstone bridge.

If we allow only one arm of the bridge to vary, *ie* a one gauge configuration with the other three gauges being substituted by fixed (high precision) resistors, the voltage output would be $V_o = 2\varepsilon_1$ if the input voltage is 4 V and $G_f = 2$. This is twofold of what we had before but we can double this again by using two strain gauges, a half–bridge configuration, giving a factor of 4 in the output. This, together with the fact that the output can easily be balanced, *eg* by using a trim-pot in parallel with one arm, makes the use of the Wheatstone bridge very effective.

Note that, in the case of the bar under analysis in this section, to use a second gauge at the opposite side of the first one, means that they need to be connected in opposite positions in the circuit so to add and not to cancel the signal. As a bonus, any bending of the bar will be compensated in this arrangement.

3.8.2 Accelerometer and load cells

Another way to build up our measurement system to probe the bar motion is to place at its end, screwed or fixed magnetically or by glue, an accelerometer. A common type of accelerometer is the piezoelectric, which consists of a quartz crystal sensor connected to a seismic mass, as shown in the next figure. This mass offers resistance to the motion of the base, so straining the sensor, which outputs proportionally a charge. Cheaper and more popular accelerometers are those based on microelectrical mechanical systems (MEMS), also shown in the next figure for the case of acceleration being proportional to the capacitance. Here, as the seismic mass senses the motion, its support changes its position, so altering the capacitance. Sensors like these are fitted in small electronic boards, together with the circuitry and other sensors, like gyroscopes. They are widely used in, for instance, passenger cars to detect a collision and, after signal processing, to trigger the airbags.

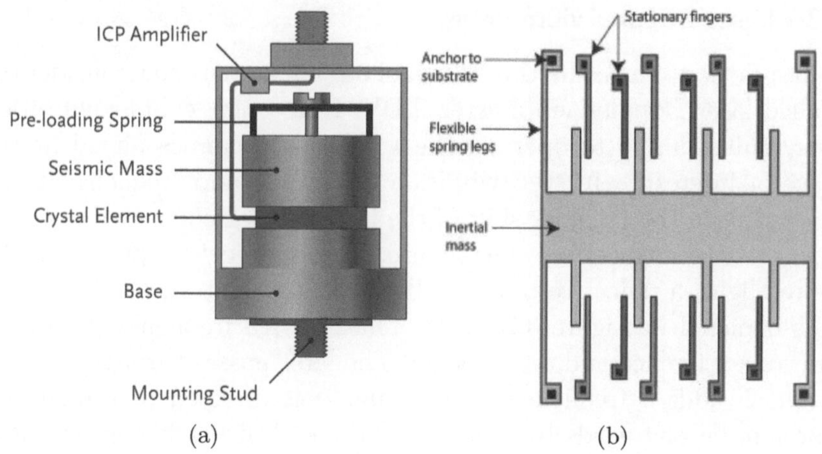

Figure 3.23 (a) A schematic view of a piezoelectric accelerometer. (b) A sketch of a MEMS accelerometer.

Accelerometers and many types of dynamic load cells can be quite rigid, meaning that they have a very high resonant frequency, but not comparable to the strain gauges. Typical values are in between 10 and 40 kHz, in any case well above the fundamental (axial) frequency of the bar studied in this section, which is some 4 kHz for a 20 cm long steel bar. The accelerometer manufacturer provides the calibration constant of the sensor, which is stable along the years. A typical value is 1 pC/g (pico Coulombs per acceleration of the gravity).

As for load cells, they can be based on strain gauges, hydraulic and on piezoelectricity. For dynamic measurement, the piezoelectric load cell is better because it has a high natural frequency. The presence of a seismic mass is required for the accelerometer but not for a piezoelectric load cell.

To understand why the use of high natural frequency measuring devices is important, consider an input signal that has a spectrum with important frequency content limited to f_c. Roughly, if the load cell has a natural frequency of $3f_c$ then the signal will not have energy to excite its natural frequency. This means that the load cell is not going to distort the signal. Hence, the higher the natural frequency of a measuring device the better and this is the case of piezoelectric load cells. Commercial load cells can be found with natural frequencies above 50 kHz, which is due to the high stiffness of the piezoelectric material, crystal or ceramic, and of the mounting cage.

3.8.3 Laser Doppler vibrometer

As a last device to measure the motion of our bars, let us now consider the so called laser Doppler vibrometer, LDV. It is a non–contact vibration sensor, quite simple to operate: the LDV laser beam is aimed at the surface of interest, with the vibration amplitude and frequency being extracted from the Doppler shift of the laser beam frequency due to the motion of the surface, see next figure. To improve the quality of the reflected light, a reflective tape can be placed in the target area.

As depicted in Figure 3.25, a laser beam with frequency f_o is split into two: a reference and a test beam. The later passes through a Bragg cell, which adds a frequency shift f_b and it is directed to the surface whose motion one needs to measure, *ie* the end of the bar in our case.

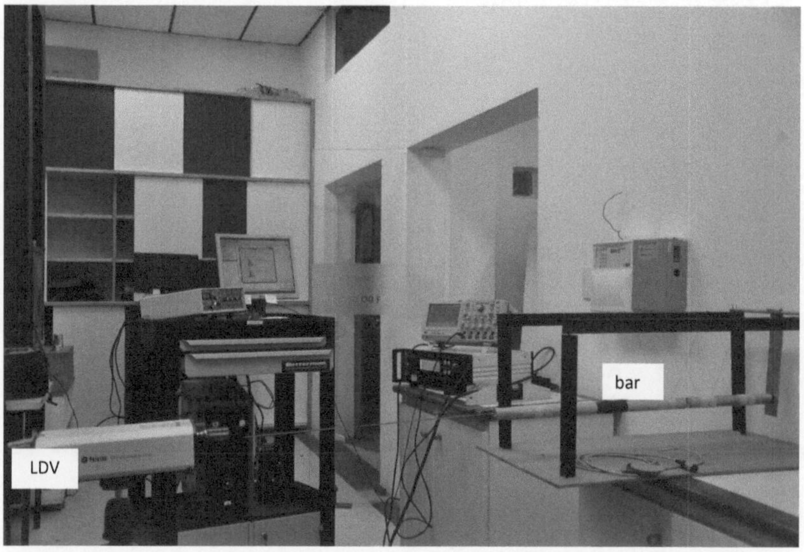

Figure 3.24 A LDV aimed at the end of a bar hanged by strings.

The motion of the bar face Doppler shifts the laser beam by

$$f_d = \frac{2v \cos\alpha}{\lambda},$$

with v being the target velocity, α the angle between the laser beam and the velocity vector and λ the laser wavelength.

Light scatters from the target in all directions, but some portion of the light is collected by the LDV and reflected by the beam splitter to the photo detector. This light has a frequency equal to $f_o + f_b + f_d$. This scattered light is combined with the reference beam at the photo–detector. The detector responds to the beat frequency between the two beams, which is $f_b + f_d$ and typically in the tens of MHz range.

Important advantages of the LDV are the ease of setting it up, the fact that it does not add extra mass to the system and that the output velocity signal is analogue. This means that the sensor continuously measures the motion, *ie* any frequency component can be measured. One is here limited only to the analogue to digital (A/D) conversion board. A LDV is capable of measuring a wide range of velocities and the signal, when time integrated by the controller or by software, gives the position in time of the object. In the case of the bar analysed here this is ideal because we can measure small vibrations of the bar end and at the same time its rigid body overall motion.

A good data acquisition board for dynamic structural and material tests should have a good response in the range of at least 1 MHz.

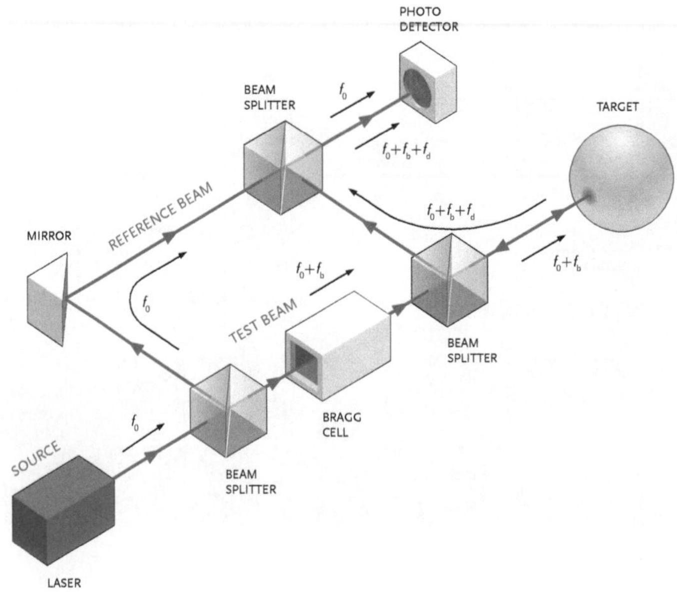

Figure 3.25 A laser Doppler vibrometer operation principle.

3.9 Closure

We have seen in some details the wave propagation phenomenon in bars. Starting from the definition of a linear elastic material, we established the concepts of wave velocity and particle velocity and that a wave is a stress–strain carrier. We also took a look at the impact of bars and the complex patterns that arise in this event.

We moved on by introducing a non–linear elastic material and the dependency of the wave velocity on frequency, leading to the concept of wave dispersion. Some basic understanding of motion sensors were also given here. Hence, terrain was gained and we can move on to more complex scenarios of impact, *ie* the transverse elastic impact in beams.

3.10 Problems

1. Two identical bars, A and B, of mass m_b, length L, and cross-sectional area A, are placed end–to–end touching one another. A third identical bar, C, moving with speed V_0 hits the two rest bars. The axes of all the bars are collinear. Describe the stress state and

the particle speed in each bar at times, $t = 0, L/c, 3L/c$ and show how at these instants momentum and energy are conserved.

2. Two bars of lengths L and $3L$, with same cross–section and material, travel against each other at a velocity V_0. Describe their motion in detail supported by the next figure.

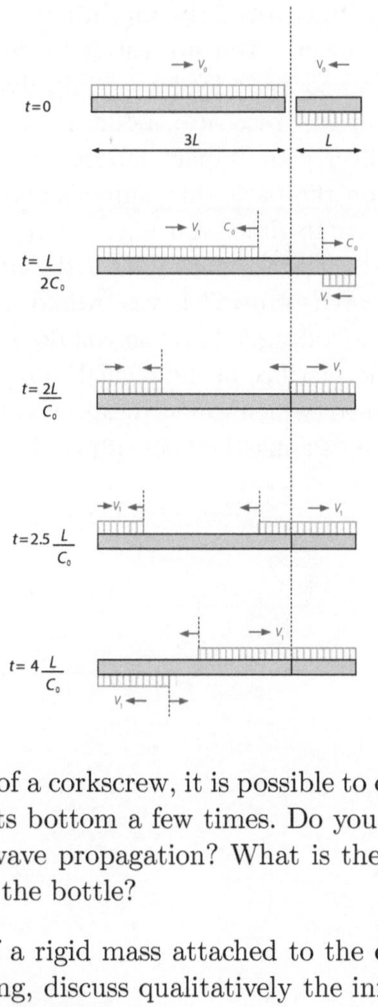

Figure 3.26 Waves in two bars of lengths L and $3L$ after impact.

3. In the absence of a corkscrew, it is possible to open a bottle of wine by impacting its bottom a few times. Do you see some important effect here of wave propagation? What is the ultimate reason for the opening of the bottle?

4. For the case of a rigid mass attached to the end of a vertical rod fixed on a ceiling, discuss qualitatively the influence of gravity on the natural frequencies.

5. What would be the gauge arrangements in a four gauges Wheatstone bridge and in which position should they be glued in a long

bar subjected to axial impact so that its dynamic Poisson ratio can be measured. Indicate the overall measurement system used.

6. What is the response of a linear elastic cantilever bar of length L, cross section A, density ρ and elastic modulus E loaded at its free extreme with a velocity V_0?

7. Some studies on fracture of the hip in humans are undertaken by using pigs as subjects. You are asked to develop a rig based on wave propagation so as to load a pig hip dynamically. As part of the contract you will receive a modern tensile machine that will definitely establish your impact laboratory. It happens that you are vegetarian on the basis that animals should not suffer for our benefit. At the same time, you have no vegetarian students who could easily undertake this task with little supervision. Should you accept this research project? If yes, would it be correct to assign it to a student or colleague because you do not want to be directly connected to the sacrifice of the pigs? If no, your laboratory could not now be formed, which would refrain you from studying vehicle safety, potentially saving the lives of people. Discuss this problem among colleagues.

CHAPTER 4

Elasto–dynamics of beams

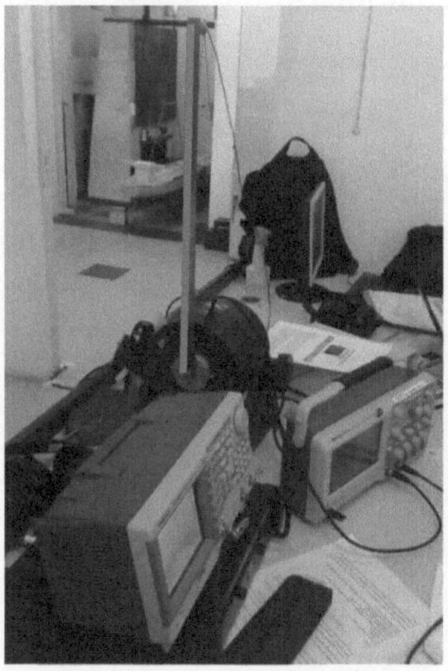

Figure 4.1 A beam mounted on a shaker for vibration tests.

After bars, beams are the simplest structural elements and have as a main characteristic the ability of sustaining transverse loads in addition to axial loads. Hence, it is important to explore this structure, particularly its dynamic response.

DOI: 10.1201/9781003485292-4

In the next sections, we shall analyse beams under transverse dynamic loads, which lead the beams to vibrate. We will explore both the cases of free and forced vibration, with some hints on on flexural waves. As for experimental mechanics, we will study digital signals, Fourier transform and very basic signal filtering concepts.

4.1 Equilibrium equation for beams

In deriving the equilibrium equations of structural members like beams, plates and shells, we avoid working with local stresses. Instead, we integrate these stresses across the transverse section, A, giving rise to the so called generalised stresses, typically, for beams, bending moment, $M = \int_A \sigma_x z \mathrm{d}A$, transverse shear force, $Q = \int_A \sigma_{xz} \mathrm{d}A$, and the normal force, $N = \int_A \sigma_x \mathrm{d}A$. These stresses and the kinematics of a beam loaded by a distributed load are shown in the next figure.

Let us start by considering the equilibrium of the beam segment. Note that we allow here for the action of membrane forces, *ie* forces acting on the direction of the main beam axis.

If we equate the axial forces to the axial inertia, it follows that

$$\frac{\partial N}{\partial x} = m\frac{\partial^2 u}{\partial t^2},$$

but this inertia term is usually small in relation to the transverse inertia of the beam, allowing us to consider the membrane force constant along the beam length. Here, m is the beam mass divided by its length.

Rotary inertia in rigid, perfectly plastic beam is discussed in N. Jones, Structural Impact, CUP, 1998, from which this figure was adapted.

Moment equilibrium on the beam element yields

$$\frac{\partial^2 M}{\partial x^2} + N\frac{\partial^2 w}{\partial x^2} + p(x,t) = m\frac{\partial^2 w}{\partial t^2},$$

use being made of the fact that the transverse force equilibrium is given by

$$Q = \frac{\partial M}{\partial x},$$

with the rotary inertia being disregarded since it has a secondary influence on the overall beam response. As for the presence of the normal force in this equation, it comes from its vertical component, as noted in the figure, using the approximation $\sin(\mathrm{d}w/\mathrm{d}x) \approx \mathrm{d}w/\mathrm{d}x$.

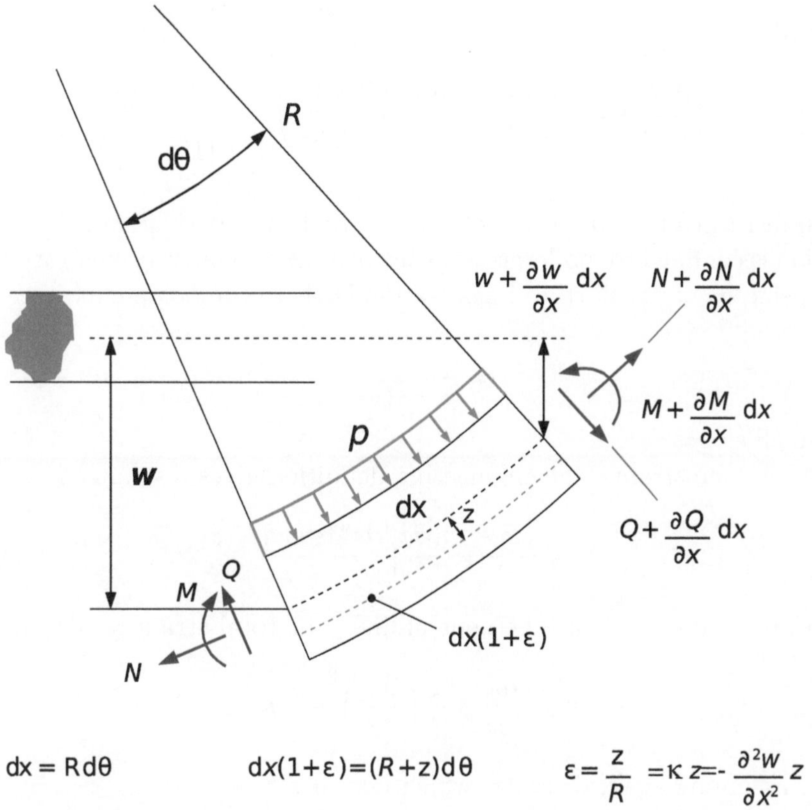

Figure 4.2 A beam element subjected to moderate transverse displacements.

The kinematic relations, *ie* how the beam transverse and lateral displacements are related to strains and curvature, can be obtained considering first that

$$\frac{\mathrm{d}s}{\mathrm{d}x} = \sqrt{\left(\frac{\mathrm{d}w}{\mathrm{d}x}\right)^2 + \left(1 + \frac{\mathrm{d}u}{\mathrm{d}x}\right)^2} \approx 1 + \frac{\mathrm{d}u}{\mathrm{d}x} + \frac{1}{2}\left(\frac{\mathrm{d}w}{\mathrm{d}x}\right)^2.$$

Now, $\tan\phi = -\frac{\frac{\mathrm{d}w}{\mathrm{d}x}}{1+\frac{\mathrm{d}u}{\mathrm{d}x}}$, and its derivative with respect to s, the length of the infinitesimal deformed element, is

$$\frac{\partial \tan\phi}{\partial s} = \sec^2\phi \frac{\partial \phi}{\partial s} = -\frac{\partial}{\partial x}\frac{\frac{\mathrm{d}w}{\mathrm{d}x}}{1+\frac{\mathrm{d}u}{\mathrm{d}x}}\frac{\mathrm{d}x}{\mathrm{d}s}$$

and using

$$\cos\phi = \left(1 + \frac{\mathrm{d}u}{\mathrm{d}x}\right)\frac{\mathrm{d}x}{\mathrm{d}s}$$

we write

$$\frac{\partial \phi}{\partial s} = -\left[\frac{\mathrm{d}^2 w}{\mathrm{d}x^2}\left(1+\frac{\mathrm{d}u}{\mathrm{d}x}\right) - \frac{\mathrm{d}w}{\mathrm{d}x}\frac{\mathrm{d}^2 u}{\mathrm{d}x^2}\right]\left(\frac{\mathrm{d}x}{\mathrm{d}s}\right)^3 \approx \frac{-\frac{\mathrm{d}^2 w}{\mathrm{d}x^2}}{\left[1+\left(\frac{\mathrm{d}w}{\mathrm{d}x}\right)^2\right]^{3/2}},$$

when disregarding cross products of u, w and their derivatives, since attention is confined to moderate displacements and relatively small strains and using $\mathrm{d}s \approx \mathrm{d}x$. Further linearisation leads to the beam curvature as

$$\kappa = -\frac{\mathrm{d}^2 w}{\mathrm{d}x^2}$$

for small slopes.

As for the strains, the engineering definition gives a strain of

$$\varepsilon^z = \frac{(R+z)(\partial\phi/\partial s)\mathrm{d}s - \mathrm{d}x}{\mathrm{d}x},$$

for a fibre z distant from the neutral line. The total strain results in

$$\varepsilon^z = \frac{\mathrm{d}u}{\mathrm{d}x} + \frac{1}{2}\left(\frac{\mathrm{d}w}{\mathrm{d}x}\right)^2 + z\kappa,$$

when assuming again $\mathrm{d}s \approx \mathrm{d}x$, which becomes

$$\varepsilon^z = z\kappa,$$

for small slope and moderate displacement.

It can be proven that these kinematic relations are consistent with the equilibrium equations. This is to say that components of the generalized forces in a given direction which were disregarded, had their kinematic counter–parties also not considered, and vice–versa.

Consistent relations in the light of the principle of virtual work is discussed in N. Jones, Structural Impact, CUP, 2012.

Note that both, the equilibrium equations and the kinematic relations, are material independent. We have then on one hand a kinematic field and, on the other hand, so to say, a force field. Both fields are material independent and the only way to bind them is by invoking the beam material behaviour via the so called material constitutive law.

Adopt now the linear elastic material law

$$\sigma = E\varepsilon$$

and calculate the bending moment across the beam section by integrating $\mathrm{d}M = z\mathrm{d}F = z\sigma\mathrm{d}A$, *ie*

$$M = EI\kappa = -EI\frac{\partial^2 w}{\partial x^2},$$

with $I = \int_A z^2\mathrm{d}A$ being the inertia moment of the beam cross–section. Introducing this expression in the equilibrium equation and disregarding the membrane force we obtain the linear governing equation of a beam as

$$EI\frac{\partial^4 w}{\partial x^4} = p(x,t) - m\frac{\partial^2 w}{\partial t^2},$$

which is a fourth order linear partial differential equation. Once the distributed load, $p(x,t)$, boundary conditions, *ie* the way the beam is held in space, and the initial conditions, *ie* load and imposed velocity fields, are known, it is possible to solve this equation and to obtain the time variation of the transverse displacement field, w, at any position of the beam span. w being known, it is possible to obtain the generalized stress and strains.

Note that the boundary conditions are four in number and they are obtained by examining the displacement, slope, transverse shear force and bending moment in the beam supports or regions of symmetry. There are also two initial conditions related to the initial displacement and velocity. These initial conditions can cause the motion of the beam, even in the absence of external forces.

4.2 Free vibration

Let us start our study on the dynamics of beams by investigating the linear elastic beam response in the absence of load, *ie* $p(x,t) = 0$. This is the case when the beam presents a free motion which, due to its repetitive nature, is called free vibration. The governing equation becomes

Note that the beam starts its motion due to some external load or initial condition, like an initial displacement, and it is then left to vibrate freely. Observe that here a has units of m^2/s and therefore does not represent a wave velocity.

$$a^2\frac{\partial^4 w}{\partial x^4} + \frac{\partial^2 w}{\partial t^2} = 0, \qquad \text{with} \qquad a = \sqrt{\frac{EI}{\rho A}},$$

which can be solved for some relatively simple, yet practical, cases.

Let us assume that the transverse displacement of the beam can be represented by the product of two functions,

$$w(x,t) = W(x)T(t),$$

which, when substituted in the governing equations gives

$$a^2 \frac{\partial^4 W(x)/\partial x^4}{W(x)} = -\frac{\partial^2 T(t)/\partial t^2}{T(t)}.$$

One side of this equation depends only on x and the other only on t, which is only possible when they are a constant, say ω^2, allowing us to write

$$\frac{\partial^4 W(x)}{\partial x^4} - \left(\frac{\omega}{a}\right)^2 W(x) = 0 \qquad \text{and} \qquad \frac{\partial^2 T(t)}{\partial t^2} + \omega^2 T(t) = 0.$$

These two ordinary differential equations have the solutions

$$W(x) = a_1 \sin \beta x + a_2 \cos \beta x + a_3 \sinh \beta x + a_4 \cosh \beta x$$

and

$$T(t) = A_1 \sin \omega t + A_2 \cos \beta t,$$

with

$$\beta^4 = \frac{\omega^2}{2}.$$

A_n and a_n are constants to be determined by the boundary and initial conditions, as the next example shows.

■ FREE VIBRATION OF A CANTILEVER

The next figure shows a shaker to which a beam is connected on its base via a clamped fixture. Suggest suitable boundary and initial conditions for the beam and obtain its natural frequencies.

Derivatives of hyperbolic sine and co–sine do not change signal. Also, $-\sinh^2 x + \cosh^2 x = 1$.

To solve this problem we need to determine the constants A_n and a_n in the above solution. We know that at the support of this cantilever beam, the displacement and curvature are zero, *ie* $W(0) = 0$ and $\mathrm{d}W(0)/\mathrm{d}x = 0$. Also, the bending moment and the transverse shear force at $x = L$ are zero, *ie* $\mathrm{d}^2W(L)/\mathrm{d}x^2 = 0$ and $\mathrm{d}^3W(L)/\mathrm{d}x^3 = 0$. Introducing these conditions in $W(x)$ gives

$$\begin{bmatrix} 0 & 1 & 0 & 1 \\ 1 & 0 & 1 & 0 \\ -\sin\beta L & -\cos\beta L & \sinh\beta L & \cosh\beta L \\ -\cos\beta L & \sin\beta L & \cosh\beta L & \sinh\beta L \end{bmatrix} \begin{Bmatrix} a_1 \\ a_2 \\ a_3 \\ a_4 \end{Bmatrix} = \begin{Bmatrix} 0 \\ 0 \\ 0 \\ 0 \end{Bmatrix}$$

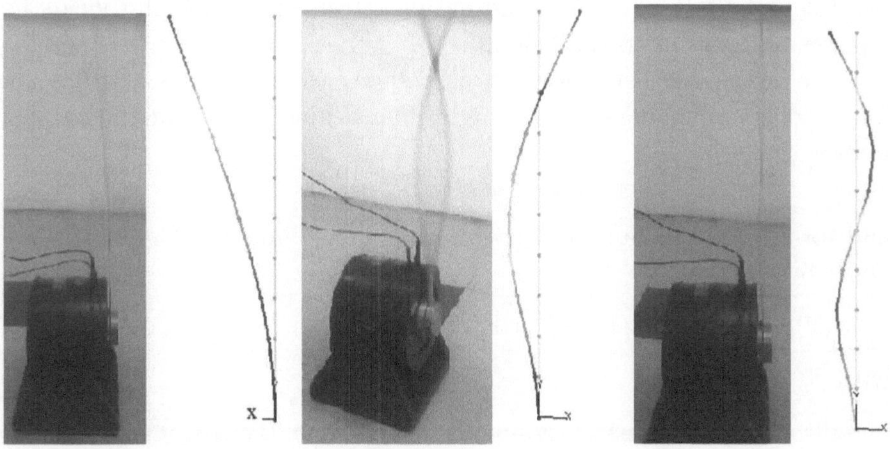

Figure 4.3 Cantilever beam undergoing free vibration in the first, second and third vibration mode. By each photo there is the FE result. The respective measured natural resonant frequencies are 3.2 Hz, 20.1 Hz and 54.0 Hz, which are in error no larger than 3.42% in comparison with the FE analysis. Theoretical values are 2.75 Hz, 17.26 Hz and 48.34 Hz.

To avoid a trivial solution to this system, its determinant must be zero, which gives

$$\cos\beta L \cosh\beta L = -1.$$

A sketch of the functions $\cos x$ and $-1/\cosh x$, together with a more refined calculation, will show that they intercept at

$$\beta_1 L = 1.875, \qquad \beta_n L \approx n\pi/2, \quad n = 3, 5, 7 \dots$$

Knowing β_n, the frequencies, called natural, $\omega = \beta^2 a$ are obtained for this beam configuration. They are compared to experimental values in the previous figure, indicating that the model performs quite well for the first modes of vibration.

Since our model has no damping, there is no distinction among resonance, natural and damped natural frequencies.

■

We further explore the problem examined above by determining an expression for $W(x)$. We need to know the constants a_n, for $n = 1, 2, \dots$, which can be obtained by expressing the constants a_2, a_3, a_4 in the above system of equation as a function of a_1, giving

$$W_n(x) = a_1 \left[\cos\beta_n x - \cosh\beta_n x - \frac{\cos\beta_n L + \cosh\beta_n L}{\sin\beta_n L + \sinh\beta_n L}(\sin\beta_n x - \sinh\beta_n x)\right].$$

These are the so called natural modes of vibration. Note that the constant a_1 can be obtained only when one imposes some condition on the beam. For instance, at the beam free end we can set it to $W(L) =$ 5mm and let the beam vibrate. Using the above expression, it is then possible to obtain a_1 and the actual displacement, which can yield the stresses and strains in the beam.

The next table lists the natural frequencies and modes of vibration for various supporting cases of single span beams. Note that the natural frequencies are

$$\omega_n = \beta_n^2 \sqrt{EI/m} \quad \text{(rad/s)}$$

and for $n \geq 5$, one can use $\lambda_n = \lambda_{n-1} + \pi$. The modes of vibration are given by

$$\phi(x)_n = C_{1n} \sin \beta_n x + C_{2n} \cos \beta_n x + C_{3n} \sinh \beta_n x + C_{4n} \cosh \beta_n x,$$

with $\beta_n = \lambda_n / L$ and

$$c_n = \left[(\sinh \lambda_n + A \sinh \lambda_n)/(\cosh \lambda_n + B \cos \lambda_n)\right]^{\alpha}.$$

mics of elastic beams to init

TABLE 4.1 Natural frequencies and vibration modes equations for various beam configurations. Adapted from lecture notes by C.A. Nunes Dias, GMSIE—USP.

beam configuration	simply sup.	clamped	free	clamped–s.sup.	cantilever
frequency equation	$\sin\lambda = 0$	$\cos\lambda \cosh\lambda = 1$	$\cos\lambda \cosh\lambda = 1$	$\tan\lambda - \tanh\lambda = 0$	$\cos\lambda \cosh\lambda = -1$
λ_1	π	4.7300	4.7300	3.9266	1.8751
λ_2	2π	7.8532	7.8532	7.0686	4.6941
λ_3	3π	10.9956	10.9956	10.2102	7.8548
λ_4	4π	14.1372	14.1372	13.3518	10.9955
modes of vibration					
α	–	-1	-1	-1	1
A	–	-1	-1	-1	-1
B	–	-1	-1	-1	1
c_1	–	1.0008	1.0008	1.0008	0.7341
c_2	–	1.0008	1.0008	1.0000	1.0185
c_3	–	0.9999	0.9999	1.0000	0.9992
c_4	–	1.0000	1.0000	1.0000	1.0000
C_{1n}	1	c_n	$-c_n$	c_n	c_n
C_{2n}	0	-1	1	-1	-1
C_{3n}	0	$-c_n$	$-c_n$	$-c_n$	$-c_n$
C_{4n}	0	1	1	1	1

4.3 Dynamics of elastic beams to initial conditions

We now state once more, as we did for rods, that the overall behaviour of a linear elastic beam can be expressed as a linear combination of its all natural vibration modes

$$w(x,t) = \sum_{n=1}^{\infty} W_n(x) T_n(t).$$

This can be substituted in the beam equation of motion, free vibration case, to give

$$\frac{EI}{\rho A}\sum_{n=1}^{\infty}\frac{\mathrm{d}^4W_n(x)}{\mathrm{d}x^4}T_n(t)+\sum_{n=1}^{\infty}W_n(x)\frac{\mathrm{d}^2T_n(t)}{\mathrm{d}t^2}=0.$$

Multiplying this equation by $W_m(x)$ and integrating along the beam length results in

$$\sum_{n=1}^{\infty}\left\{\frac{EI}{\rho A}\int_0^L W_m(x)\frac{\mathrm{d}^4W_n(x)}{\mathrm{d}x^4}\mathrm{d}x\right\}T_n(t)+$$
$$\sum_{n=1}^{\infty}\left\{\int_0^L W_m(x)W_n(x)\mathrm{d}x\right\}\frac{\mathrm{d}^2T_n(t)}{\mathrm{d}t^2}=0.$$

We shall see in the sequence that the so called orthogonality conditions lead the above equation to

$$\frac{\mathrm{d}^2T_n(t)}{\mathrm{d}t^2}+\omega_nT_n(t)=0,\qquad n=1,2,\ldots$$

whose solution is

$$T_n(t)=A_n\cos\omega_nt+B_n\sin\omega_nt\qquad n=1,2,\ldots$$

being necessary to know the A_n, B_n constants from the initial conditions $T_n(0)$ and $\mathrm{d}T_n(0)/\mathrm{d}t$. If the initial conditions of the actual problem are specified, *ie*

$$w(x,0)=\sum_{n=1}^{\infty}W_n(x)T_n(0)\quad\text{and}\quad\frac{\partial w(x,0)}{\partial t}=\sum_{n=1}^{\infty}W_n(x)\frac{\mathrm{d}T_n(0)}{\mathrm{d}t}$$

and if they are multiplied by ρAW_m and integrated along the beam length, it follows that

$$T_n(0)=\rho A\int_0^L W_n(x)w(x,0)\mathrm{d}x$$

and

$$\frac{\mathrm{d}T_n(0)}{\mathrm{d}t}=\rho A\int_0^L W_n(x)\dot{w}(x,0)\mathrm{d}x,\qquad n=1,2,\ldots$$

Note here how each vibration mode has a weight on the time response of the beam.

Substitute now these initial conditions into $T_n(t) = A_n \cos\omega_n t + B_n \sin\omega_n t$ to finally obtain

$$w(x,t) = \sum_{n=1}^{\infty} W_n(x)\left[T_n(0)\cos\omega_n t + \frac{1}{\omega_n}\frac{\mathrm{d}T_n(0)}{\mathrm{d}t}\sin\omega_n t\right].$$

This general solution allow us to solve a series of interesting impact related problems, as illustrated in the next section. Before though we present the orthogonality conditions used above and also in the previous chapter, proven now to be valid.

For details see S.S. Rao, Vibration of Continuous Systems, Wiley, 2007.

For that, assume that the beam governing equation has the harmonic solution $w(x,t) = W(x)\sin\omega t$. Upon substitution in the equilibrium beam equation we obtain

$$a^2\frac{\mathrm{d}^4W_n(x)}{\mathrm{d}x^4} = \omega_n^2 W_n(x),$$

for a given eigenfrequency ω_n and eigenfunction W_n. We now multiply this equation by W_m and integrating along the beam length,

$$a^2\int_0^L W_m(x)\frac{\mathrm{d}^4W_n(x)}{\mathrm{d}x^4}\mathrm{d}x = \omega_n^2\int_0^L W_m(x)W_n(x)\mathrm{d}x.$$

Integration by parts of its left hand to lower the fourth order derivative results in

$$a^2\int_0^L W_m(x)\frac{\mathrm{d}^4W_n(x)}{\mathrm{d}x^4}\mathrm{d}x = c^2\int_0^L \frac{\mathrm{d}^2W_m(x)}{\mathrm{d}x^2}\frac{\mathrm{d}^2W_n(x)}{\mathrm{d}x^2}\mathrm{d}x.$$

This gives

$$a^2\int_0^L \frac{\mathrm{d}^2W_m(x)}{\mathrm{d}x^2}\frac{\mathrm{d}^2W_n(x)}{\mathrm{d}x^2}\mathrm{d}x = \omega_n^2\int_0^L W_n(x)W_m(x)\mathrm{d}x.$$

A similar procedure, but now multiplying by $W_n(x)$ the solution written in terms of $W_m(x)$, yields

$$a^2\int_0^L \frac{\mathrm{d}^2W_n(x)}{\mathrm{d}x^2}\frac{\mathrm{d}^2W_m(x)}{\mathrm{d}x^2}\mathrm{d}x = \omega_m^2\int_0^L W_m(x)W_n(x)\mathrm{d}x,$$

which can be subtracted from the previous equation to give

$$(\omega_n^2 - \omega_m^2)\int_0^L W_n(x)W_m(x)\mathrm{d}x = 0,$$

or

$$\int_0^L W_n(x)W_m(x)\mathrm{d}x = 0, \quad n, m = 1, 2, 3\ldots \quad \omega_n \neq \omega_m.$$

Now, for $\omega_n = \omega_m$, we can normalize the nth mode as

$$\int_0^L \rho A W_n^2(x)\mathrm{d}x = 1,$$

then being valid, using the Kronecker delta notation,

$$\int_0^L \rho A W_n(x)W_m(x)\mathrm{d}x = \delta_{ij}$$

and

$$\int_0^L EIW_m(x)\frac{\mathrm{d}^4 W_n(x)}{\mathrm{d}x^4}\mathrm{d}x = \omega_n^2\delta_{ij}.$$

■ A BEAM SUDDENLY UNLOADED

Consider the next figure, where a simply supported beam is loaded by a mass via a string attached to its centre. Describe the motion of the beam after the string is suddenly cut.

Figure 4.4 A simply supported beam under a central load of mass, which is suddenly removed.

We should first notice that the presence of the mass causes the beam to be initially deflected, as indicated in the figure. The initial displacement of a beam of length L with a central load due to a mass G is given by

$$w_0(x) = \frac{Gg}{12EI}x(x^2 - 3L^2/4), \quad \text{for} \quad 0 \le x \le L/2$$

and

$$w_0(x) = \frac{Gg}{12EI}(L-x)(x^2+L^2/4-2Lx), \quad \text{for} \quad L/2 \leq x \leq L,$$

with the beams own weight being disregarded.

The initial velocity is zero throughout the whole beam span and therefore $\mathrm{d}T(0)/\mathrm{d}t = 0$. The normal modes for this simple supported beam are

$$W_n(c) = C_n \sin(n\pi x/L)$$

and from

$$\int_0^L \rho A W_n^2 \mathrm{d}x = 1$$

we obtain $C_n = \sqrt{2/\rho AL}$.

Now,

$$T_n(0) = \int_0^L \rho A W_n(x) w_0(x) \mathrm{d}x = \rho A \sqrt{\frac{2}{\rho AL}} \frac{Gg}{12EI} \times$$

$$\int_0^{\frac{L}{2}} \sin\frac{n\pi x}{L} x \left(x^2 - \frac{3L^2}{4}\right) \mathrm{d}x +$$

$$\int_{\frac{L}{2}}^{L} \sin\frac{n\pi x}{L}(L-x)\left(x^2 + \frac{L^2}{4} - 2Lx\right) \mathrm{d}x,$$

or

$$T_n(0) = \sqrt{\frac{2\rho A}{L}} \frac{Gg}{EI} \left(\frac{L}{n\pi}\right)^4 \sin\frac{n\pi}{2}.$$

The transverse displacement of the beam after the string is cut becomes

$$w(x,t) = \sum_{n=1}^{\infty} W_n(x) T_n(0) \cos\omega_n t,$$

$$w(x,t) = \sum_{n=1}^{\infty} \frac{2Gg}{EIL}\left(\frac{L}{n\pi}\right)^4 \sin\frac{n\pi}{2} \sin\frac{n\pi x}{L} \cos\left[\left(\frac{n\pi}{L}\right)^2 \sqrt{\frac{EI}{\rho A}} t\right],$$

since

$$\omega_n = \left(\frac{n\pi}{L}\right)^2 \sqrt{\frac{EI}{\rho A}}.$$

This solution is dominated by the first vibration mode, with the even modes being zero and the odd modes having little significance in the response, except for the first one. Note also that damping was not considered in the solution.

A RELEASED BEAM

A beam of length L and cross–section A has its end, at $x = L$, released from a height H, so rotating at its left support, at $x = 0$, see next figure. Obtain the ensued motion of the beam considering that it does not bounce back after contact with the support is established.

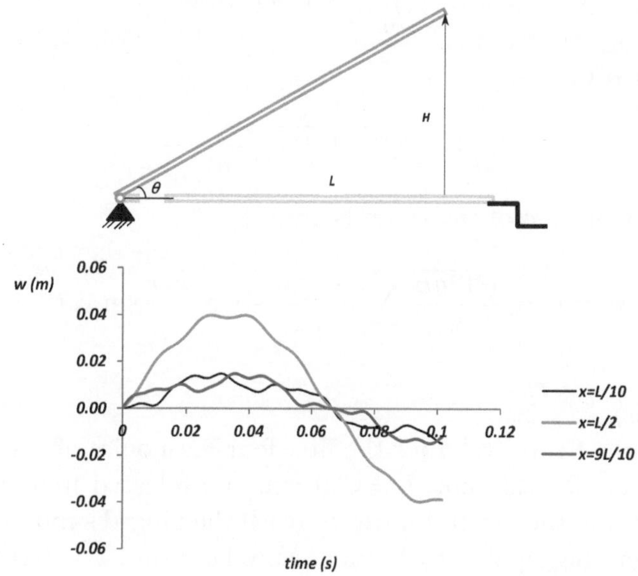

Figure 4.5 (a) A falling beam and (b) its time–transverse displacement along time.

We know that a general expression for the transverse displacement of a simply supported beam is

$$w(x,t) = \sum_{n=1}^{\infty} \sqrt{\frac{2}{\rho AL}} \sin \frac{n\pi x}{L} \left[T_n(0) \cos \omega_n t + \frac{\dot{T}_n(0)}{\omega_n} \sin \omega_n t \right],$$

with $T_n(0) = \int \rho A W_n(x) 0 \mathrm{d}x = 0$.

Equating the kinetic and potential energy of the beam leads to

$$\frac{1}{2}\bar{I}\omega^2 = \int_0^L m\mathrm{d}xg\frac{Hx}{L},$$

so that the beam velocity reads

$$\dot{w}(x,0) = \omega x = \sqrt{3gH}\frac{x}{L},$$

since the moment of inertia around the support is $\bar{I} = m_b L^2/3$, with $m_b = mL$.

We have now that

$$\dot{T}_n(0) = \int_0^L \rho A W_n \dot{w}(x,0)\mathrm{d}x,$$

which resolves to

$$\dot{T}_n(0) = \sqrt{6\rho AgHL}\frac{-\cos n\pi}{n\pi}$$

so the final vibration of the beam is given by

$$w(x,t) = \frac{\sqrt{12gH}}{\pi}\sum_{n=1}^{\infty}\frac{-\cos n\pi}{n\omega_n}\sin\frac{n\pi x}{L}\sin\omega_n t,$$

with $\omega_n = \left(\frac{n\pi}{L}\right)^2\sqrt{\frac{EI}{\rho A}}$.

Note in the figure, valid for the first four harmonics of a steel beam of $H = 1$ mm, $B = 25$ mm, $L = 560$ mm and released from $H_f = 300$ mm, the delay in the flexural waves to reach the hinged support at $x = 0$ when disregarding rigid body motion. Note here that $I = BH^3/12$.

■

Another interesting application of the free vibration of beams is the analysis of the impact of a mass on it. It seems at first glance that the problem requires the forced term $f(x,t)$ in the governing equations. On the contrary, we can solve this problem with the forced term equal to zero. We introduce the impact velocity as an initial condition and the impact mass as a boundary condition.

4.4 Impact of a mass on a beam

This problem falls in the category of the previous section, where the initial condition, the impact velocity, is given. Together with the impact mass, one can solve this problem, but the solution is not so straightforward. Also, given the importance of this topic in the context of this book, this solution warrants a section of its own.

We deal here with the problem of a simple supported beam of length L subjected to the impact of a mass G hitting its mid span at a velocity V_0. To obtain the ensued motion of the beam we note that the transverse beam deflection is given by

This solution is presented in Théorie de L'Élasticité des Corps Solides, De Clebsch, translated from German by Barré de Saint–Venant and Flamant, 1883, Dunod Éditeur, Paris and further explored in W. Goldsmith, Impact: the theory and physical behaviour of colliding solids, Dover edition, 2001.

$$w(x,t) = \sum_{n=1}^{\infty}(A_n \sin\beta_n x + B_n \cos\beta_n x + C_n \sinh\beta_n x + D_n \cosh\beta_n x)$$

$$(E_n \sin\omega_n t + F_n \cos\omega_n t),$$

which equals to

$$w(x,t) = \sum_{n=1}^{\infty}\left(A_n \sin\beta_n x - A_n \frac{\cos\beta_n \frac{L}{2}}{\cosh\beta_n \frac{L}{2}} \sinh\beta_n x\right)(E_n \sin\omega_n t),$$

when considering the solution symmetrical in $x = L/2$ and valid from the support to the midspan.

Applying now the boundary conditions for this simply supported beam, symmetrically loaded,

$$w(0,t) = 0, \quad \partial^2 w(0,t)/\partial x^2 = 0, \quad \partial w(L/2,t)/\partial x = 0 \quad w(x,0) = 0,$$

the above equation becomes

$$w(x,t) = \sum_{n=1}^{\infty} \frac{1}{a^2\beta_n^2} G_n X_n \sin\beta_n^2 a^2 t,$$

with

$$G_n = a^2\beta_n^2 E_n A_n \cos\beta_n L/2$$

and

$$X_n = \frac{\sin\beta_n x}{\cos\beta_n L/2} - \frac{\sinh\beta_n x}{\cosh\beta_n L/2}.$$

G_n is a function of the integration constants and they are uniquely determined since the impact mass and velocity are given. To obtain it, we introduce first the impact mass by equating the transverse shear force to the (half) inertia force, *ie*

$$EI\frac{\partial^3 w(L/2,t)}{\partial x^3} = \frac{G}{2}\frac{\partial^2 w(L/2,t)}{\partial t^2},$$

from which we obtain the characteristic equation

$$\phi_n(\tan\phi_n - \tanh\phi_n) = 2m_r,$$

with $\phi_n = \beta_n L/2$ and $m_r = m_b/G$, the beam to impact mass ratio.

We know that

$$\frac{\partial w(x,0)}{\partial t} = \sum_{n=1}^{\infty} G_n X_n = \psi(x).$$

Multiplying it by X_m and integrating with respect to the total beam and striker masses, dq, it follows that

$$\int \sum_{n=1}^{\infty} G_n X_n X_m \mathrm{d}q = \int \psi(x) X_m \mathrm{d}q,$$

where $q = G + m_b$, $\mathrm{d}q = m_b \mathrm{d}x/L$.

Now, for $n = m$,

$$G_n = \frac{\int \psi(x) X_n \mathrm{d}q}{\int {X_n}^2 \mathrm{d}q}.$$

Since, $\psi(L/2) = V_0$ and zero for $x \neq L/2$ (here enters the initial condition of our problem), we have that

$$\int \psi(x) X_n \mathrm{d}q = \int_0^{L/2} X_n (m_b/L) 2\mathrm{d}x + G X_{x=L/2} V_0 = G X_{x=L/2} V_0.$$

As for the denominator of G_n we have

$$\int X_n^2 \mathrm{d}q = \int_0^{L/2} X_n^2 2(m_b/L)\mathrm{d}x + G X_n (L/2)^2,$$

with the first term being

$$\frac{2m_b}{L}\int_0^{L/2}\left(\frac{\sin\beta_n x}{\cos\beta_n L/2} - \frac{\sinh\beta_n x}{\cosh\beta_n L/2}\right)^2 \mathrm{d}x =$$

$$\frac{2m_b}{L}\left\{\frac{L}{4}\left(\frac{1}{\cos^2\beta_n L/2} - \frac{1}{\cosh^2\beta_n L/2}\right) - \frac{3L}{4\beta_n}(\tan\beta_n L/2 - \tanh\beta_n L/2)\right\}$$

and the last

$$X_{L/2} = \tan \beta_n L/2 - \tanh \beta_n L/2 = \frac{4m_b}{\beta_n LG},$$

so that

$$G_n = \frac{4V_0}{\frac{\beta_n L}{2}\left(\frac{1}{\cos^2 \beta_n L} - \frac{1}{\cosh^2 \beta_n L/2}\right) + \frac{2m_r}{\beta_n L/2}}.$$

This resolves to

$$G_n = \frac{4V_0}{\phi_n \left(\frac{1}{\cos^2 \phi_n} - \frac{1}{\cosh^2 \phi_n}\right) + \frac{2m_r}{\phi_n}},$$

so that the transverse displacement of the beam in time and space becomes

$$w(x,t) = \frac{L^2 V_0}{a^2} \sum_{n=1}^{\infty} \frac{1}{\phi_n^3} \frac{\frac{\sin \frac{2\phi_n x}{L}}{\cos \phi_n} - \frac{\sinh \frac{2\phi_n x}{L}}{\cosh \phi_n}}{\frac{1}{\cos^2 \phi_n} - \frac{1}{\cosh^2 \phi_n} + \frac{2m_r}{\phi_n^2}} \sin \frac{4\phi_n^2 a^2}{L^2} t.$$

The next figure presents some results for a simply supported steel beam of length 500 mm and $B = 25$ mm, $H = 1$ mm, struck at its centre a velocity of $V_0 = 1$ m/s, impact mass of $G = 97.5$ g. This mass equals to the beam mass, so that $m_r = 1$ and the four first roots of the frequency equation are 1.19161, 4.11972, 7.18994, 10.29839, 13.42093.

As a complement to this section, we present the deflection and frequency equation for the following cases (from W. Goldsmith book):

- Clamped beam struck at the centre

  Deflection

$$w(x,t) = \frac{L^2 V_0}{2a^2} \times$$

$$\sum_{n=1}^{\infty} \frac{1}{\phi_n^2} \frac{(1 - \cos \phi_n \cosh \phi_n)(\cos \phi_n - \cosh \phi_n)(\sin \phi_n + \sinh \phi_n)}{(1 - \cos \phi_n \cosh \phi_n)^2 + m_r(\cos \phi - \cosh \phi_n)^2}$$

$$\left[\frac{\sinh \frac{2\phi_n x}{L} - \sin \frac{2\phi_n x}{L}}{\cosh \phi_n - \cos \phi_n} - \frac{\cosh \frac{2\phi_n x}{L} - \cos \frac{2\phi_n x}{L}}{\sinh \phi_n + \sin \phi_n}\right] \sin \frac{4\phi_n^2 a^2}{L^2} t.$$

  Frequency equation

$$m_r = \frac{\phi_n (1 - \cos \phi_n \cosh \phi_n)}{\sin \phi_n \cosh \phi_n + \cos \phi_n \sinh \phi_n}$$

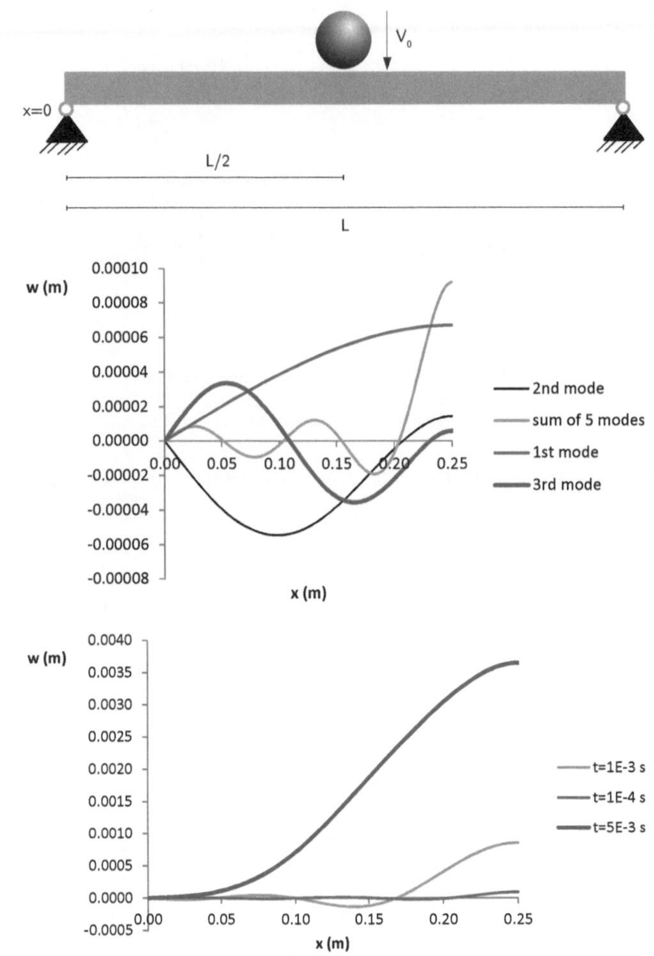

Figure 4.6 (a) A simply supported beam hit by a mass. (b) The contribution of the various modes at $t = 1E - 4$ and (c) its time–transverse displacement.

- Cantilever beam struck at its end $x = L$

  Deflection

$$w(x,t) = \frac{2L^2 V_0}{a^2} \times$$

$$\sum_{n=1}^{\infty} \frac{1}{\phi_n^2} \frac{(\sin\phi_n \cosh\phi_n - \cos\phi_n \sinh\phi_n)(\sin\phi_n + \sinh\phi_n)(\cos\phi_n + \cosh\phi_n)}{(\sin\phi_n \cosh\phi_n - \cos\phi_n \sinh\phi_n)^2 + m_r(\sin\phi + \sinh\phi_n)^2}$$

$$\left[\frac{\cosh\frac{\phi_n x}{L} - \cos\frac{\phi_n x}{L}}{\cosh\phi_n + \cos\phi_n} - \frac{\sinh\frac{\phi_n x}{L} - \sin\frac{\phi_n x}{L}}{\sinh\phi_n + \sin\phi_n}\right] \sin\frac{\phi_n^2 a^2}{L^2} t.$$

Frequency equation

$$m_r = \frac{\phi_n(\sin\phi_n \cosh\phi_n - \cos\phi_n \sinh\phi_n)}{1+\cos\phi_n \cosh\phi_n}$$

4.5 Dynamic response of beams to a force

When a beam is subjected to a load, the equation of motion is no longer homogeneous, having now the forced term $p(x,t)$. For the linear elastic response, we can still use the combination of normal modes to obtain

$$EI\sum_{n=1}^{\infty}\frac{\mathrm{d}^4W_n(x)}{\mathrm{d}x^4}T_n(t)+\rho A\sum_{n=1}^{\infty}W_n(x)\frac{\mathrm{d}^2T_n(t)}{\mathrm{d}t^2}=p(x,t),$$

which can be written, following the same previous development for the free response of beams, as

$$\frac{\mathrm{d}^2T_n(t)}{\mathrm{d}t^2}+\omega_n^2T_n(t)=q_n(t), \qquad n=1,2,\ldots$$

whose solution is

$$T_n(t)=A_n\cos\omega_nt+B_n\sin\omega_nt+\frac{1}{\omega_n}\int_0^t q_n(\tau)\sin\omega_n(t-\tau)\mathrm{d}\tau,$$

where

$$q_n(t)=\int_0^L W_n(x)p(x,t)\mathrm{d}x \qquad n=1,2,\ldots$$

It follows then that the final solution for the transverse beam displacement is

$$w(x,t)=\sum_{n=1}^{\infty}\left[a_n\cos\omega_nt+b_n\sin\omega_nt+\frac{1}{\omega_n}\int_0^t q_n(\tau)\sin\omega_n(t-\tau)\mathrm{d}\tau\right]W_n(x).$$

■ A BEAM SUDDENLY LOADED BY A MASS

The contact between a mass G and a simply supported beam is prevented by a string holding the mass, as shown in the next figure. The string is suddenly cut and it is required to know the ensued motion of the beam.

The problem can be seen as a step–function force of intensity Gg applied at some position ξ of the beam

$$p(x,t)=Gg\delta(x-\xi).$$

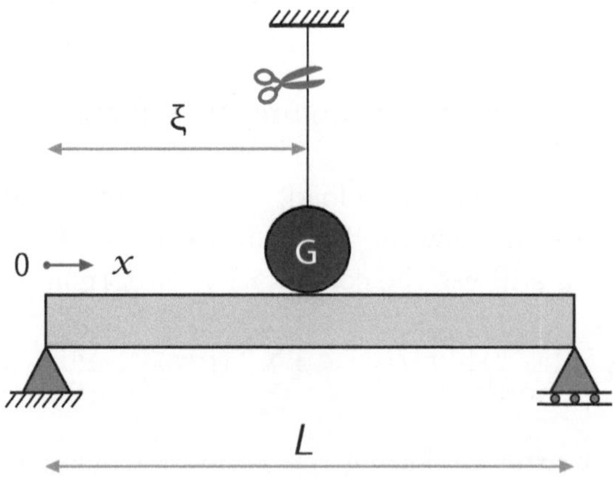

Figure 4.7 A beam suddenly loaded by a mass G.

We have seen that the normalized modes for this beam are

$$W_n(x) = \sqrt{\frac{2}{\rho AL}} \sin \frac{n\pi x}{L},$$

and the natural frequencies are

$$\omega_n = \frac{n^2\pi^2}{L^2}\sqrt{\frac{EI}{\rho A}}$$

so that the generalized force is

$$q_n(t) = \int_0^L W_n(x)p(x,t)\mathrm{d}x = \int_0^L \sqrt{\frac{2}{\rho AL}} \sin \frac{n\pi x}{L} Gg\delta(x-\xi)\mathrm{d}x =$$

$$\sqrt{\frac{2}{\rho AL}} Gg \sin \frac{n\pi\xi}{L},$$

with the time solution being

$$T(t) = A_n \cos \omega_n t + B_n \sin \omega_n t +$$

$$\sqrt{\frac{2}{\rho AL}} Gg \sin \frac{n\pi\xi}{L} \frac{1}{\omega_n} \int_0^T \sin \omega_n (t-\tau)\mathrm{d}\tau.$$

Since the initial conditions for this problem are zero, $A_n = B_n = 0$ and the beam response to this step load reads

$$w(x,t) = \frac{2GgL^3}{\pi^4 EI}\sum_{n=1}^{\infty}\frac{1}{n^4}\sin\frac{n\pi x}{L}\sin\frac{n\pi\xi}{L}(1-\cos\omega_n t).$$

■

4.6 Flexural waves

The governing equation for a beam seen in this chapter resembles the equation that rules a propagation of a disturbance in a bar. Therefore, it is interesting to investigate how a continuous sequence of transverse displacement in a beam would propagate along its length.

If we try to substitute the general wave solution for the bar wave equation, $w(x,t) = f(x-ct) + g(x+ct)$, in the governing equation for beams we will see that it is not a solution. This should not be a surprise since the constant a in the beam governing equation does not have dimensions of velocity.

Let us now investigate how a harmonic pulse would propagate in a beam, *ie* would cause the portion of the beam reached by the pulse to move up and down. Assume that a harmonic wave of amplitude A, wave length of λ, and period T travels in the positive x direction with a velocity v. This can be represented by, for instance, the function

$$w(x,t) = f(x-ct) = Af_1\left(\frac{x}{\lambda/2\pi} - \frac{vt}{\lambda/2\pi}\right) = A\cos\left(kx - \omega t\right),$$

with $\omega = 2\pi f = 2\pi/T$ being, as before, the wave frequency, T the wave period and

$$k = \frac{2\pi}{\lambda}$$

the wave number.

Substituting this harmonic motion in the beam governing equation we have

$$v = \frac{2\pi}{\lambda}\sqrt{\frac{EI}{\rho A}},$$

which has units $[LT^{-1}]$ and it is called flexural wave velocity in beams. It depends on the material parameters, beam cross–section and wave length.

Recall now from Chapter 3 that a given general disturbance can be represented by a series of harmonic waves, each with a given frequency. Each of these harmonic waves has their own wavelength, λ, and, therefore, according to the above equation, travel with their own velocity. If they have a common origin in a point of the beam, after some time, the low frequency harmonics will have lagged behind the high frequency ones. This leads again to the phenomenon of wave dispersion, as we have already studied for bars. Hence, beams are a dispersive medium.

When a beam is loaded, say, by a pulse load or by the impact of an object, the generated disturbance can be decomposed in a packet of waves. This group of waves travel in the beam at a velocity $v_g \neq v$, called group velocity.

A relation between the group and phase velocities can be obtained by considering two harmonic waves of same amplitude A and slightly different frequencies, $\omega_1 = \omega + \Delta\omega$ and $\omega_2 = \omega - \Delta\omega$ that travel in space. The wave packet, with $\Delta k = (k_1 - k_2)/2$ and $\Delta\omega = (\omega_1 - \omega_2)/2$, is

According to K.F. Graff, Wave Motion in Elastic Solids, this argument was first given by G.G. Stokes in 1876.

$$\begin{aligned} w_p = w_1 + w_2 &= A[\cos(k_1 x - \omega_1 t) + \cos(k_2 x - \omega_2 t)] \\ &= 2A\cos(\Delta k x - \Delta\omega t)\cos(kx - \omega t), \end{aligned}$$

which is depicted in the next figure. We observe from the above equation that there are two waves of low (amplitude modulated wave) and high (carrier wave) frequency, travelling at velocities

$$v_g = \frac{\Delta\omega}{\Delta k} \qquad \text{and} \qquad v = \frac{\omega}{k},$$

respectively. This envelope moving at the group velocity is responsible for carrying energy.

We can write the group velocity as

$$v_g = \lim_{\Delta k \to 0} \frac{\Delta\omega}{\Delta k} = \frac{\mathrm{d}(kv)}{\mathrm{d}k} = v + k\frac{\mathrm{d}v}{\mathrm{d}k} = v - \frac{\mathrm{d}v}{\mathrm{d}\lambda},$$

which, for the propagation of flexural waves in a beam, equals

$$v_g = \frac{4\pi}{\lambda}\sqrt{\frac{EI}{\rho A}},$$

ie twice the wave velocity.

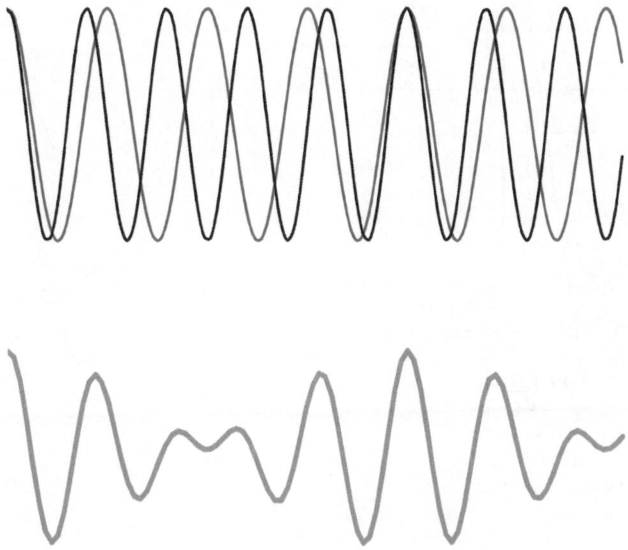

Figure 4.8 Two waves of similar frequencies, when added give an amplitude–modulated carrier. Phase and group velocity concepts.

4.6.1 Rayleigh and Timoshenko beam theories

It is clear from the previous expression that, as the wave length decreases, the group velocity increases boundless, which is physically unsound. Here, we must realize that the flexural wave, when reaching some portion of the beam, makes its cross–section to rotate since the wave displaces the beam transversely. This requires energy and its consideration leads to the so called Rayleigh theory for beams. Also, the transverse shearing stress makes the beam cross–section not to remain perpendicular to the centroid axis. Indeed, the slope of the centroid axis, $\partial w/\partial x$ is now made up of the contribution of bending, $\partial w_b/\partial x$, and shear $\partial w_s/\partial x$, see next figure, and such a consideration leads to the so called Timoshenko beam theory.

The dynamics of Timohenko beams is explored in C.A.N. Dias and M. Alves, A method to solve the nonlinear eigenvalue problem of Timoshenko plane frames with rigid offsets and end releases, Journal of Sound and Vibration, Volume 332, p. 1372–1387, 2013.

In deriving the governing equations for this new beam theory, we still use the bending moment—curvature relation

$$M = -EI\frac{\partial \phi}{\partial x}.$$

The transverse shear force is

$$Q = \int_A \tau \mathrm{d}A = G\int_A \gamma \mathrm{d}A = \bar{k}G\gamma_0 A = \bar{k}AG\left(\frac{\partial w}{\partial x} - \phi\right),$$

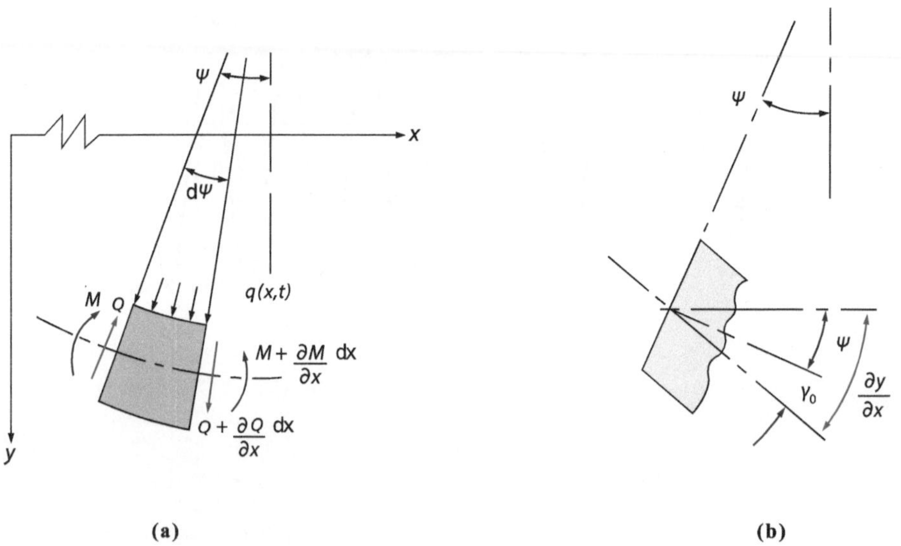

Figure 4.9 Deformation in a generic beam due to bending and shear stresses.

with γ being the shear strain and $\bar{k}$ a coefficient equal, for instance, to 0.833 and 1.111 for rectangular and circular cross–sections, respectively, which gives Q from the single value of the shear slope at the centroid of the cross–section.

The equation of motion comes from the vertical equilibrium and from bending moment summation,

$$\frac{\partial Q}{\partial x} + p = \rho A \frac{\partial^2 w}{\partial t^2} \qquad Q - \frac{\partial M}{\partial x} = \rho I \frac{\partial^2 \phi}{\partial t^2}.$$

Taking the x derivative of the latter and substituting it in the former, it follows, for the free vibration case, that

$$\frac{EI}{\rho A}\frac{\partial w^4}{\partial x^4} - \underbrace{\frac{I}{A}\frac{\partial^4 w}{\partial x^2 \partial t^2}}_{\text{Rayleigh term}} + \underbrace{\frac{E}{\bar{k}G}\frac{\partial^4 w}{\partial x^2 \partial t^2} + \frac{\rho I}{\bar{k}GA}\frac{\partial^4 w}{\partial t^4}}_{\text{Timoshenko terms}} + \frac{\partial^2 w}{\partial t^2} = 0,$$

where a linear elastic material is assumed.

This equation reduces to $\frac{EI}{\rho A}k^4 - \frac{I}{A}k^4c^2 - k^2c^2 = 0$ when only the Rayleigh term is retained.

If we now use as a trial solution for the above equation the harmonic transverse displacement

$$w = A\cos(\omega t - kx),$$

we obtain

$$\frac{EI}{\rho A}k^4 - \frac{I}{A}\left(1 + \frac{E}{\bar{k}G}\right)k^4c^2 - k^2c^2 + \frac{\rho I}{\bar{k}GA}k^4c^4 = 0,$$

with $c = \omega/k$.

For very large wave numbers (very small λ), the limit of the above equation for $k \to \infty$ gives the roots $\pm\sqrt{E/\rho}$ and $\sqrt{kG/\rho}$, *ie* the flexural wave is bounded to a finite value.

The solution to the above equation for cylindrical steel and aluminium bars is plotted in the next figure against frequency.

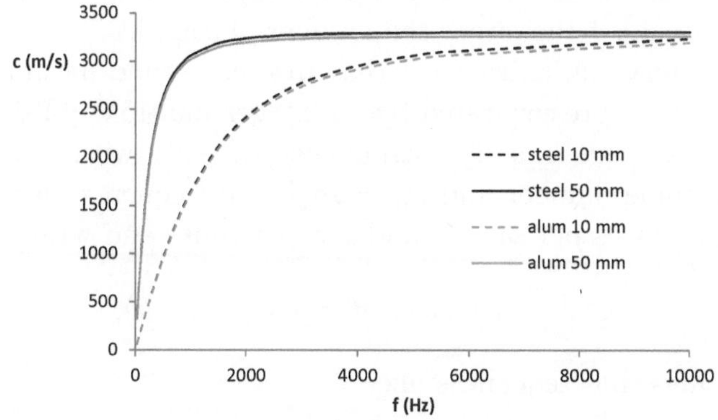

Figure 4.10 Flexural wave speed in aluminium and steel cylindrical bars versus frequency (diameter as indicated).

The elementary theory for flexural waves given here, when applied to a cylindrical rod of radius r, gives reasonable results for $r/\lambda < 0.1$. For a steel rod with radius 5 mm, a limit frequency is 100 kHz, approximately.

4.7 Digital signals

The measurement of a signal is usually performed by electronic boards connected to a computer. There are variants to this arrangement, like the use of oscilloscopes. In any case, it is always necessary at some stage to have the measured data in the computer memory so it can be processed. A digital board performs the task of digitizing the analogue signal coming from the sensor, allowing software to read the data. Because the board is digital, it samples the signal, *ie* it reads one item of information every given time interval. This sample rate, N, can be chosen by the experimentalist and is limited by the maximum sample rate of a given channel of the board. Boards with sample rates as high as 1 MHz per channel and with 12 bits amplitude resolution are common. The signal resolution of 12 bits means that the full voltage range of the channel, commonly in between -10 V and $+10$V, can be divided in 2E12 = 4096 parts. For a 16 bit amplitude resolution the signal is divided by 65536. A digital signal is a discrete–time signal having a set of discrete (quantized) values.

Let us consider the simple discretized harmonic signal

$$x(n) = \sin(\omega n) = \sin(2\pi \bar{f} n),$$

with $\bar{f}$ being cycles per sample. If we digitize this function by 4 samples in its period, it means that $\bar{f} = 1/4$. For 8 samples it follows that $\bar{f} = 1/8$, so the smaller the $\bar{f}$ the larger the number of samples.

Consider now the affirmative that discrete—time harmonic signals whose frequencies are separated by an integer multiple of 2π are in fact identical. To see this, take the signal $\cos(n\omega_0 + \theta)$, with $-\infty < n < \infty$ being the sample number and $\omega_0 = 2\pi \bar{f}_0$ the frequency in radians per sample ($\bar{f}_0$ is cycles per sample and not per time) and write

See, for instance, Introduction to Digital Signal Processing, J.G. Proakis and D. G. Manolakis, 1988.

$$\cos[n(\omega_0 + 2\pi) + \theta] = \cos(n\omega_0 + \theta),$$

which indicates that sequences like

$$x_1(n) = A\cos(n\omega_1 + \theta),$$

$$x_2(n) = A\cos(n\omega_2 + \theta) \qquad \dots$$

or, in general,

$$x_k(n) = A\cos(n\omega_k + \theta) \qquad k = 0, 1, 2 \dots$$

are identical, with

$$\omega_k = \omega_0 + 2\pi k, \qquad -\pi \le \omega_0 \le \pi, \qquad \frac{-1}{2} \le \bar{f}_0 \le \frac{1}{2}.$$

Here, the fundamental range could well be from 0 to 2π but instead we adopt $-\pi \le \omega_0 \le \pi$. Note that, although frequency is a positive physical quantity, we can use negative frequencies if convenient, which is the case here. These negative frequencies follow from the Euler identity, $e^{\pm i\alpha} = \cos\alpha \pm i \sin\alpha$, so that

$$2\cos(\omega t + \theta) = e^{i(\omega t + \theta)} + e^{i(-\omega t - \theta)}.$$

This result points to the uniqueness of discrete–time signals in the frequency range of $-\pi \le \omega \le \pi$, *ie* $\frac{-1}{2} \le \bar{f} \le \frac{1}{2}$. Signals, s_a, whose frequency falls out of this interval, *ie* $|\omega| > \pi$ are identical to signals, s, with frequency $|\omega| < \pi$. We refer to s_a as an alias of s.

Next, consider an analogue signal with frequency F,

$$x_a(t) = \cos(2\pi F t).$$

This signal is sampled every T seconds so that the sample frequency is

$$F_s = \frac{1}{T},$$

with the resulting discrete–time signal being

$$x(n) = \cos(2\pi F n T) = \cos(2\pi F n \frac{1}{F_s}) = \cos(2\pi \bar{f} n).$$

Clearly, the sampling, $\bar{f}$, the analogue, F_s, and the digital, F, signal frequencies are related by

$$\bar{f} = \frac{F}{F_s},$$

as illustrated next.

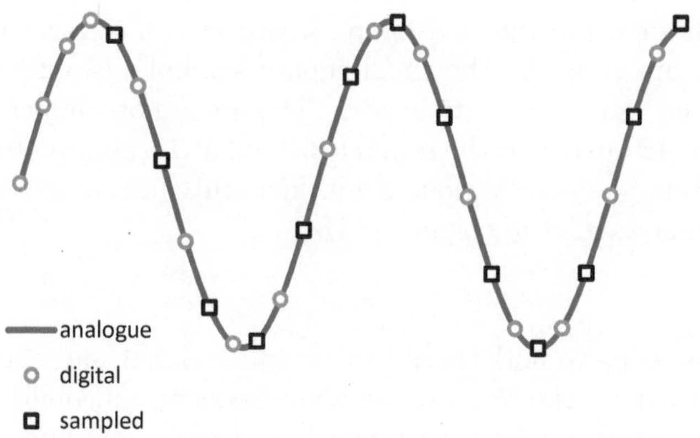

Figure 4.11 An analogue, digital and digital sampled signal.

Remember now that $\frac{-1}{2} \leq \bar{f} \leq \frac{1}{2}$, which leads to

$$\frac{-F_s}{2} \leq F \leq \frac{F_s}{2}.$$

Hence, sampling of a continuous signal is likely to map its infinity frequency range into the finite range $|\bar{f}| \leq \frac{1}{2}$, meaning that, if a signal to be measured has an expected harmonic of frequency F_{max}, the sample rate should be

$$F_s \geq F_{max}.$$

This minimum limit for the sample rate is famously known as the Nyquist rate. In other words, F_s is the minimum sampling rate that avoids aliasing of a signal. This explains the factor 2 used in the calculation of the sampling frequency.

Note that the total window time for a signal, say t, is related to the number of samples, n, by $t = nT$, with T, the sampling period, being the inverse of the sampling frequency. Hence, if the sampling rate is set too high, for the same number of samples, we are covering a smaller window frame. In this case, we need to increase the number of readings, which means that more memory has to be used.

■ SAMPLING RATES

Use different sample rates to acquire a harmonic signal of amplitude 1 and frequency 10 kHz.

Let us sample this signal using two frequencies: one of 15 kHz, which does not obey the Nyquist criterion, and one of 50 kHz, which does. The result is shown in the next figure, where we can see that the higher sampling, represented by the small square symbols, captured the amplitude of the signal along time well. This does not happen with the low sampling frequency. If the signal captured at 15 kHz, represented by the open circles, had the various points smoothly interpolated, it would indicate a sine wave of frequency 5 kHz.

■

It is interesting to note that, if we sample a signal, say, $\sin 1000\pi t$ at 1000 Hz, we satisfy the Nyquist sampling frequency theorem. However, the sampled signal will be $\sin \pi n$, which is zero for any integer n. This means we are missing the signal altogether because it happened that the sampling occurred at the zero–crossing points. A simple remedy to this situation is to slightly increase the sampling frequency. But let us keep in mind that in the real world a signal encompasses many harmonics and it is unlikely that such a coincidence will happen.

A signal can be seen as formed by an array of harmonics and this is best understood using the concept of Fourier series, as now seen.

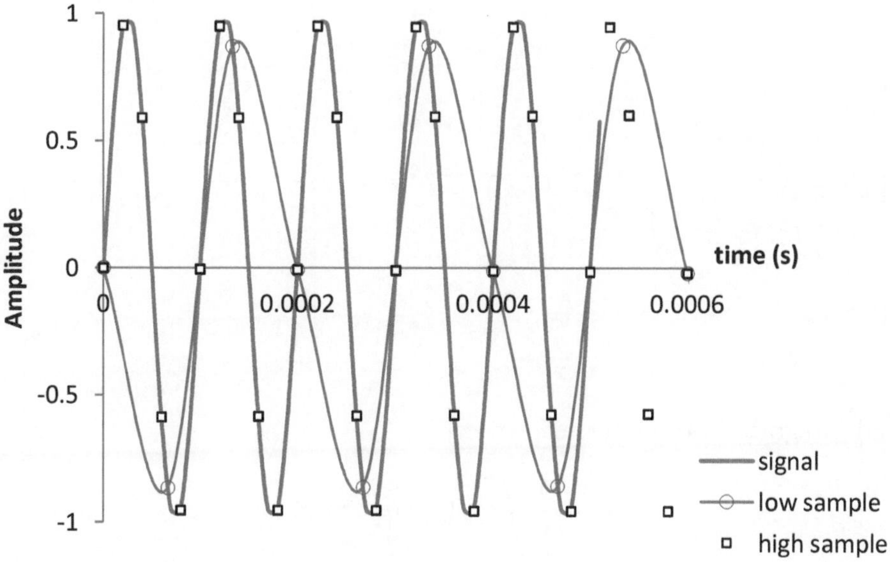

Figure 4.12 Sampling of a harmonic signal.

4.7.1 Fourier series and Fourier transform

Let us plot the function $\sin(\omega t)$ in the interval, say, [0..1] with $\omega = 4\pi$, as shown in the next figure. Now, on the same plot we add a function, say $0.8\cos(5\omega t) + 0.5\sin(2\omega t)$ and we observe that the period of both waves is the same. We conclude with similar examples that we can add as many trigonometric functions as we wish, of any frequency larger than ω and yet such a composed function repeats itself every ωt.

We can modify slightly this reasoning and plot the function $0.406\cos(\pi x) + 0.045\cos(3\pi x) + 0.016\cos(5\pi x)$ and the line $0.5 - x$, as in the figure, and be convinced that the straight line can be approximated by a sum of trigonometric functions well.

This was anticipated by J. Fourier as long ago as 1807 and the idea can be generalized to

$$f(x) = \sum_{n=0}^{\infty} (a_n \cos nx + b_n \sin nx),$$

ie any periodic function can be represented by a sum of harmonic waves, each one with an amplitude given by the coefficients of this series. To determine these coefficients, multiply the expression above first by $\cos nx$

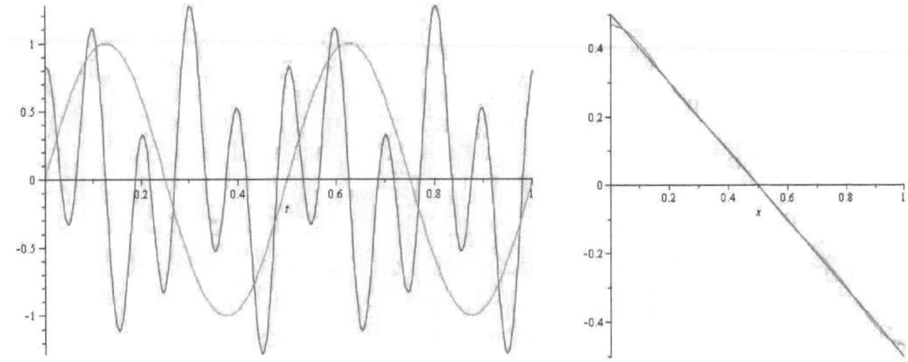

Figure 4.13 (a) A pure harmonic signal (green) and a sum of harmonics (red). (b) A straight line approximated by a sum of harmonics.

Use the identities $2\sin A\cos B = \sin(A-B)+\sin(A+B)$ and $2\cos A\cos B = \cos(A-B)+\cos(A+B)$.

and integrate in the interval $[-\pi:\pi]$,

$$a_n = \frac{1}{\pi}\int_{-\pi}^{\pi} f(x)\cos nx\,\mathrm{d}x,$$

since $\int_{-\pi}^{\pi}\sin nx\cos nx = 0$. The same procedure, but now with the function $\sin nx$, gives

$$b_n = \frac{1}{\pi}\int_{-\pi}^{\pi} f(x)\sin nx\,\mathrm{d}x.$$

Note that

$$a_0 = \frac{1}{\pi}\int_{-\pi}^{\pi} f(x)\mathrm{d}x,$$

represents the average value of $f(x)$.

FOURIER SERIES OF A SQUARE FUNCTION

Represent a square function of amplitude $A=1$ and period 2π by a Fourier series.

The coefficients a_n are all zero since

$$a_n = \frac{1}{\pi}\int_{0}^{\pi} A\cos nx\mathrm{d}x + \frac{1}{\pi}\int_{-\pi}^{0} -A\cos nx\mathrm{d}x =$$

$$\frac{-1}{n\pi}\sin nx\,|_0^{\pi} + \frac{1}{n\pi}\sin nx\,|_{-\pi}^{0} = 0$$

and

$$b_n = \frac{1}{\pi}\int_{0}^{\pi} A\sin nt\mathrm{d}x + \frac{1}{\pi}\int_{-\pi}^{0} -A\sin nx\mathrm{d}x = \frac{4}{n\pi}, \qquad n = 1,3,5\ldots$$

which allow us to represent $f(x)$ by

$$f(x) = \frac{4}{\pi}\left(\frac{\sin 1x}{1} + \frac{\sin 3x}{3} + \frac{\sin 5x}{5} + \dots\right)$$

as graphically represented in the next figure, which also includes other functions and their harmonics.

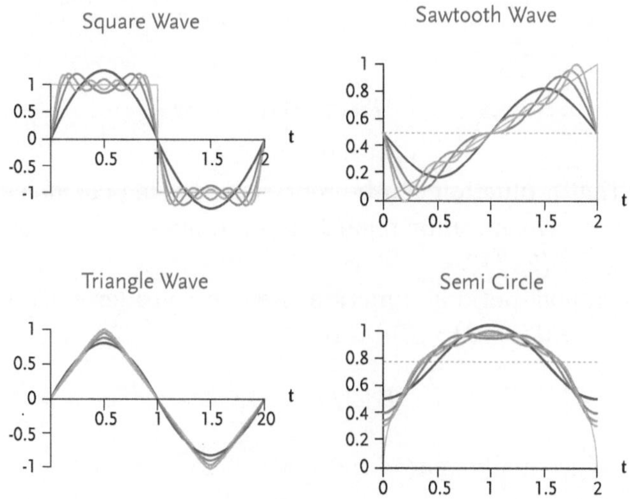

Figure 4.14 Various periodic signals (in light grey) and their representation by a sum of harmonics (first harmonic in red).

■

The period of these waves is not necessarily 2π. We only need to make the change $x = (2\pi/T_p)t = \omega_0 t$ to work in time, t, and frequency, ω_0 (rad/s). The Fourier series then becomes

$$f(t) = \frac{a_0}{2} + \sum_{n=1}^{\infty} (a_n \cos n\omega_0 t + b_n \sin n\omega_0 t),$$

which will serve as a basis for the introduction of the Fourier transform.

As already seen, the Euler formula allows us to relate the sine and cosine of an argument θ according to $e^{i\theta} = \cos\theta + i\sin\theta$, so that

$$\cos n\omega_0 t = \frac{e^{in\omega_0 t} + e^{-in\omega_0 t}}{2} \quad \text{and} \quad \sin n\omega_0 t = \frac{e^{in\omega_0 t} - e^{-in\omega_0 t}}{2i}$$

which, when substituting in the series expansion for $f(t)$, allows us to approximate a periodic signal by

$$f(t) = c_0 + \sum_{n=1}^{\infty} \left(c_n e^{in\omega_0 t} + c_{-n} e^{-in\omega_0 t}\right) = \sum_{n=-\infty}^{\infty} c_n e^{in\omega_0 t},$$

with

$$c_0 = \frac{a_0}{2}, \quad c_n = \frac{a_n - ib_n}{2}, \quad c_{-n} = \frac{a_n + ib_n}{2},$$

or

$$c_n = \frac{1}{T_p} \int_{-T_p/2}^{T_p/2} f(t) e^{-in\omega_0 t} \mathrm{d}t.$$

This shows that a function can be represented using coefficients ranging from $-\infty$ to ∞, which can be regarded as a function of n or as a function of ω_n since $\omega_n = (2\pi/T)n$.

An argument by L. Vas in her Mathematical Methods for the Physical Sciences classes.

Now, for a non–periodic function, we can consider that its period is $T_p \to \infty$, *ie* $\omega \to \mathrm{d}\omega = (2\pi/T)\mathrm{d}n$, or

$$T\mathrm{d}\omega/2\pi = 1,$$

since $\mathrm{d}n = 1$ for equally spaced intervals. It follows that

$$T_p c(\omega) = \int_{-T_p/2}^{T_p/2} f(t) e^{-in\omega t} \mathrm{d}t,$$

leading to

$$f(t) = \sum_{n=-\infty}^{\infty} c_n e^{i\omega t} = \sum_{n=-\infty}^{\infty} \frac{T_p \mathrm{d}\omega}{2\pi} c(\omega) e^{i\omega t},$$

or, at the limit,

$$f(t) = \frac{1}{2\pi} \int_{-\infty}^{\infty} \underbrace{\left(\int_{-\infty}^{\infty} f(t) e^{-i\omega t} \mathrm{d}t \right)}_{F(\omega),\ \text{Fourier Transform}} e^{i\omega t} \mathrm{d}\omega.$$

We can then write our function $f(t)$ as

$$f(t) = \frac{1}{2\pi} \int_{-\infty}^{\infty} F(\omega) e^{i\omega t} \mathrm{d}\omega,$$

with

$$F(\omega) = \int_{-\infty}^{\infty} f(t)e^{-i\omega t}\mathrm{d}t$$

being the Fourier transform of our function, $f(t)$.

This equation has the important characteristic of representing a function in the frequency domain rather then in time, allowing us to obtain the spectrum of a signal, *ie* its representation in terms of frequency. Of course we measure a signal in time, but with the Fourier transformation we readily represent it in the frequency domain. This is useful in many ways since the frequency content of a signal can disclose interesting aspects that are not evident in the time domain.

Observe that the Fourier transform is a complex number. Accordingly, we write

$$F(\omega) = \int_{-\infty}^{\infty} f(t)e^{-i\omega t}\mathrm{d}t = \int_{-\infty}^{\infty} f(t)\cos\omega t\mathrm{d}t + i\int_{-\infty}^{\infty} -f(t)\sin\omega t\mathrm{d}t,$$

or

$$F(\omega) = A(\omega) + iB(\omega),$$

with

$$\underbrace{|F(\omega)| = \sqrt{A(\omega)^2 + B(\omega)^2}}_{modulus} \qquad \underbrace{\phi(\omega) = \tan^{-1}\frac{B(\omega)}{A(\omega)},}_{phase}$$

which are the modulus and the phase of the transformed signal at the given frequency.

Note that in the case of Fourier series of a periodic signal, we obtain the amplitude of the various harmonic that are associated with discrete values of frequency. In the case of the Fourier transform of a non–periodic signal, the spectrum is defined for all real values of frequency.

■ FOURIER TRANSFORM AND SERIES

Obtain the Fourier transform and the Fourier series coefficients for the function depicted in the next figure.

The Fourier transform (of the non–periodic function) is calculated from

$$F(\omega) = \int_{-1}^{0} (t+1)e^{-i\omega t}\mathrm{d}t + \int_{0}^{1} (-t+1)e^{-i\omega t}\mathrm{d}t =$$
$$\frac{2}{\omega^2}(1-\cos\omega) = \frac{2}{\omega^2}2\sin^2\frac{\omega}{2},$$

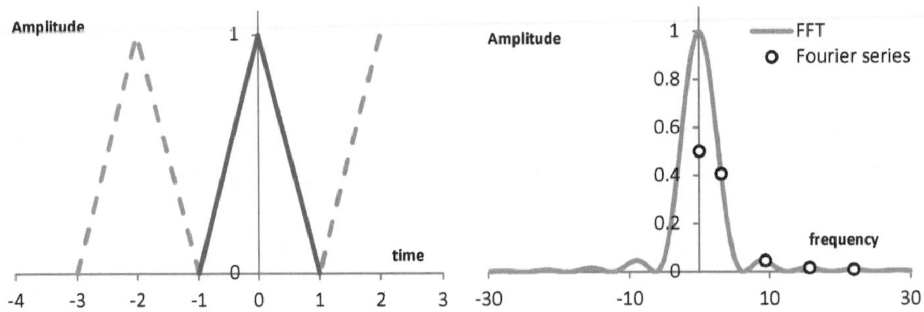

Figure 4.15 A periodic (dashed) and non–periodic (continuous) function.

which gives

$$F(\omega) = \mathrm{sinc}^2\left(\frac{\omega}{2}\right),$$

with $\mathrm{sinc}x = \sin x/x$.

The Fourier series representation of the periodic function (dashed lines in the figure) is given by

$$f(t) = \frac{1}{2} + \frac{4}{\pi^2}\left(\cos \pi t + \frac{1}{3^2}\cos 3\pi t + \frac{1}{5^2}\cos 5\pi t \ldots\right).$$

The above figure plots both the Fourier transform and the Fourier coefficients. It is evident that the former is continuous on the frequency domain, while the Fourier series coefficients are discrete. Note also that, for this particular function, the Fourier transform has only real values and that its maximum amplitude is 1 since $\lim_{x\to 0}\mathrm{sinc}x = 0$. This maximum amplitude is different from the first term of the Fourier series for this function.

■

■ FOURIER TRANSFORM OF A TOP HAT FUNCTION

Obtain the Fourier transform of the top hat function in the figure, exploring the case when its period tends to zero.

The very definition of the Fourier transform gives

$$F(\omega) = \int_{-a}^{a} Ae^{-i\omega t}\mathrm{d}t = \frac{A}{i\omega}\left(e^{ia\omega} - e^{-ia\omega}\right)$$
$$= \frac{2A}{\omega}\sin a\omega = 2Aa\mathrm{sinc}(a\omega),$$

which is plotted in the figure for various values of a and for $A = 1/2$.

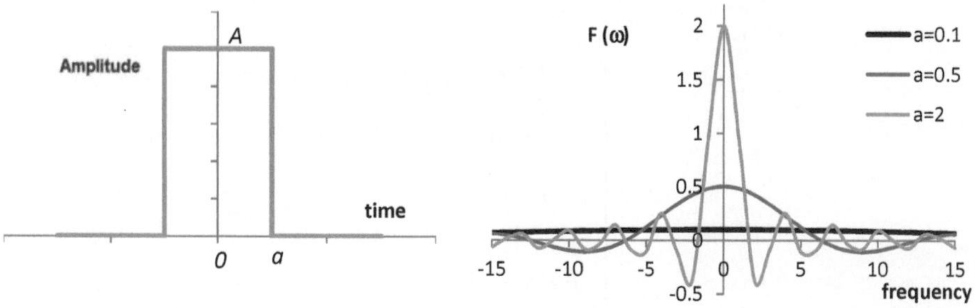

Figure 4.16 The top hat function on time and frequency.

Observe the important fact that, for small values of a, *ie* very sharp input, we obtain a spectrum that is spread flat and contain many frequencies. This is the basis for one of the methods of experimental modal analysis, where one hammers the structure and excite many vibration modes. It is also interesting to note that we have in the figure negative intensities of the Fourier transform. We can calculate the modulus and phase and we will note that this function is such that it has only the real component, with the phase having the value of either 0 or π rad and the amplitude being $|F(\omega)|$.

■

4.7.2 Discrete Fourier transform

The knowledge we have gained so far on Fourier transform must be enhanced so as to take into account that in modern systems the reading data are digital. We deal with a set of points equally spaced in time and whose amplitude is proportional to the measured variable.

Let then our signal $f(t)$ be represented by N samples, each recorded within an interval T. In the Fourier integral

$$F(\omega) = \int_{-\infty}^{\infty} f(t) e^{-i\omega t} \mathrm{d}t$$

we use

$$\omega t = \frac{2\pi}{N} t = \frac{2\pi}{N} nT = \frac{2\pi}{N} nk$$

and $f(t) = f(k)$, so we have

$$F(n) = \sum_{k=0}^{N-1} f(k) e^{-i\frac{2\pi}{N}nk}, \qquad n = [0..N-1],$$

or, in matrix form,

$$\begin{bmatrix} F(0) \\ F(1) \\ F(2) \\ F(3) \\ \vdots \\ F(N-1) \end{bmatrix} = \begin{bmatrix} 1 & 1 & 1 & 1 & \dots & 1 \\ 1 & A & A^2 & A^3 & \dots & A^{1(N-1)} \\ 1 & A^2 & A^4 & A^6 & \dots & A^{2(N-1)} \\ 1 & A^3 & A^6 & A^9 & \dots & A^{3(N-1)} \\ \vdots & & & & & \\ 1 & A^{N-1} & A^{N-2} & A^{N-3} & \dots & A \end{bmatrix} \begin{bmatrix} f(0) \\ f(1) \\ f(2) \\ f(3) \\ \vdots \\ f(N-1) \end{bmatrix},$$

with $A = e^{-i2\pi/N}$ and noting that $A^{3(N-1)} = A^{(N-3)}$ and so on.

Note that $F[n]$ has complex coefficients, while $f[k]$ has real ones. Having in mind that the spectrum is symmetric about $N/2$, the terms $F[n]$ and $F[N-n]$ produce two frequency components: the higher, at $n > N/2$ being the aliasing of the lower, this later having physical meaning $(n \leq N/2)$.

■ FOURIER TRANSFORM CALCULATION

Obtain the Fourier transform of the signal in the figure and give a detailed interpretation to the calculations.

This example and sub–section are drawn, with permission, from a lecture text by S. Roberts, Department of Engineering Science, University of Oxford.

Our signal is represented by

$$f(t) = 5 + 2\cos(2\pi t - \pi/2) + 3\cos 4\pi t,$$

meaning that it has a constant term of amplitude 5 and two harmonic terms, the first at 1 Hz with a phase shift of 90° and another at 2 Hz.

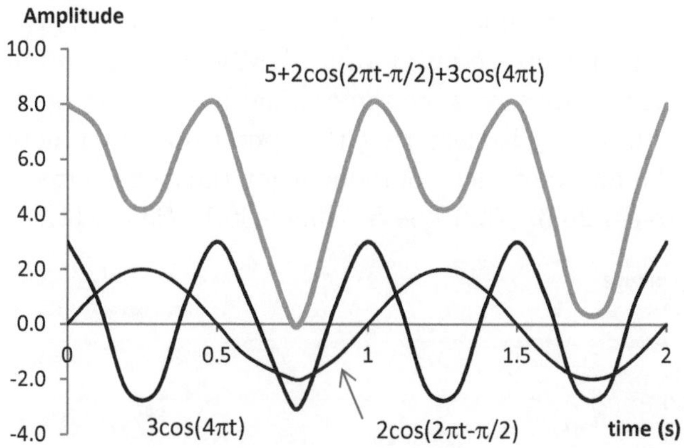

Figure 4.17 A signal formed by two harmonics and one DC component.

We first need to decide how many samples of the signal we want to deal with. In order to meet the Nyquist criterion, we need at least a sample frequency of $2 \times 2 = 4$ Hz, so we sample

$$f[k] = 5 + 2\cos(\pi k/2 - \pi/2) + 3\cos\pi k,$$

since $t = kT_s = k/4$, which gives $f[0] = 8, f[1] = 4, f[2] = 8, f[3] = 0$. It follows, from the very definition of the Fourier Transform that

$$F[n] = \sum_{k=0}^{3} f[k]e^{-ink\pi/2} = \sum_{k=0}^{3} f[k](-i)^{nk},$$

or, in matrix form,

$$\begin{bmatrix} F[0] \\ F[1] \\ F[2] \\ F[3] \end{bmatrix} = \begin{bmatrix} 1 & 1 & 1 & 1 \\ 1 & -i & -1 & i \\ 1 & -1 & 1 & -1 \\ 1 & i & -1 & -i \end{bmatrix} \begin{bmatrix} f[0] \\ f[1] \\ f[2] \\ f[3] \end{bmatrix} = \begin{bmatrix} 20 \\ -i4 \\ 12 \\ i4 \end{bmatrix}.$$

For $n = 0$, note that $F[0] = \sum f[k]$, so that $f_0[k] = F[0]/N$, *ie* $f[0]$ is the average of all the samples, which in our case gives the d.c. value of $F[0]/N = 20/4 = 5$, which agrees with our original signal.

As for the first harmonic, we have that $F[1] = -i4 = F^*[3]$ and the peak amplitude is $\frac{2}{N}|F[1]| = 2$, while the phase is $\arg F[1] = -90°$. Finally, $F[2] = 12$ and this gives

$$f_2[k] = \frac{1}{N}F[2]e^{i\frac{2\pi}{N}2k} = F[2]e^{i\pi k}/4 = 3\cos(\pi k).$$

The resulting amplitude of each harmonic, of 2 and 3, are sometimes shown in a root—mean—square format, RMS, *ie* the amplitude divided by $\sqrt{2}$. The next figure shows the same signal but with 16 points sampled at 10 Hz and 20 Hz. It is clear that the spectrum is symmetrical about position 8. The phase of each component for the sample frequency of 20 Hz is (in degrees) 26.6, -70.0, -58.8, -46.9, -35.1, -23.3, -11.7.

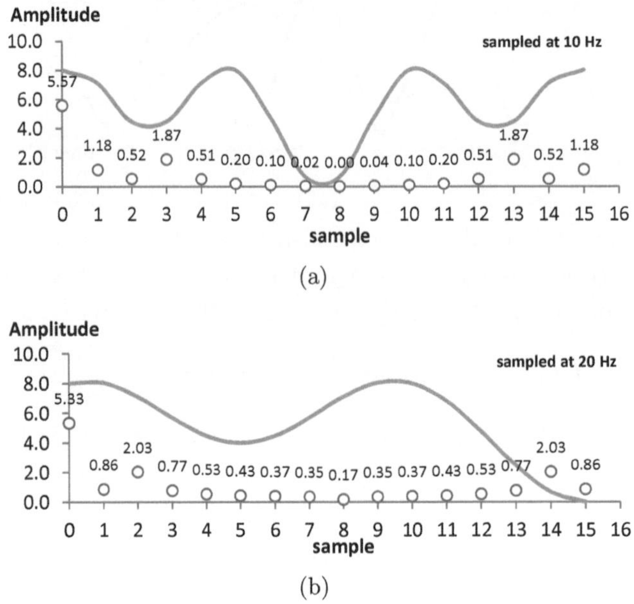

Figure 4.18 The signal sampled at (a) 10 Hz and at (b) 20 Hz.

■

4.7.3 Filters

One problem when reading signals like the ones generated by impacting a bar is the presence of noise. Noise not only makes the reading of a signal difficult but also may hide important information of a given phenomenon. A filter is used to remove noise from a signal. There are different types of filters, some basic ones being exemplified in the next figure.

Analogue filters are based on hardware, with the cut–off frequency being set by operating keys. Usually, an analogue filter is placed in between the amplifiers and the acquisition board.

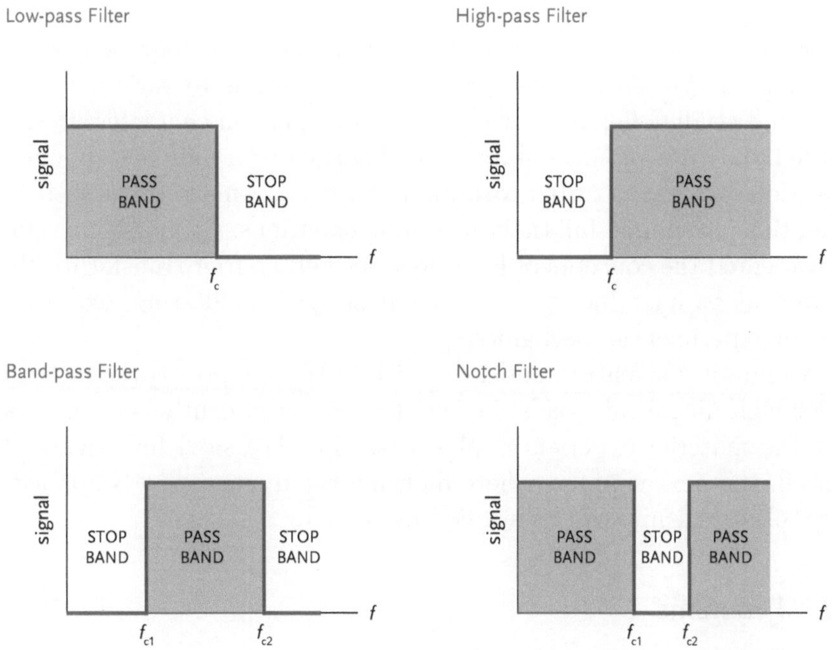

Figure 4.19 Basic types of ideal filters.

There are also digital filters, which operate in the digitalised signal to output a more noisy free signal. Note that we can acquire a signal to the computer's memory and perform a spectral analysis, *ie* FFT the signal, remove the unwanted supposedly known contents, and apply the inverse of FFT to obtain in the time domain a cleaner signal. This is a valid procedure but one should exercise care, as already commented in Chapter 1 and as now indicated.

If one operates the signal at a digital level, it means that all the frequency content up to a sample rate of f_s of a given signal is loaded to the computer memory. When using a FFT tool to analyse the signal, components of it with frequency content larger than $f_s/2$ will be presented in the spectrum but not in the right position, due to the mirrored aspect of the Fourier transform. This, of course, disturbs the read signal. This is to say that it is always preferable to use an analogue filter, if filtering is required at all, before sampling the signal.

4.8 Closure

We have seen various aspects of the dynamic behaviour of beams in this chapter. We investigate the response of beams to various types of loads and studied flexural waves, with its dispersion characteristics. We studied the case of mass impact on beams under different boundary conditions. It was possible to obtain the displacement of these beams along time, solving in full their governing equations. By analysing signals we presented the concepts of Fourier series and Fourier transform, which allowed us to introduce the important subject of filtering, extensively used in experimental mechanics.

An important feature not dealt with in this chapter is the fact that, under high magnitude loads, beams deform permanently. This is to say that the material experiences plastic strains. We shall investigate this topic in the next chapter, where high intense dynamic loads applied to visco–plastic beams will be studied in details.

4.9 Problems

1. A cantilever of length L, transverse inertia I, cross-section area A, elastic modulus E and density ρ, has on its free extreme a string that holds a mass G. This string is suddenly cut. Determine the ensued motion of the beam.

2. A beam with the beam properties of the previous exercise has a length L and falls horizontally from a height H, impacting two supports distant apart from L. Obtain the motion after the collision, assuming that the beam sticks at the simple supported condition.

3. The beam depicted in the next figure is released from rest at an angle θ. Obtain the resulting motion of the beam considering the beam properties of the previous exercise. In particular, derive an expression for the maximum stress in the beam.

4. From the governing equation for a Timoshenko beam

$$\frac{EI}{\rho A}\frac{\partial w^4}{\partial x^4} - \underbrace{\frac{I}{A}\frac{\partial^4 w}{\partial x^2 \partial t^2}}_{\text{Rayleigh term}} + \underbrace{\frac{E}{kG}\frac{\partial^4 w}{\partial x^2 \partial t^2} + \frac{\rho I}{kGA}\frac{\partial^4 w}{\partial t^4}}_{\text{Timoshenko terms}} + \frac{\partial^2 w}{\partial t^2} = 0,$$

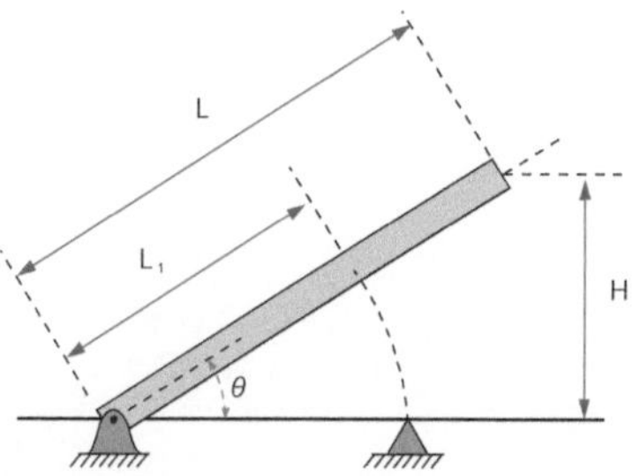

Figure 4.20 A beam hitting a support.

when only the Rayleigh term is retained, prove that

$$\frac{EI}{\rho A}k^4 - \frac{I}{A}k^4c^2 - k^2c^2 = 0$$

5. Prove that the displacement of a simply supported beam hit at its middle by an impulse load $P\delta(x - L/2)\delta(t)$ is

$$w(x,t) = \frac{2P}{\rho AL}\sum_{n=1,3,5,...}^{\infty}\frac{(-1)^{\frac{n-1}{2}})\sin\beta_n x \sin\omega_n t}{\omega_n}$$

6. Show that the beam curvature can be approximated by $\kappa = -\frac{\partial^2 w}{\partial x^2}$ for small transverse displacements.

7. Use a spread sheet to generate a sinusoidal signal of unity amplitude and frequency f. Study the resulting sampled signal by using different sampling frequencies.

8. For the discrete Fourier transform, prove that $F[0] = \sum f[k]$ and that $f_0[k] = F[0]/N$, *ie* $f[0]$ is the average of all the sample, the d.c. value of the signal.

9. In most countries, the use of seat belts in cars is compulsory for at least the front drive and passenger. At the same time, no protection device is required for motorcyclists, except helmets. Do you think that the legislation should change so as to impose on motorcyclists a similar level of protection as required for a car occupant? What devices would you recommend to be compulsory for motorcyclists?

10. Discuss the fact that most cars are capable of travelling well above legal speed limits. Should a control device be installed in cars so they are restrained from speeding over these limits?

CHAPTER 5

Visco–plastic dynamics of beams and plates

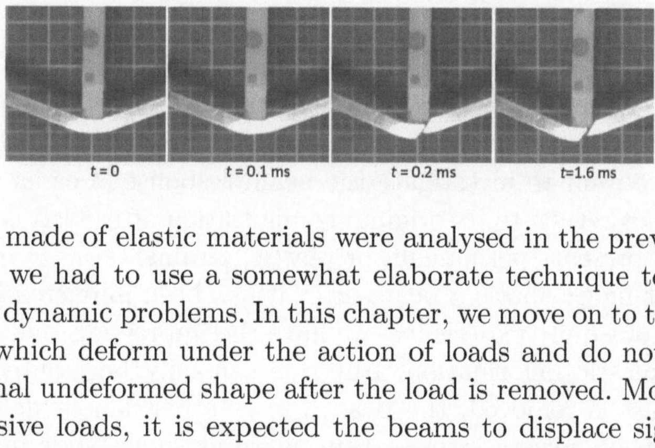

Figure 5.1 A beam failing under the transverse impact of a drop mass.

Beams made of elastic materials were analysed in the previous chapter, where we had to use a somewhat elaborate technique to solve the associated dynamic problems. In this chapter, we move on to the analysis of beams which deform under the action of loads and do not return to their original undeformed shape after the load is removed. Moreover, for very intensive loads, it is expected the beams to displace significantly, which require in the analysis the consideration of non–linear terms in the governing equation. Hence, we enter into two realms of non–linearity: material and geometrical, both taking place along time, so characterizing a non–linear dynamic analysis.

To start our study on beam impact, we need to use the equilibrium equations for beams derived in the last chapter. We then set ourselves

DOI: 10.1201/9781003485292-5

the task to solve these equations and we will see at once that we need to define the material behaviour and to introduce the concept of yield surface. We move on by solving the impact problem of a mass falling on a beam and of a beam loaded by a pressure pulse. We introduce the concept of equivalent pülse and add a section on transverse shear effects on the beam response, now loaded by an initial velocity. We take care also of large displacements of beams, obtaining expressions for strain and strain rate, considering its influence on the material flow stress, so the use of the term visco–plastic for this chapter title.

As to enrich our knowledge on dynamics of more complex structures, we also study plates here. They have a different stress and strain field and their plastic regime is governed by a tridimensional yield surface. We deal here with the problems of a circular plate being loaded by a blast load and by the drop of a mass. These two cases cover practical applications that are important to be explored. We also go further by studying high velocity impact on plates, finishing this chapter with a section on experiments with beams and fiber laminated plates under transverse loads.

5.1 Plastic behaviour of beams

5.1.1 Perfectly plastic material model

As we have seen in the previous chapters, to relate the generalized stresses with the strains we need to invoke the material constitutive law. Here, we want to investigate the beam response in cases where the beam does not return to its original configuration after load is removed, *ie* the beam presents permanent, or plastic, strains.

The next figure shows a beam after it has been impacted by a drop mass. From our daily experience, we have the impression that the beam underwent plastic deformation. But this can only be assured if, after the drop mass is removed, the beam presents permanent deformation. The figure also shows a stress–strain curve at some point in the span of the beam. As indicated, after load removal, a permanent deformation takes place. By unloading, one removes the elastic strains and what is left are plastic strains. This is the reason why plastic strains are called an internal variable.

This is not strictly correct. Even an unloaded structure can lock elastic strains within it.

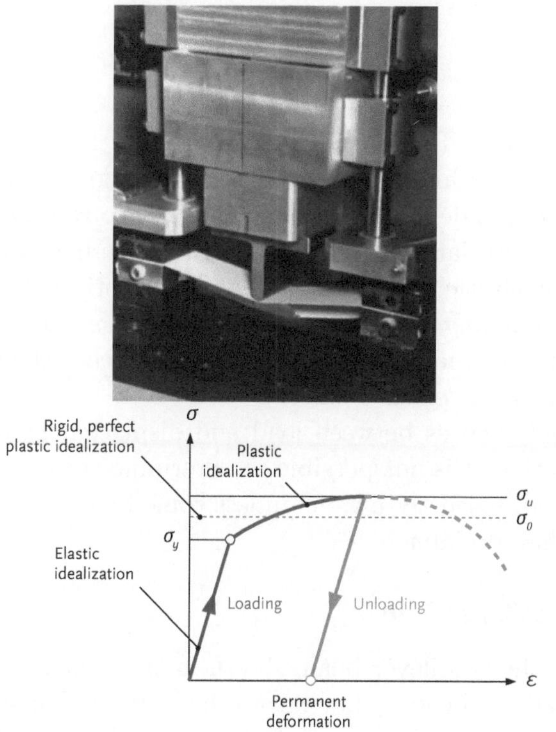

Figure 5.2 A beam undergoing large displacements due to a transverse impact and a typical stress–strain diagram indicating permanent deformation.

We have interest in the plastic response of beams; we add that typical values for the elastic region in the stress–strain diagram for a common steel are $\varepsilon^e = 0.2\%$, $\sigma_y = 300\text{MPa}$, $\sigma_u = 400\text{MPa}$, and that a beam slightly permanently bent may easily present strains of say 5%, *ie* many times greater than the elastic limit strain. This is to say that, in an approximate analysis of a plastically deformed beam, the elastic cooperation to the overall beam response may be disregarded.

Moreover, for many metals, and also for other less conventional materials, the difference between the elastic limit stress, σ_y, and the ultimate stress, σ_u, is not too large. This indicates that the material behaviour can be represented by a single stress parameter, σ_0, which can be given by an expression like

$$\sigma_0 = \frac{\int_0^{\varepsilon_r} \sigma \mathrm{d}\varepsilon}{\varepsilon_r},$$

as depicted in the figure above, with ε_r being the rupture strain. σ_0 is called the flow stress and sometimes can be set as the yielding stress, as the ultimate stress, or even as

$$\sigma_0 = \sqrt{\sigma_y \sigma_u}.$$

We arrive then at the so called perfectly plastic material model or constitutive law, to which we add the adjective rigid to convey that no elastic return is considered. Hence, we can chose to represent the plastic behaviour of a material by a single parameter, the flow stress σ_0, as we do in the linear elastic realm when using the elastic modulus. Also, in the case of a beam, any load below a critical value, say P_c, will not cause any displacement in the beam. When the load reaches the value P_c only then the beam will displace at once.

A rigid–perfectly plastic idealization implies no vibration, no elastic return.

This non–uniqueness between load and displacement (or stress and strain) implies that it is not possible to determine the strains in a beam idealized as rigid, perfectly plastic, unless some further assumptions are made, as we shall do later.

5.1.2 Beam collapse load

Let us analyse the cantilever beam shown in the next figure and obtain the collapse load, *ie* the load that causes the beam to displace. Again, for loads below the collapse load, a perfectly plastic beam will not displace since there is no elasticity in the assumed material model.

For now, let us restrict ourselves to a quasi–static tip load so that the inertia terms in the equilibrium equations are set to zero. We also consider small displacements, hence the membrane forces are too small, leading, from the last chapter, to the equilibrium equation

$$\frac{\partial^2 M}{\partial x^2} = 0,$$

since the distributed load is zero (the concentrated load is considered in the boundary condition). The solution of the above differential equation is

$$M = P(L - x),$$

because $M(L) = 0$. We seek the load P_{c_l} which collapses the beam, *ie* the collapse load that causes every point in a given cross–section (here, in the support, $x = 0$) to go plastic. From the above equation it follows that

$$P_{c_l} = \frac{M_0}{L},$$

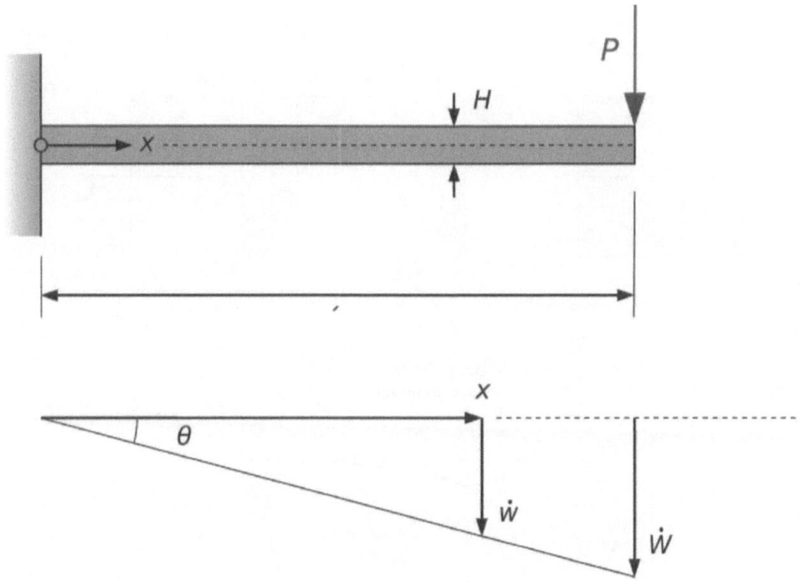

Figure 5.3 A cantilever beam and the associated kinematic field.

where M_0 is the plastic collapse bending moment given by

$$M = M_0 = 2\int_0^{H/2} \sigma_0 B y \mathrm{d}y = \frac{\sigma_0 B H^2}{4}$$

for a rectangular cross–section beam, according to the next figure.

The above collapse load was obtained by solving the equilibrium equation. Alternatively, we can simply equate the external power,

In impact analyses it is better to work with power rather than work.

$$P_{c_u}\dot{W} = P_{c_u}L\dot{\theta}$$

to the internal one,

$$M_0\dot{\theta}$$

to obtain

$$P_{c_u} = \frac{M_0}{L},$$

which happens to be equal to the collapse load obtained by solving the equilibrium equation.

Some remarks are in place here. The subscripts l and u in these equations refer to the lower and upper bound theorems of plasticity. Note that the lower bound solution comes from the equilibrium equation while the upper one from energy balance. When the upper and lower

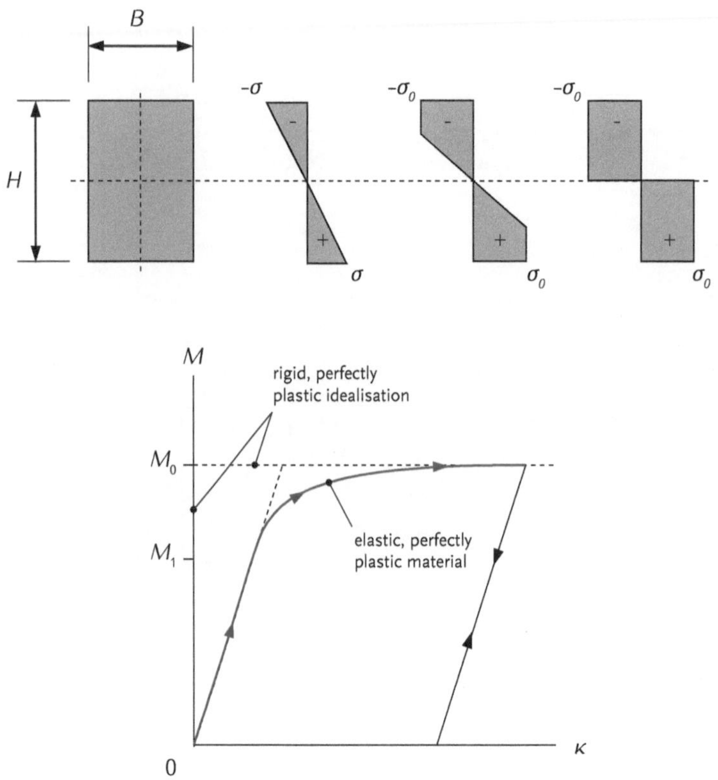

Figure 5.4 Cross–section of a beam becoming fully plastic and the associate bending moment versus curvature.

bound solutions give the same result, we say that the solution is exact. It is important to bear in mind that usually upper bound solutions are somewhat easier to be obtained. Also, it is not always possible to obtain both solutions and they even may not be equal, having as a consequence that the solution is not exact. Finally, to obtain an upper bound solution, the analyst needs to devise a kinematic field. For the present problem, the kinematic field, a bent straight line, is quite obvious but this is not always the case.

■ COLLAPSE LOAD FOR A CLAMPED–SIMPLY SUPPORTED BEAM

Obtain the collapse load for the rectangular cross–section clamped–simply supported beam submitted to a static distributed load p, as depicted in the next figure.

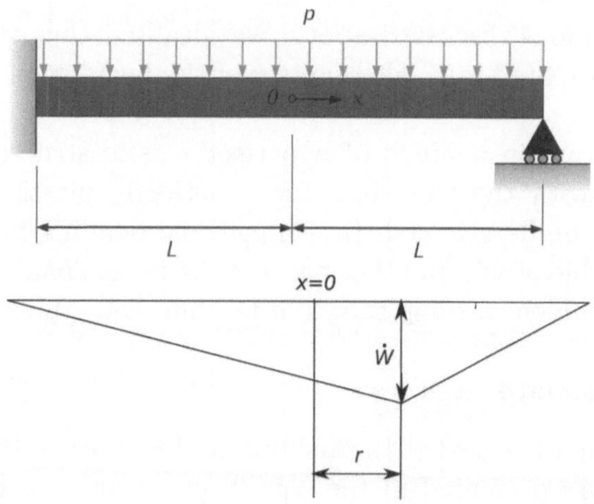

Figure 5.5 (a) A clamped, simply supported beam under distributed load (b) and the associated kinematic field.

We start by obtaining, from equilibrium, the bending moment distribution as

$$M(x) = \frac{p(L-x)^2}{2} - \left(\frac{M_0}{2L} + \frac{pL}{2}\right)(L-x),$$

whose maximum value occurs at $(\mathrm{d}M/\mathrm{d}x = 0)$

$$r = (3 - 2\sqrt{2})L.$$

Since $M(-L) = M_0$, it follows that

$$p_{c_l} = \left(\frac{3}{2} + \sqrt{2}\right)\frac{M_0}{L^2}.$$

Now, energy equilibrium can be written as

$$\int_{-L}^{r} p\frac{x+L}{r+L}\dot{W}\mathrm{d}x + \int_{r}^{L} p\frac{L-x}{L-r}\dot{W}\mathrm{d}x =$$

$$M_0\frac{\dot{W}}{L+r} + M_0\frac{\dot{W}}{L+r} + M_0\frac{\dot{W}}{L-r},$$

which gives

$$p_{c_u} = \left(\frac{3}{2} + \sqrt{2}\right)\frac{M_0}{L^2}.$$

■

In the solutions discussed above, we idealised that the mechanical energy was dissipated in single points x of the beam span, giving rise to hinges.

The curvature in a hinge of a perfect plastic structure is infinite, in agreement with the fact that, for a perfectly plastic material, the strains cannot be determined. In a hinge, the bending moment attains the collapse value of M_0. In other parts of the beam span with no hinges, the bending moment attains values in the interval $-M_0 < M < M_0$.

For plates, we refer to a hinge line instead of a hinge point.

5.1.3 Yield surface

We have so far disregarded the membrane force acting in a beam, although they are presented in the equilibrium equations derived in the previous chapter. We will now take a closer look at the interaction of the membrane forces with the bending moment. By doing so, we can solve more complex beam problems, when both bending moments and normal forces are present.

Membrane forces arise in the presence of moderate or large transverse displacements of the beam. In general, the bending moment dominates the response for small transverse displacement, but decreases in importance as the transverse displacement increases. Membrane effects may not be relevant, even for large displacements, if the boundary condition allows in–plane movement, as indicated in the next figure. However, for the pinned boundary condition shown in the same figure there is no bending moment at the pinned support and yet the membrane stresses become dominant as soon as some visually detected transverse displacement takes place.

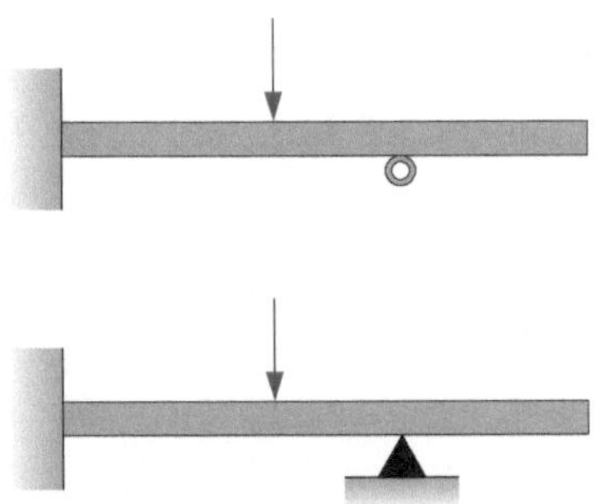

Figure 5.6 Free in–plane sliding and free rotation at the support.

Bending moment and membrane stresses cause yielding in a perfectly plastic beam cross–section, but not independently. They obey the so called yield condition. Considering the cross–section of a rectangular beam in the next figure, we can write

$$M = \sigma_0 B(H - \eta H)\eta H \rightarrow \frac{M}{M_0} = 4\eta\,(1 - \eta)$$

and

$$N = \sigma_0 B[H - 2(H - \eta H)] \rightarrow \frac{N}{N_0} = 2\eta - 1,$$

where $N_0 = \sigma_0 BH$. Combining both equations gives

$$\frac{M}{M_0} + \left(\frac{N}{N_0}\right)^2 = 1,$$

which is a yield surface, depicted in the next figure, relating bending moment and membrane force in a beam made of a perfectly plastic material.

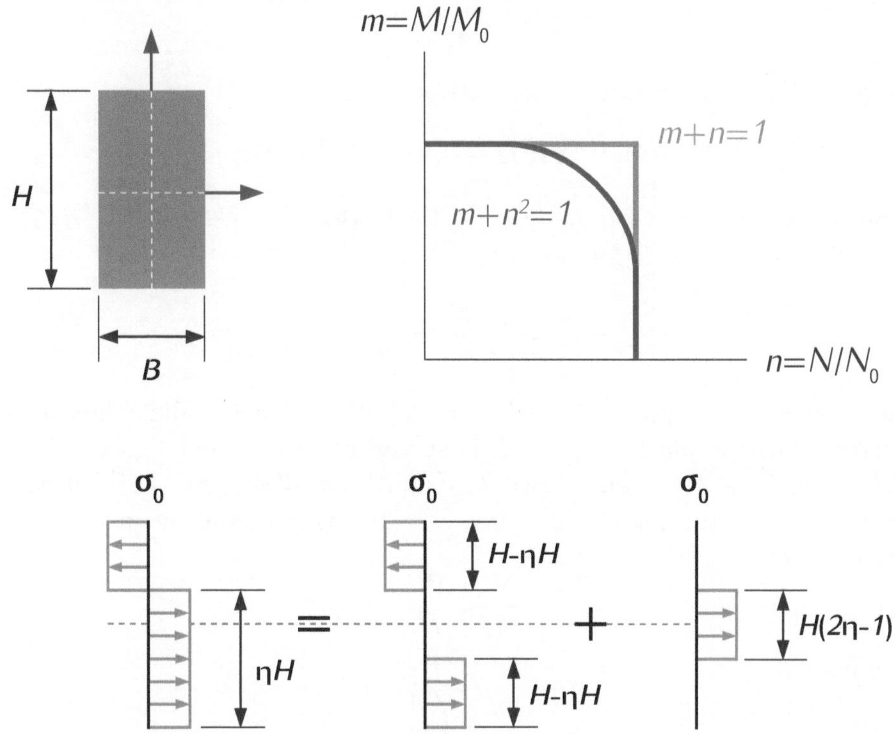

Figure 5.7 Bending moment and membrane stresses in a beam.

We can at once apply this yield surface to the problem of a clamped beam centrally loaded. The equilibrium equation to be solved for this quasi–static problem is

$$\frac{\partial^2 M}{\partial x^2} + N\frac{\partial^2 w}{\partial x^2} = 0.$$

The kinematic (displacement) field is described by $w = W(1 - x/L)$, so the above equation becomes $\partial^2 M/\partial x^2 = 0$, whose solution is

$$M = a_1 x + a_2,$$

a_1 and a_2 being constants to be determined. Note that, at the beam support, the bending moment is not the collapse one since we have to consider the contribution of the membrane force. But we know that the bending moment has the same magnitude at the support and at the mid-span so that

$$M = a_1\left(x - \frac{L}{2}\right) \rightarrow a_1 = \frac{-2a_2}{L}.$$

Note that $P - 2N\sin(\partial w/\partial x) + 2Q\cos(\partial w/\partial x) = 0,\ \partial w/\partial x \ll 1.$

Now, vertical equilibrium at the load point gives

$$P/2 = -a_1 + NW/L \rightarrow A = -P/2 + NW/L,$$

since $Q = \partial M/\partial x = a_1, \partial w/\partial x = -W/L$. Because $\dot{\varepsilon} = \dot{\kappa}(\eta H - H/2)$ and $N/N_0 = 2\eta - 1$, it follows that

$$\frac{\dot{\varepsilon}}{\dot{\kappa}} = \frac{H}{2}\frac{N}{N_0} = \frac{W}{2}.$$

Here we use the approximation $\varepsilon = \Delta L/4l$, which is valid when assuming that the plastic region, l, is spread as in the next figure, with $\Delta L = 2\sqrt{L^2 + W^2} - 2L \cong L(W/L)^2$ and $\dot{\varepsilon} = W\dot{W}/2Ll$. This allows us to quantify the membrane force as a function of the beam central displacement,

$$\frac{N}{N_0} = \frac{W}{H},$$

leading to

$$\frac{P}{P_{c_l}} = 1 + \left(\frac{W}{H}\right)^2,$$

where $P_{c_l} = 4M_0 L$ is the collapse load for a central loaded beam with no membrane force.

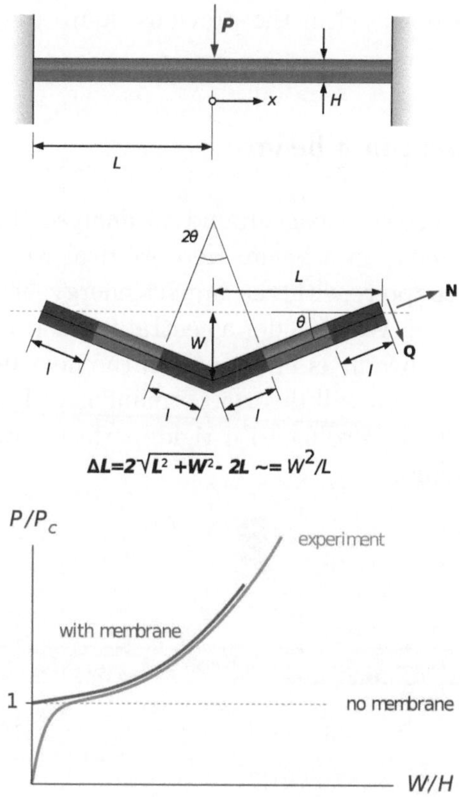

Figure 5.8 Clamped beam centrally loaded and the theoretical *versus* experimental force–displacement diagram.

It should be noted in this solution that, if $W > H$ then $N > N_0$, which is not possible since our material model does not strain harden, *ie* does not increase its strength with straining, and therefore the load point cannot lie outside the yield surface. This means that our previous solution is valid in the interval $0 < W \leq H$ and we need to seek for another one when the mid–displacement becomes larger than the beam height. By noting that

$$\frac{M}{M_0} = \left(1 - \frac{W^2}{H^2}\right)\left(1 - \frac{2x}{L}\right),$$

the bending moment is zero for $W = H$ so that $P = -2Q - 2N\partial w/\partial x = 2N_0 W/L$, giving

$$\frac{P}{P_c} = \frac{2W}{H}$$

for $W > H$.

This solution is depicted in the previous figure together with some experimental results.

5.2 Central impact in a beam

We now have the necessary background to analyse the transverse impact of a dropping mass on a beam. Here we deal with a different case from the previous chapter, with the impact energy being large enough to cause plastic strains. We consider a central impact in a clamped rectangular cross–section beam, as illustrated in the next figure. Due to the impact loading, the beam will deform permanently. The beam material is assumed to behave according to a rigid–perfectly plastic law, so no elastic vibration occurs.

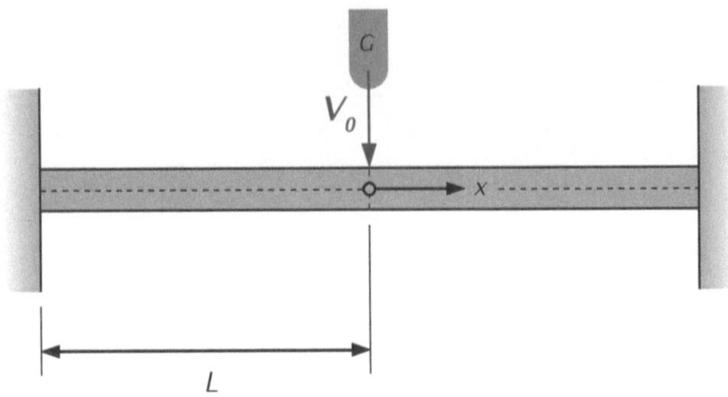

Figure 5.9 A clamped beam under transverse impact.

The first obvious approach to solve this problem, meaning to obtain the beam mid–displacement, is to equate the impact energy with the energy absorbed by the hinges, giving

$$\frac{GV_0^2}{2} = \frac{4M_0}{L}W \quad \rightarrow \quad W = \frac{GV_0^2 L}{8M_0}.$$

Note that this solution does not consider membrane forces, so it is restricted to small transverse displacements of the beam. Observing this solution, we note that it carries information of the beam material, via M_0, of the impact mass, G, and velocity, V_0, and of the beam geometry, L, B, H. The beam density does not play any role here.

If the mass of the beam is not being considered, it means that the beam inertia contribution to the solution is altogether disregarded. This

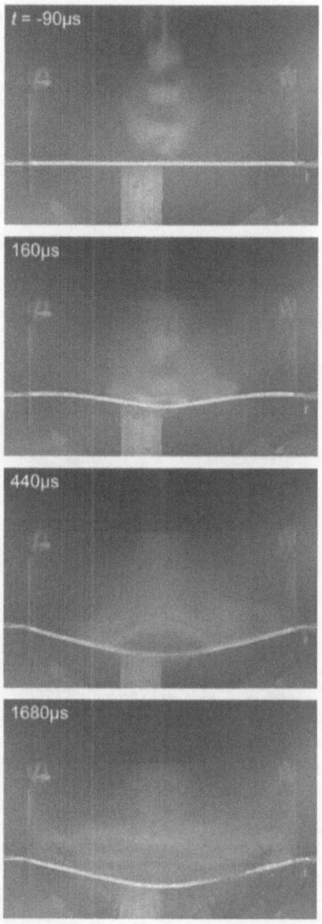

Figure 5.10 Response of an aluminium beam subjected to the impact of a wet sand slug at $V_0 = 94$ m/s. From T. Uth and VS Deshpande, Response of clamped sandwich beams subjected to high velocity impact by sand slugs, International Journal of Impact Engineering, 69, p. 165—181, 2014.

is quite reasonable if the impact velocity is not too high, say lower than 50 m/s.

There are structural problems when even relative low impact velocities trigger important inertia effects, as in the buckling of tubes.

Another aspect of this solution is the fact that the beam transverse displacement and velocity profiles are linear since we did not consider any extra energy which could alter the beam curvature other than at the support and impact point.

Experimental investigations, as shown in the previous figure, indicate, in contrast with the above solution, that the beam profile does alter its initial straight shape if the impact velocity is sufficiently high. The reason is that the impact energy, suddenly applied to the beam, would require it to displace very fast. But the beam, lazily, counter reacts to this request because it has mass. Therefore, the more dense a beam the more distorted it becomes, which prompts us to seek a more refined solution.

The matter of moving hinge is a debatable one. See S.R. Reid and X.G. Gui, On the elastic–plastic deformation of a cantilever, Int J Impact Eng, 6, 109–127, 1987 and D. Shu, W.J. Stronge and T.X. Yu, Oblique impact at the tip of a cantilever, Int J Impact Eng, 12, 37–47, 1992.

We have studied before the concept of plastic hinge. In light of our previous studies on wave propagation, we know that a disturbance does not propagate instantaneously. Consider for a moment a cantilever beam impacted in its extreme by a mass, as indicated in the next figure. When the mass hits the end of the beam, the support does not feel this impact. It requires some time until the local hinge reaches the support, so we refer to plastic hinge propagation. We have seen this phenomenon when studying the elastic impact in beams.

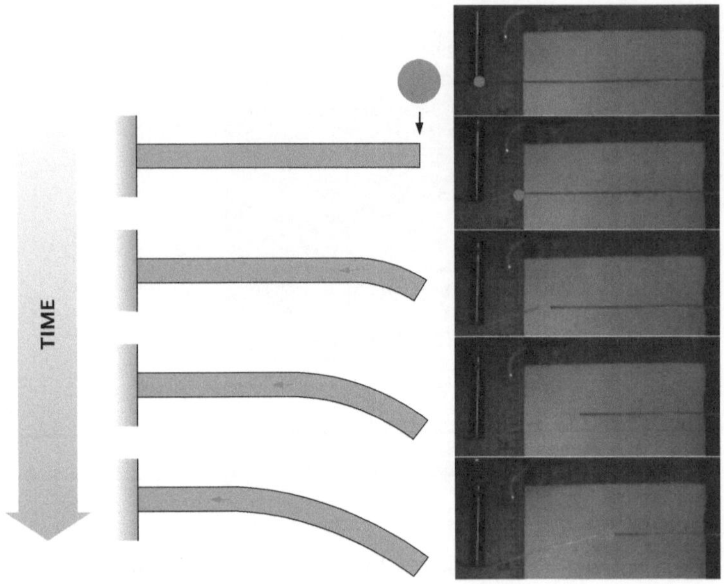

(a) (b)

Figure 5.11 (a) Illustration of plastic hinge propagation in a cantilever beam under transverse impact. (b) a carbon fiber beam impacted at 99.6 m/s. Pictures are taken at 30 μs time interval (from Impact on carbon fiber composite: Ballistic tests, material tests, and computer simulations, by S Chocrona, AJ Carpenter, NL Scott, RP Bigger and K Warren, Int J of Impact Eng, 131, p 39–56, 2019.)

A moving hinge absorbs energy as it propagates and causes the bending moment to reach its maximum absolute value of M_0. This means that the beam span covered by the hinge propagation is under a bending moment of M_0. In any case, the concept of moving plastic hinge is a consequence of an exact dynamic solution and experimental and numerical studies have been performed on this topic. This transient phase is rather short since, roughly, the hinge speed is some 150–400 m/s for metal beams.

Let us postulate for our problem of central beam impact that the impact mass causes the formation of two symmetric hinges in the same position around the impact point, as in the next figure. We can also postulate that these two hinges will move to the supports, so changing locally the beam curvature as they travel.

Various types of impact loading on beams, plates and shells are detailed analysed in N. Jones, Structural Impact, CUP, 2011, from where parts of this Chapter are based.

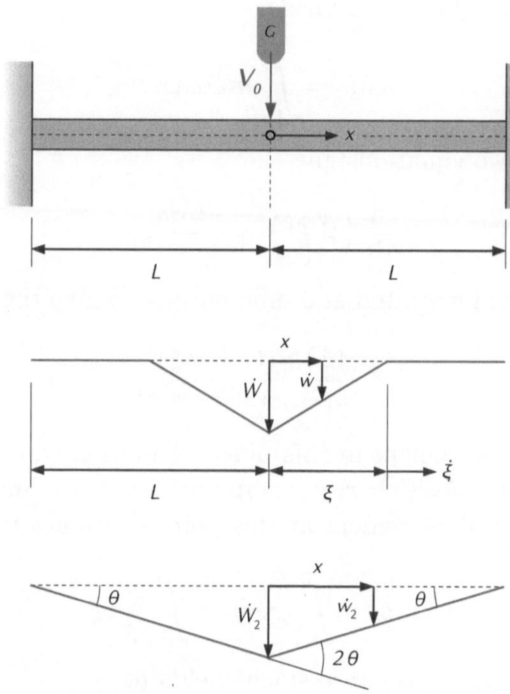

Figure 5.12 The kinematics of the central impact of a mass in a clamped beam.

The transverse velocity profile is

$$\begin{aligned} \dot{w} &= \dot{W}(1 - x/\xi), \quad 0 < x < \xi \\ \dot{w} &= 0, \qquad\qquad\qquad \xi < x < L \end{aligned}$$

and vertical equilibrium in the interval $0 < x < \xi$ gives

$$G\ddot{W} = -2\int_0^{\xi} m\frac{\mathrm{d}\dot{w}}{\mathrm{d}t} =$$

$$-2\int_0^{\xi} m\left\{\ddot{W}(1 - x/\xi) + \frac{\dot{W}\dot{\xi}x}{\xi^2}\right\}\mathrm{d}x = 0,$$

or

$$G\ddot{W} + m(\ddot{W}\xi + \dot{W}\dot{\xi}) = 0$$

and since $\mathrm{d}(\dot{W}\xi)/\mathrm{d}t = \ddot{W}\xi + \dot{W}\dot{\xi}$ it follows that

$$\dot{W} = \frac{V_0}{1 + \frac{m}{G}\xi},$$

where the hinge position varies with time, *ie* $\xi = \xi(t)$.

Now, moment equilibrium yields

$$2M_0 = \int_0^{\xi} m\ddot{w}x\mathrm{d}x,$$

so the previous two equations give

$$\frac{\mathrm{d}}{\mathrm{d}t}\left(\dot{W}\xi^2\right) = \frac{12M_0}{m},$$

which can be time integrated and differentiate to give the hinge velocity

$$\dot{\xi} = \frac{12M_0\,(G + m\xi)^2}{mGV_0\xi(2G + m\xi)}.$$

The beam displacement in this phase of motion can be obtained by observing that, at a position x, it is zero until the travelling hinge reaches x. After that, the displacement at this point continues to increase such that

$$w(x,t) = \int_{t(x)}^{t} \dot{w}\mathrm{d}t = \int_x^{\xi} \frac{\dot{w}}{\dot{\xi}}\mathrm{d}\xi,$$

which, using the previous expressions, yields to

$$w = \int_x^{\xi} \dot{w}\frac{\mathrm{d}\xi}{\dot{\xi}} = \frac{GV^2}{24mM_0}\left\{\frac{1+\beta}{(1+\alpha)^2} - \frac{1+2\beta}{1+\beta} + \frac{2\beta}{1+\alpha} + 2\ln\frac{1+\alpha}{1+\beta}\right\},$$

where $\alpha = m\xi/G, \beta = mx/G$.

When the hinges reach the support, the beam impact point has a velocity of $\dot{W} = V_0/(1 + mL/G)$ and the beam total kinetic energy of

$$K = \frac{1}{2}\frac{V_0^2}{(1 + mL/G)^2}\left(G + \frac{2mL}{3}\right)$$

needs to be dissipated by the hinges in the support and in the beam centre. This additional displacement, w_2, is obtained from

$$4M_0\theta = 4M_0\frac{w_2}{L-x} = \frac{1}{2}\frac{V_0^2}{(1 + mL/G)^2}\left(G + \frac{2mL}{3}\right),$$

so the final beam profile is

$$w = \frac{GV_0^2}{24mM_0}\left\{\frac{\bar{\alpha}-\beta}{(1+\bar{\alpha})(1+\beta)} + 2\ln\frac{1+\bar{\alpha}}{1+\beta}\right\},$$

where $\bar{\alpha} = mL/G$.

Note that, by disregarding the first phase of motion, *ie* transient phase of hinge propagation, we obtain the final displacement

$$w_f = \frac{GV_0^2 L(1 - x/L)}{8M_0},$$

which coincides with our initial approach to the problem and it does not contain information of the beam mass. On the other hand, when most of the energy is dissipated during the hinge propagation, case of a light mass, the second phase of motion can be disregarded and the final beam displacement becomes

$$w = \frac{GV_0^2 L}{12\bar{\alpha}M_0}\ln\frac{\bar{\alpha}}{1+\beta}.$$

We have managed so far to obtain a complete solution to our problem of a mass dropping on a beam but remember that no consideration was given to the membrane forces. In a practical situation, it is likely that they will develop due to the large displacements resulting from the impact event. As already seen, in the case of the membrane forces, we need to use in our solution a yield surface relating bending moment and membrane force, as the one developed previously in this chapter.

5.3 Beam loaded by a pressure pulse

A pressure pulse, or blast load, is a somewhat common dynamic load. Take the case, for instance, of an explosion. The sudden change in the air pressure imparts energy to surrounding objects and structures, with disastrous consequences.

In modelling a beam or other structures subject to blast loads, the solution of the differential governing equations are made easier if the pressure pulse is constant, as it is the case of a rectangular pulse of intensity p and duration τ. We will show later that actual complex pressure pulses can be represented by an equivalent rectangular one so the solutions we will now develop can be used directly to time varying loading.

Blast loads on beams is the subject of the Ph.D. Thesis by A.A. Nassr, Experimental and Analytical Study of The Dynamic Response of Steel Beams and Columns to Blast Loading, McMaster University, 2012.

Consider the rectangular cross–section clamped beam in the next figure made of a rigid, perfectly plastic material and loaded with the pressure pulse shown. When the pulse ceases to impart energy, the beam of course still moves so at least two phases of motion are expected, one for $0 < t < \tau$ and the other for $t \geq \tau$. We anticipate that there is a third phase of motion due to the hinge propagation phase.

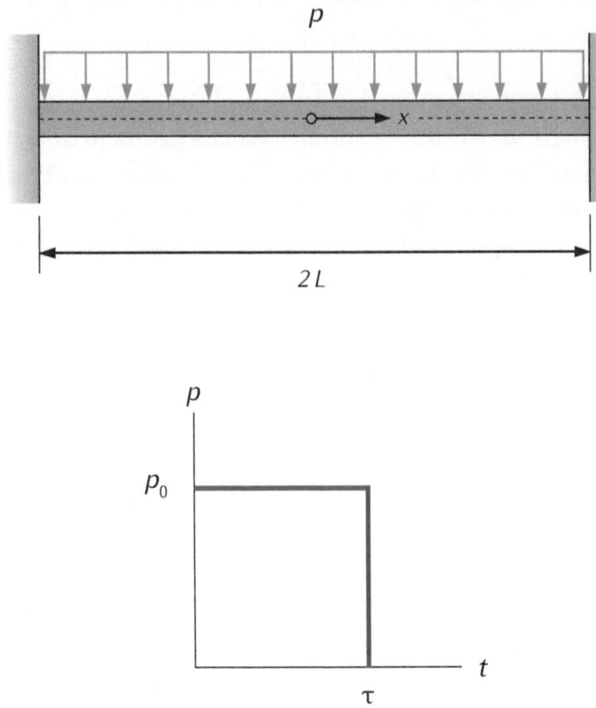

Figure 5.13 A clamped beam loaded with a rectangular pressure pulse.

The beam above can be analysed by solving the equilibrium equation

$$\frac{\partial^2 M}{\partial x^2} + p = m\frac{\partial^2 w}{\partial t^2},$$

where the influence of membrane force is disregarded. This is a limitation of our present solution since beams under blast loads may suffer severe strains and displacement. We will turn later to a more complete solution.

We solved this equation in Chapter 4 by expressing the bending moment as a function of the curvature and by using the method of separation of variables. Here, since our material model is perfectly plastic, there is no unique relation between the bending moment and the curvature.

To develop a solution to this problem let us use a kinematic field for the beam transverse velocity and perform the integration of the equation using appropriate initial and boundary conditions. The kinematic field in the present case is suggested to be described by the velocity profile

$$\dot{w} = \frac{L - x}{L - \xi_0}\dot{W}_1, \quad \text{for} \quad x > \xi_0$$

and

$$\dot{w} = \dot{W}_1, \quad \text{for} \quad x \leq \xi_0,$$

valid for the first phase of motion. In this phase of motion, we assume that the hinges are stationary at $x = \xi = \xi_0$. This velocity fields are as indicated in the next figure.

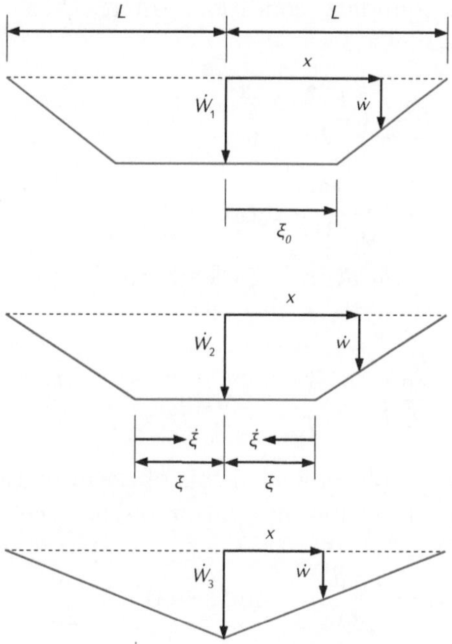

Figure 5.14 Velocity fields for a clamped beam under a pressure pulse with $\eta > 3$.

Using this velocity profile in the equilibrium equations we have

$$\frac{\partial^2 M}{\partial x^2} = -p_0 + m\frac{\partial^2 W_1}{\partial t^2}, \qquad \text{for} \qquad 0 \leq x \leq \xi_0,$$

which can be space integrated to yield an acceleration of the $2\xi_0$ beam span of

$$\ddot{W}_1 = \frac{p_0}{m},$$

since $Q(0) = Q(\xi_0) = 0$, with the consequence that $M = M_0$ in $-\xi_0 \leq x \leq \xi_0$.

It is worth commenting that in the above equation the acceleration is positive, which is expected in this phase of motion since the pressure pulse is still being applied. After the pulse ends, we shall see that the beam will start to decelerate. Note that we were able to solve a partial differential equation expeditiously because the pulse load is time independent and because the velocity profile yields a constant velocity in the span between the plastic hinges. The boundary conditions used for the determination of the integration constants refer to the fact that the bending moment always attains the collapse value, M_0, in the hinge and in the mid–span.

In the interval $\xi_0 \leq x \leq L$, double integration of the equilibrium equation, with the boundary conditions $\mathrm{d}M(\xi_0)/\mathrm{d}x = Q(\xi_0) = 0$ and $M(L) = -M_0$ gives

$$M = -\frac{-p_0 x^2}{2} + \frac{p_0}{L-\xi_0}\left(\frac{Lx^2}{2} - \frac{x^3}{6}\right) + p_0\xi_0 x - \frac{p_0}{L-\xi_0}\xi_0\left(L - \frac{\xi_0}{2}\right)x +$$
$$\frac{p_0 L^2}{2} - \frac{p_0}{L-\xi_0}\frac{L^3}{3} - p_0\xi_0 L + \frac{p_0\xi_0}{L-\xi_0}\left(L - \frac{\xi_0}{2}\right)L.$$

Knowing that $M = M_0$ at $x = \xi_0$ leads to the initial position of the plastic hinge

$$\left(1 - \frac{\xi_0}{L}\right)^2 = \frac{3}{\eta}, \qquad \text{with} \qquad \eta = \frac{p_0}{p_c} > 3.$$

The velocity and displacement in this phase of motion, which finishes at $t = \tau$, comes from the time integration of the acceleration, giving

$$\dot{W}_1 = \frac{p_0 t}{m} \qquad \text{and} \qquad W_1 = \frac{p_0 t^2}{2m}.$$

These two equations are now used as the initial conditions for the next phase of motion, which starts when the pressure pulse is removed.

In the second phase of motion, the former stationary hinges now move towards the centre so that the new velocity profile is

$$\dot{w} = \frac{L-x}{L-\xi}\dot{W}_2,$$

where now $\xi = \xi(t)$. The equilibrium equation for $x \leq \xi$ becomes

$$\frac{\partial^2 M}{\partial x^2} = \ddot{W}_2,$$

whose integration in x and the boundary condition $\mathrm{d}M(x = \xi_0)/\mathrm{d}x = 0$ leads to

$$\ddot{W}_2 = 0 \quad \text{and} \quad \dot{W}_2 = \frac{p_0 \tau}{m} \quad \text{and} \quad W_2 = \frac{p_0 \tau}{m} t - \frac{p_0 \tau^2}{2m},$$

when using the initial conditions of the previous phase of motion.

Now, for $x > \xi$, the equilibrium equation give

$$\frac{\mathrm{d}^2 M}{\mathrm{d}x^2} = p_0 \tau \frac{L - x}{(L - \xi)^2} \dot{\xi},$$

whose successive integrations in x leads to

$$M = \frac{p_0 \tau \dot{\xi}}{(L - \xi)^2} \left(\frac{Lx^2}{2} - \frac{x^3}{6} - L\xi x + \frac{\xi^2 x}{2} - \frac{L^3}{3} + L^2 \xi - \frac{L\xi^2}{2} \right) - M_0,$$

when using the boundary conditions $M(L) = -M_0$ and $\frac{\mathrm{d}M(\xi)}{\mathrm{d}x} = 0$.

This bending moment can be equated to M_0 at $x = \xi$ to give

$$M_0 = \frac{p_0 \tau}{6} (\xi - L) \dot{\xi},$$

whose integration allows us to determine the position of the plastic hinge with time as

$$L\xi - \frac{\xi^2}{2} - L^2 \xi_0 + \frac{L^2 \xi_0^2}{2} = \frac{6 M_0 (\tau - t)}{p_0 \tau}.$$

When setting $\xi = 0$ we obtain the duration of this phase of motion as

$$t_1 = \frac{\eta \tau}{3}.$$

When the hinges reach the mid–spam, the transverse beam velocity is not yet zero and there is still some kinetic energy to be dissipated in a third phase of motion. In this case, the transverse velocity profile is

$$\dot{w} = \frac{L - x}{L} \dot{W}_3$$

and the bending moment is

$$M = \frac{m}{6}\left(\frac{Lx^2}{2} - \frac{x^3}{6} - \frac{L^3}{3}\right)\ddot{W}_3 - M_0,$$

so that the acceleration amounts to

$$\ddot{W}_3 = \frac{-6M_0}{mL^2},$$

which can be time integrated to give the beam mid–spam velocity of

$$\dot{W}_3 = \frac{-6M_0}{mL^2}t + \frac{p_0\tau}{m} + \frac{6M_0}{mL^2}t_1.$$

Motion ends when this velocity is zero, at

$$t_f = \frac{2\eta\tau}{3},$$

when then the central beam displacement is

$$W_f = \frac{p_0\tau^2(3\eta - 2)}{4m}.$$

This solution is valid for $\eta > 3$.

It is opportune to note that the moment distribution in the beam should be such that, at any time and at any point of the span, its value is no greater than $|M_0|$. This is so because our material model, being perfectly plastic, does not allow for strain hardening, which limits the maximum stress to σ_0 or, equivalently, the maximum bending moment to the collapse one, M_0. But the magnitude of the bending moment is not a sufficient condition to investigate in order to check the correctness of a solution. As one learns from maximum and minimum of a function, if the second derivative of a given function is negative at a given point then there is a maximum.It can be appreciated in the next figure that the bending moment distribution for a simply supported beam reaches the required value of M_0 at the beam centre and zero at the support and even so the solution is incorrect for $\eta = 10$ for this simply supported case.

To investigate the bending moment distribution, in time and in space, is in fact a search for the solution correctness, more precisely, on the lower bound solution. But this is not enough to assure a solution is correct, as we have already seen. We should also investigate the kinematic aspects of a solution, which leads to the critical appraisal of the upper bound solution. Only then we can conclude on the exactness of a solution. The kinematic field should satisfy some basic requirements, like continuity of transverse displacement, slope and velocity across a hinge, stationary or dynamic, but these aspects are not considered here. Note that the velocity profile comes out from the expertise of the analyst and by trial and error.

See Q.M. Li, Continuity conditions at bending and shearing interfaces of rigid, perfectly plastic structural elements, International Journal of Solids and Structures, 37, p. 3651–3665, 2000. For an example of proof of the exactness of a solution see N. Jones and M. Alves, Post–severance analysis of impulsively loaded beams, International Journal of Solids and Structures, 41:6441–6463, 2004.

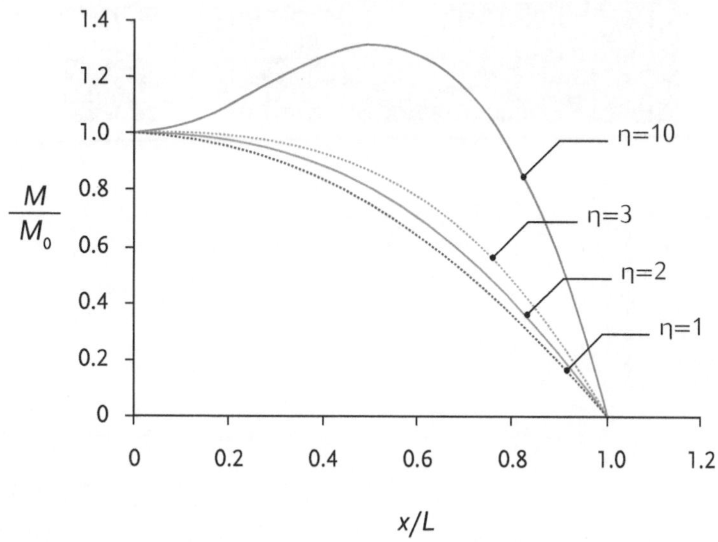

Figure 5.15 Bending moment distribution of a simply supported beam for different dimensionless pulse pressures.

5.4 Equivalent pulse

The next figure presents a rig capable of producing different shapes and intensity of pulse loads. The figure also presents a typical pulse, very close to a triangular shape. The experimental procedure involves separating the two halves of the chamber by the panel test sample, creating two volumes which are then subjected to equal values of air pressure. This does not displace the panel to any side. At the opposite end of the pannel, one of the chambers has a kind of inspection window covered with a a soft sheet of a paper like material. An electrical resistance can easily tear this paper and pressurized air is suddenly released to the environment. Since the other chamber is still pressurized this generates a load difference in the panel, which responds eventually with permanent deformation.

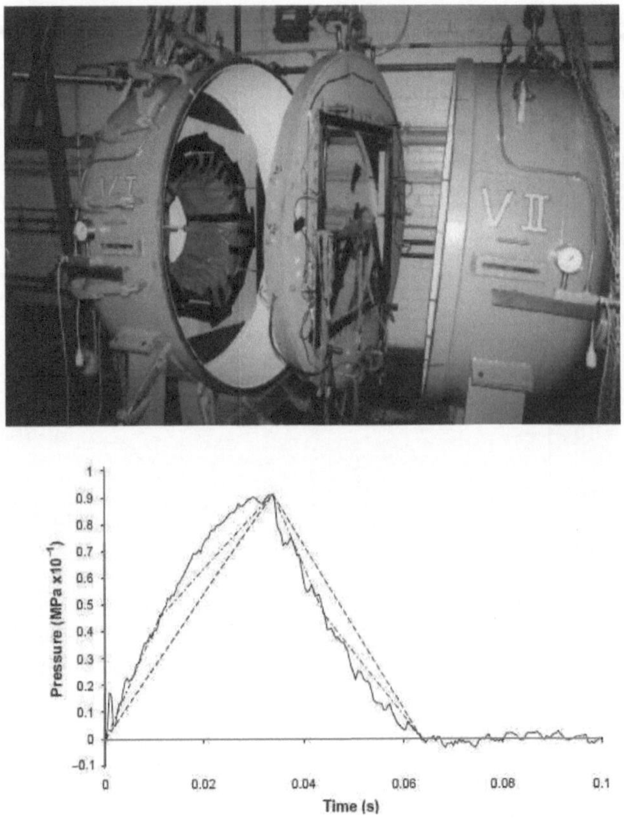

Figure 5.16 (a) A rig to apply a pressure pulse in plates and (b) an actual pressure pulse. From G.K. Schleyer, S.S. Hsu, M.D. White and R.S. Birch, Pulse pressure loading of clamped mild steel plates, International Journal of Impact Engineering, 28, p. 223–247, 2003.

As evident from the figure, this pressure pulse does not resemble the rectangular one used in the analyses of the previous sections. Had we used it, the integrations of the equations of motion would be further complicated. Hence,it would be very useful to transform a real pulse into an equivalent rectangular one. To this end, consider the definition of impulse

This was developed by C.K. Youngdall in Correlation parameters for eliminating the effect of pulse shape on dynamic plastic deformation, Journal of Applied Mechanics, 37, 744-752, 1970. See also R. L. Azevedo and M. Alves, A numerical investigation on the visco–plastic response of structures to different pulse loading shapes, Engineering Structures, 30:258–267, 2008.

$$I = \int_{t_y}^{t_f} F(t)dt$$

and of the mean time of a pulse

$$t_{mean} = \frac{1}{I}\int_{t_y}^{t_f} (t - t_y)F(t)\mathrm{d}t,$$

where t_y is the time when yielding first occurs and t_f is the time when plastic deformation finishes. t_f is not known a priori, but it can be approximate as

$$t_f = \frac{I}{p_c} + t_y,$$

with p_c being the the static collapse load of the structure under analysis.

These definitions allow us to define the effective loading according to

$$p_e = \frac{1}{2}\frac{I}{t_{mean}} = \frac{I}{t_e},$$

which represents the amplitude of a rectangular pulse acting during $2t_{mean}$. In this way, the pulse quantified by p_e and t_e could be used in an analysis as an equivalent one in substitution for the actual loading pressure pulse. Clearly, these effective values involve only integrals of the loading and do not depend on details of the loading–time profile.

As an illustration, consider a clamped beam subjected to different pulses shapes; see next figure. A finite element simulation of this problem leads to the calculation of the final mid–span displacement of the various pulses. If an equivalent pulse is used as load, the response to the various pulses practically collapses in a single curve.

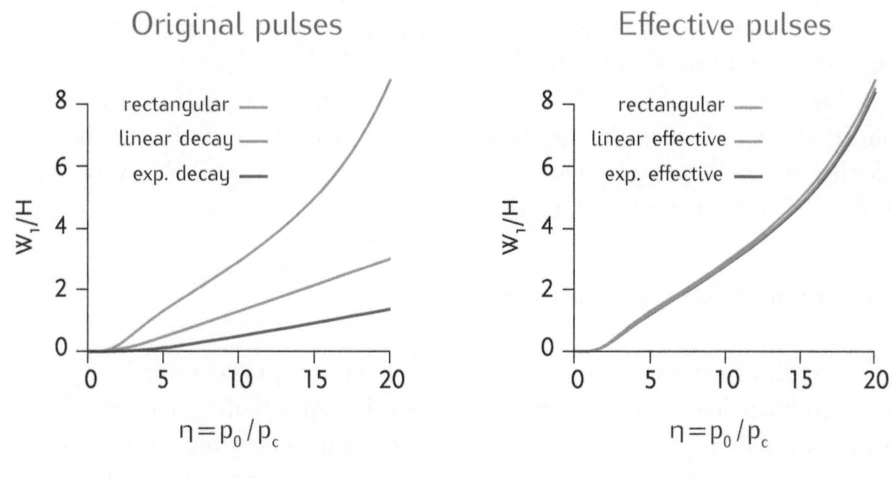

Figure 5.17 The response of a clamped beam to various pulse shapes.

Let us now briefly investigate how the final mid–spam displacement of a clamped beam behaves as the pulse load intensity increases. This is shown in the next figure.

We see that, as the pulse increases in magnitude, the final displacement tends to a single value. If we equate the impulse to momentum

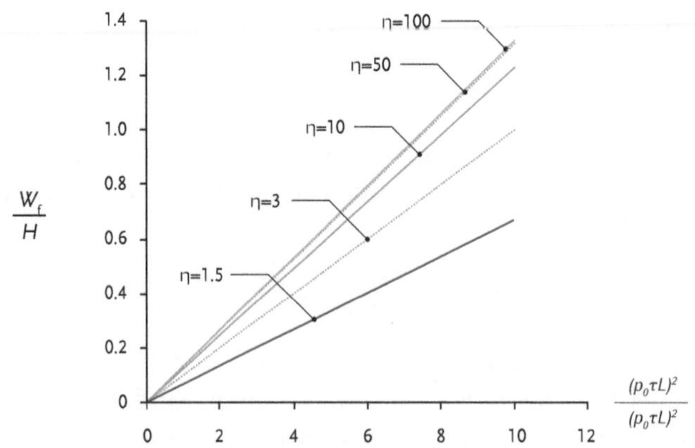

Figure 5.18 Dimensionless displacement of a clamped beam for different pulse load intensities.

applied to the whole beam it follows that

$$\int F\mathrm{d}t = m_b V \rightarrow p_0 2L\tau = 2mLV_0,$$

or

$$V_0 = \frac{p_0\tau}{m}.$$

This is to say that there is a direct equivalence between (high) pulse magnitude and initial velocity. The analysis of a blast load in a structure then becomes a problem where the load, p_0, can be applied as an initial condition, V_0. Hence, the analysis of a beam motion under an initial velocity is an important case that deserves more detailed study. We perform it next, together with consideration of shear effects.

5.5 Transverse shear effects

The next figure shows various clamped beams and plates after being exposed to blast loads of different intensities. It is interesting to note that, as the impulse increases so does the final displacement, but only up to a certain threshold impulse. Beyond it, a further increase of the pulse load leads to a decrease of the final plate displacement. This is due to the fact that the energy provided by the explosive is used to shear off the beams and plates from its supports. Such a remark reveals the importance of the transverse shear force, even for these slender structures.

Let us consider a simply supported beam subjected to a pressure pulse along its span. It can be shown that the bending moment

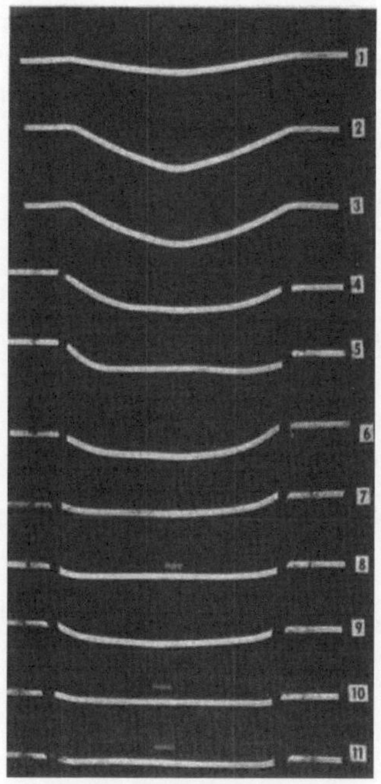

(a) Beams

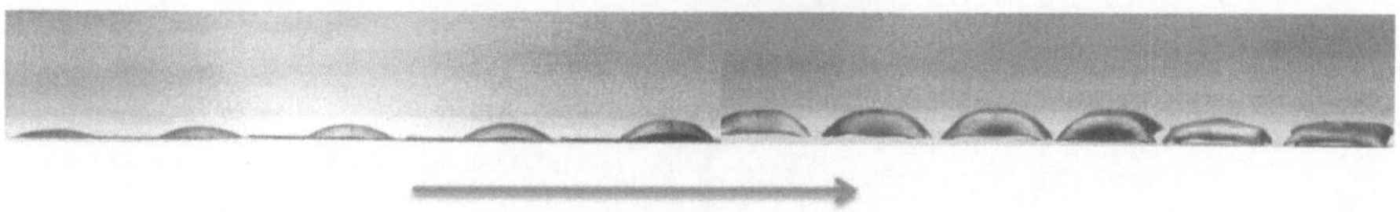

(b) Square plates

Figure 5.19 Side view of various clamped (a) beams and (b) plates subjected to impulsive loads of growing intensity as indicated by the arrows. The results for beams are from S.B. Menkes and H.J. Opat, Tearing and shear failures in explosively loaded broken beams, Experimental Mechanics, 13, pp. 480–486, 1973 and for plates from G.N. Nurick, M.E. Gelman and N.S. Marshall, Tearing of blast loaded plates with clamped boundary conditions, International Journal of Impact Engineering, 18, pp. 803–827, 1996.

distribution in the first phase of motion is given, with $\bar{\xi}_0 = \xi_0/L$ and $\bar{x} = x/L$, by

As arguments N. Jones in Structural Impact, CUP, 2012.

$$\frac{M}{M_0} = \frac{\eta}{3(1-\bar{\xi}_0)} \left\{3\bar{\xi}_0\bar{x}^2 - \bar{x}^3 - 3\bar{\xi}_0\bar{x} + 1 - 3\bar{\xi}_0 + 3\bar{\xi}_0^2\right\},$$

whose derivative gives the transverse shear force at the support as

$$\frac{Q_d}{Q_0} = \frac{-\eta(1-\bar{\xi}_0)}{L}\frac{H}{2},$$

Here, Q_0, similar to M_0 and N_0, is the transverse shear force which causes yielding, independently of the bending moment intensity or of the membrane force, in the whole cross section of a beam made of a perfectly plastic material with shear flow stress τ_0. It is given by

$$Q_0 = \tau_0 BH \approx \frac{\sigma_0}{2} BH.$$

This simple analysis indicates the harmfulness of the transverse shear force in dynamic events since it may achieve extreme high values for high pulse loads ($\eta \to \infty$) and highlights the importance of considering the transverse shear force in the yield surface.

When considering membrane forces in the analysis of beams undergoing large displacements, we used a yield surface $M \times N$. Likewise, the simplest way to consider the interaction of the bending moment and transverse shear stress is by using the square yield surface depicted in the next figure and quantified by

$$M = \pm M_0 \quad \text{for} \quad Q = 0 \quad \text{and} \quad Q = \pm Q_0 \quad \text{for} \quad M = 0.$$

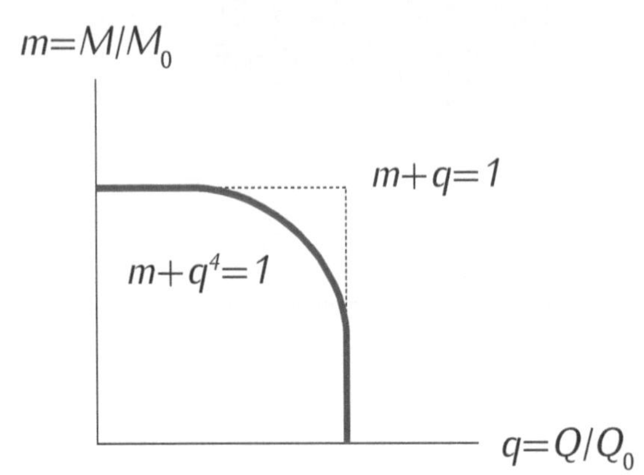

Figure 5.20 Yield surface relating bending moment and transverse shear force.

Let us now outline the solution to the problem of a simply supported beam under an initial velocity, V_0, considering the cooperation of transverse shear force and bending moment to the yielding of the beam cross–section. The next figure shows the beam and the various transverse velocity profiles assumed for this solution.

See N. Jones and M. Alves, Post-severance analysis of impulsively loaded beams, Int J of Solids and Structures, 41: 6441-–6463, 2004.

In the first phase of motion, the velocity profile is set to represent the new feature of a shear hinge at the supports, described by

$$\dot{w} = V_0 \quad \text{for} \quad 0 \le x \le \xi_0 \quad \text{and}$$

$$\dot{w} = \dot{W}_s + (V_0 - \dot{W}_s)\frac{L - x}{L - \xi_0} \quad \text{for} \quad \xi_0 \le x \le L.$$

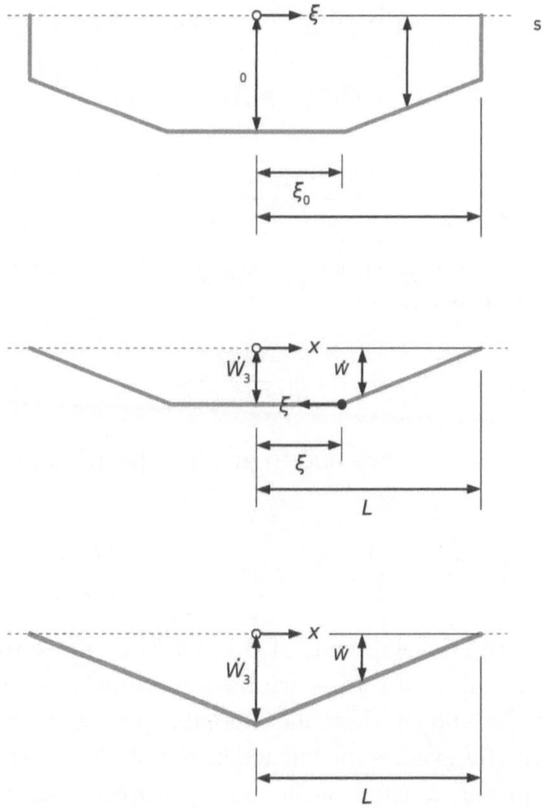

Figure 5.21 The transverse velocity profile for the various phases of motion of a simply supported beam subjected to an initial velocity .

For the interval $0 \leq x \leq \xi_0$, the spatial integration of the equilibrium equation, $\partial Q/\partial x = -p + m\partial^2 w/\partial t^2$, gives $Q(x) = 0$, with the boundary condition at the hinge, $M(\xi_0) = M_0$. For the interval $\xi_0 \leq x \leq L$, the transverse shear force distribution is

$$Q = m\ddot{W}_s(x - \xi_0)\left\{1 + \frac{L/2 + \xi_0/2 - L}{L - \xi_0}\right\},$$

which can be used to determine the acceleration at the support by noting that $Q(L) = Q_0$,

$$\ddot{W}_s = \frac{-2Q_0}{m(L - \xi_0)}.$$

When we used above $Q(L) = Q_0$, we introduced in the solution the yield condition. We see here the advantage of using a square dimensionless yield surface. Had we used a more complex surface the problem would be coupled and have been much more difficult to solve.

The bending moment distribution in this part of the beam is

$$M = \int Q\mathrm{d}x = m\ddot{W}_s(L-x)\times$$

$$\left\{\xi_0 - \frac{L}{2} - \frac{x}{2} + \frac{L^2/3 + Lx/3 - x^2/6 - \xi_0 L + \xi_0^2/2}{L-\xi_0}\right\}$$

and knowing that, at the support, $M(L) = 0$ we can obtain another expression for the acceleration,

$$\ddot{W}_s = \frac{-6M_0}{m(L-\xi_0)^2},$$

to be equated to the previous one to give the position of the stationary plastic hinge,

$$1 - \frac{\xi_0}{L} = \frac{3}{2\nu}, \qquad \text{with} \qquad \nu = \frac{Q_0}{M_0}\frac{L}{2}.$$

See W.Q. Shen and N. Jones, Dynamic plastic response and failure of a clamped beam struck transversely by a mass, International Journal of Solids and Structures, 30: 1631–1648, 1993 and J. Liu and N. Jones, Dynamic response of a rigid plastic clamped beam struck by a mass at any point on the span, International Journal of Solids and Structures, 24: 251–270, 1988.

We stop here the development of this solution, since the next phases of motion are similar to the ones we have already developed along this chapter. It can be shown that the velocity profile used in the above solution leads to the exact solution when $\nu > 3/2$. For smaller ν, *ie* for beams weaker in shear, there is no hinge propagation and the hinges become concentrated in the supports only.

The above solution dealt with an interaction between transverse shear force and bending moment. In the presence of large displacements, it is also necessary to consider the effects of membrane forces, as already anticipated when solving the static problem of a beam under concentrated load. Theoretical solutions for beams when using a cubic $M \times N \times Q$ yield surface or numerical solutions when using $M/M_0 + (N/N_0)^2 + (Q/Q_0)^4 = 1$ are available.

5.6 Large displacements effects

It is evident that, upon the transverse impact of a mass, a beam is likely to displace significantly if enough energy is imparted to it. This large transverse displacement gives rise to membrane forces that need to be considered in the (non–linear) analysis.

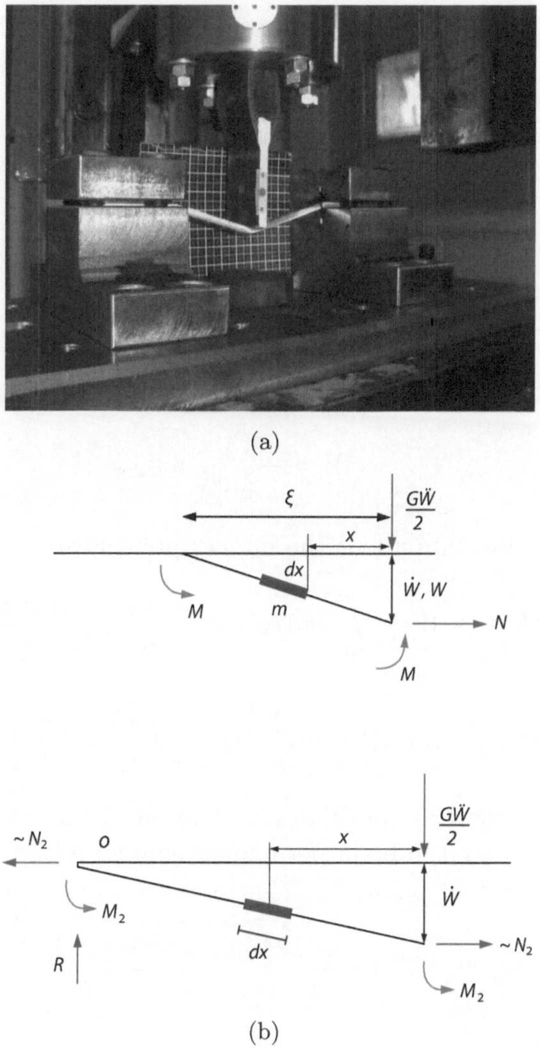

Figure 5.22 (a) A clamped beam subjected to the impact of a dropped mass and (b) details of the analysis.

Consider the clamped beam subjected to a drop mass depicted in the above figure.

In this problem there are two distinct phases of motion. The first deals with the formation of two symmetric hinges just when the impacting mass hits the beam. These hinges propagate towards the support and it is assumed that the transverse velocity profile is linear, *ie*

$$\dot{w} = \dot{W}(1 - \xi(t)/L).$$

At time t_1, the moving hinges reach the supports and the beam displaces further in a second phase of motion. The transverse displacement profile comes from the integration of the above equation and it is non–linear. Now, vertical equilibrium gives

$$\frac{G\ddot{W}}{2} + \int_0^{\xi} m\mathrm{d}x\ddot{w} = 0,$$

or

$$\frac{G\ddot{W}}{m} + \ddot{W}\xi + \dot{W}\dot{\xi} = 0 \rightarrow \frac{G\ddot{W}}{m} + \frac{\mathrm{d}}{\mathrm{d}t}(\dot{W}\xi) = 0,$$

whose time integration with the initial conditions $\dot{W}(0) = V_0$ gives

$$\dot{W} = \frac{V_0}{1 + \frac{m\xi}{G}}.$$

This solution is presented by W.Q. Shen and N. Jones in A comment on the low speed impact of a clamped beam by a heavy striker, Mech. Struct. & Mach., 19, p. 527–549, 1991.

Moment equilibrium about the impact point gives

$$2M_1 + N_1 W = \int_0^{\xi} m\mathrm{d}x\ddot{w}x = \frac{m\ddot{W}\xi^2}{6} + \frac{m\dot{W}\dot{\xi}\xi}{3},$$

or

$$12M_1 + 6N_1 W = m\frac{\mathrm{d}}{\mathrm{d}t}\left[\dot{W}\frac{G^2}{m^2}\left(\frac{V_0}{\dot{W}} - 1\right)^2\right],$$

from which it follows the positive acceleration of the beam (and of the drop mass, considered to be in permanent contact with the beam upper surface) at the impact point,

$$\ddot{W} = \frac{m}{G^2}\frac{12M_0(1+\bar{w}^2)}{1 - \frac{V_0^2}{\dot{W}^2}},$$

with $\bar{w} = W/H$. Here, use is made of the yield condition $M/M_0 + (N/N_0)^2 = 1$, together with $N_1/N_0 = W/H$, since the displacement in this phase of motion is small.

We can integrate the above equation to obtain

$$\int \frac{\mathrm{d}\dot{W}}{\mathrm{d}t}\left(1 - \frac{V_0^2}{\dot{W}^2}\right)\mathrm{d}t = \int \frac{12mM_0}{G^2}\left(1+\bar{w}^2\right)\mathrm{d}t,$$

or

$$\int \left(\dot{W} - \frac{V_0^2}{\dot{W}}\right)\mathrm{d}\dot{W} = \int \frac{12mM_0}{G^2}\left(1+\bar{w}^2\right)\mathrm{d}W,$$

since $\mathrm{d}t = \mathrm{d}W/\dot{W}$. This results, with $W(0) = 0$ and $\dot{W}(0) = V_0$, in

$$\left(\frac{\dot{W}}{V_0}\right)^2 - \ln\left(\frac{\dot{W}}{V_0}\right)^2 - 1 = \frac{24HmM_0}{V_0^2G^2}\left(\bar{w} + \frac{1}{3}\bar{w}^3\right)$$

and setting $\xi = L$ we obtain the beam transverse displacement when this first phase of motion finishes, *ie*

$$\frac{24HmM_0}{V_0^2G^2}\left(\bar{w} + \frac{1}{3}\bar{w}^3\right) = \left(\frac{1}{1+\frac{mL}{G}}\right)^2 - \ln\left(\frac{1}{1+\frac{mL}{G}}\right)^2 - 1$$

since, in this case, $\dot{W} = V_0/(1 + mL/G)$.

On the second phase of motion, the beam is led to a new configuration, whose moment equilibrium around o gives

$$G\ddot{W}L/2 + M_2 + M_2 + N_2W = \int_0^L m\mathrm{d}x\ddot{W}(1 - x/L)(L - x),$$

or

$$\ddot{W} = -\frac{12M_2 + 6N_2W}{2mL^2 + 3GL}.$$

The term $12M_2 + 6N_2W$ is evaluated to $12M_0(1 + \bar{w}^2)$ or to $24M_0\bar{w}$, according to the transverse displacement being smaller or larger than H, respectively.

For low impact energy, $W \leq H$ and we integrate the above equation (multiply it by $\dot{W}$, use $\ddot{W} = \mathrm{d}\dot{W}/\mathrm{d}t$ and the velocity at the end of the first phase of motion as an initial condition) to obtain

$$\begin{aligned}\dot{W}^2 = -\frac{24HM_0}{3GL + 2mL^2}\left(H\bar{w} + \frac{H\bar{w}}{3}\right) + \\ \frac{G^2V_0^2(G + 2mL)}{(G + mL)(3G + 2mL)mL} - \\ \frac{V_0^2G^2}{(3GL + 2mL^2)m}\left[1 + \ln(1 + mL/G)^2\right].\end{aligned}$$

By setting $\dot{W} = 0$ in the above equation we can obtain the displacement of the central impact point for this case of $W \leq H$.

On the other hand, when the displacement equals the beam height, the velocity is

$$\dot{W}_H^2 = -\frac{32M_0H}{3GL+2mL^2} + \frac{G^2V_0^2(G+2mL)}{(G+mL)L(3G+2mL)m} - \frac{G^2V_0^2}{L(3G+2mL)m}\left[1+\ln(1+mL/G)^2\right],$$

which is used as an initial velocity in the integration of the (de)acceleration equation

$$\ddot{W} = -\frac{24M_0}{2mL^2+3GL}\bar{w},$$

to give

$$\dot{W}^2 = \dot{W}_H^2 - \frac{24HM_0}{L(3G+2mL)}(\bar{w}^2-1).$$

When $\dot{W} = 0$, we obtain the final transverse mid–displacement of the clamped beam for $W > H$ as

$$W_f = \sqrt{H\left(H + \frac{3LG+2mL^2}{24M_0}\dot{W}_H^2\right)}.$$

In developing the above solution, we consider the whole dynamics of the beam, including the energy absorbed by the propagating plastic hinges. But knowing that the later is small, we can obtain a solution for this impact problem by equating the kinetic energy of the drop mass to the energy absorbed by the concentrated hinges located at the supports and at the impact point, together with the energy absorbed in stretching the whole beam span. Accordingly, we write

$$K_{\text{int}} = K_{ext} \rightarrow \frac{GV_0^2}{2} = \int_0^W P\text{d}W$$

and using $P/P_c = 1 + (W/H)^2$ gives

$$\frac{W}{H} + \frac{1}{3}\left(\frac{W}{H}\right)^3 = \frac{GV_0^2L}{8H^2M_0}, \qquad \text{for} \qquad W \leq H,$$

or

$$\frac{W}{H} = \frac{A}{2} - \frac{2}{A} \quad \text{for} \quad W \le H,$$

with

$$A = \sqrt[3]{12\lambda + 4\sqrt{4 + 9\lambda^2}}, \qquad \lambda = \frac{GV_0^2 L}{8HM_0}.$$

For larger transverse displacements, $W > H$, we have

$$\frac{GV_0^2}{2} = \int_0^H P_c \left[1 + \left(\frac{W}{H}\right)^2\right] \mathrm{d}W + \int_H^W P_c \left[2\left(\frac{W}{H}\right)\right] \mathrm{d}W,$$

from which it follows that

$$\frac{W}{H} = \sqrt{\frac{3GL^2V_0^2}{8HM_0}}.$$

5.7 Strains and strain rate in beams

When a beam is dynamically loaded, its change in shape is globally quantified by the transverse displacement. At a local level, we can use the strains as a measure of the new beam configuration. These strains are only approximately known for beams idealized by perfectly plastic material models.

For impact loading, the strains in a beam, or any other structure, will change rapidly and we refer to strain rate as a measure of how much the strains alter its magnitude along time. Obviously that the strain and its rate will not be constant and it is important to evaluate both. The chief reason to know the strains in a beam is that its magnitude is an indication of material failure.

The reason for knowing the strain rates is that there are various materials which increase their strength as they are loaded more rapidly, as we shall see in more details in Chapter 7. For a rigid–perfectly plastic material model, this is to say that the flow stress will increase with the strain rate. Hence, we have the situation where the flow stress is changing as the structure is loaded, *ie* the material becomes stronger due to these dynamic effects.

We notice at once that the generalized stresses for beams, M_0, N_0, Q_0 are dependent on the flow stress σ_0. Clearly, if we correct the flow stress for the strain rate effects we can also correct the generalised stresses and so our previous solutions can still be used within some restrictions. The yield surfaces will increase in size and we refer to strain rate (isotropic) hardening.

To quantify the material strain rate sensitivity we should develop a material constitutive law. In the realm of the strength of materials science, the constitutive laws proposed are of phenomenological nature, *ie* no attention is given to metallurgical reasons for the strain rate effect on the material strength; only the global response to the material is measured. Once various strength curves for a given material are known, a mathematical law is suggested that fits better the material behaviour. These constitutive laws will be discussed in Chapter 7 and for now we will work with the simple equation

$$\sigma_d = \sigma_0 \left(\frac{\dot{\varepsilon}}{\dot{\varepsilon}_0} \right)^n ,$$

This is the so called Norton–Hoffman equation and it is referred to as a viscous relation.

where n is a material constant, the subscript d stands for dynamic and $\dot{\varepsilon}_0$ is a reference strain rate which took place during the measurement of the quasi–stain flow stress σ_0. To use this equation we need to evaluate the strain rate in a beam and this depends on the specific problem, according to the boundary and initial conditions and type of loading. Of course that the strain rate depends also on the strain itself, whose calculation is of our concern now.

5.7.1 Strains

Further details of the calculation of bending, shear and membrane quasi–static and dynamic hinge lengths are discussed in M. Alves and N. Jones, Impact failure of beams using damage mechanics: Part I — Analytical model, International Journal of Impact Engineering, 27, p. 837–861, 2002.

The approach adopted in this chapter is based on a rigid, perfect plastically material model. When using such a model, a given structure will displace thanks to the formation of hinges, where it is assumed the deformation is concentrated. For a clamped beam under a central point load, for example, the hinges are at the load and at the supports and all deformation is assumed to be concentrated there. If the analysis takes into account finite displacements, then the membrane strains could well be considered to spread along the whole span.

In order to evaluate the strains, it is necessary to define a region, rather than a point, where the strains spread. Some authors suggest a plastic hinge of value between H and $2H$ but a more rational way to evaluate the hinge length and the strains would be better. Based on the next figure, we can admit that the hinge length is the distance associated with the bending moments M_y, which causes yielding at the out most fibre, and M_0, which is the beam collapse moment. Hence, depending on the boundary condition and the beam cross–section shape we will have different bending moment distribution and, therefore, hinge lengths, which are listed in the next table for some supporting cases and loads.

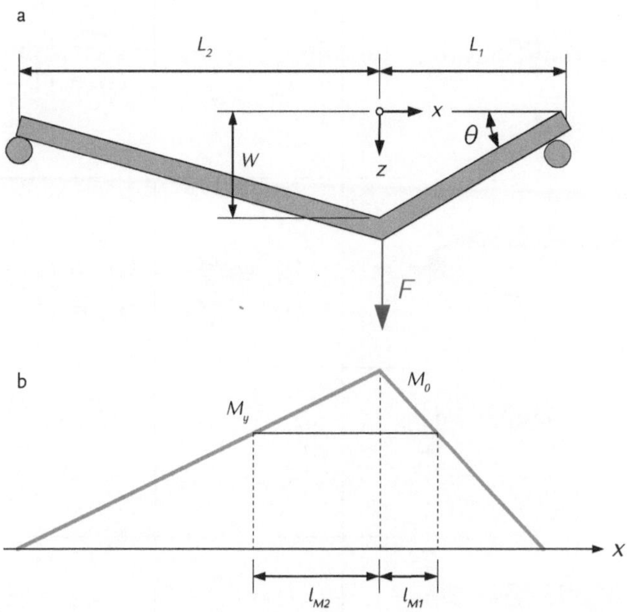

Figure 5.23 (a) A beam loaded off-centre showing the bending hinge length definition. (b) Associated bending moment distribution. The beam displacement is exaggerated.

The bending strain, at the point of the maximum beam displacement, is

$$\varepsilon_M = \frac{1}{2}\left(\frac{W}{L_1 l_{M_1}}z + \frac{W}{L_2 l_{M_2}}z\right),$$

when taking an average value of the strains at the left- and right-hand sides of the load point.

We can specialize this expression, using the hinge values in the table, for a clamped beam subjected to an off centre concentrated load, *ie* $l_{M_1} = L_1/6$, $l_{M_2} = L_2/6$, and at the point of maximum bending strain, *ie* $z = H/2$. In this case, the bending strain is

$$\varepsilon_M = 3h^2(1 + l^2)\bar{w}/2,$$

with $h = H/L_1$, $\bar{w} = W/H$ and $l = L_1/L_2$ ($L_1 \leq L_2$).

TABLE 5.1 Bending hinge length for various beam configurations. $s_f = M_0/M_y$ is the shape factor used in plasticity and equal 3/2 for a uniform rectangular cross–section beam.

beam configuration	bending hinge length	
	rectangular cross-section	general cross-section
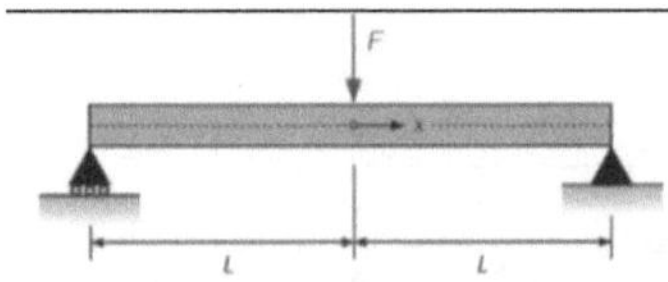	$l_M = \frac{L}{3}$	$l_M = (1 - s_f^{-1})L$
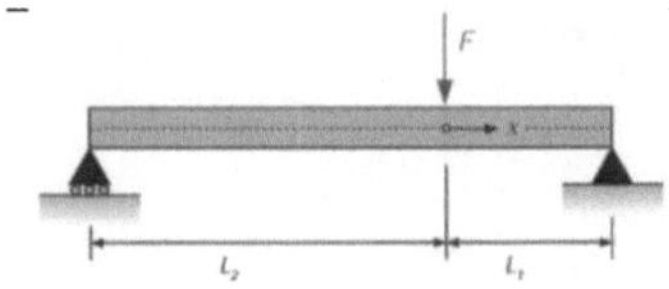	$l_{M_1} = \frac{L_1}{3}$ $l_{M_2} = \frac{L_2}{3}$	$l_{M_1} = (1 - s_f^{-1})L_1$ $l_{M_2} = (1 - s_f^{-1})L_2$
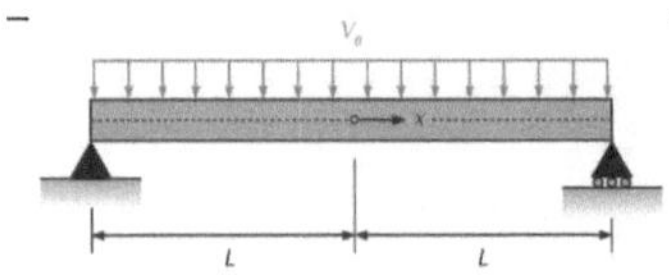	$l_M = \frac{L}{\sqrt{3}}$	$l_M = \frac{L}{\sqrt{1-s_f^{-1}}}$
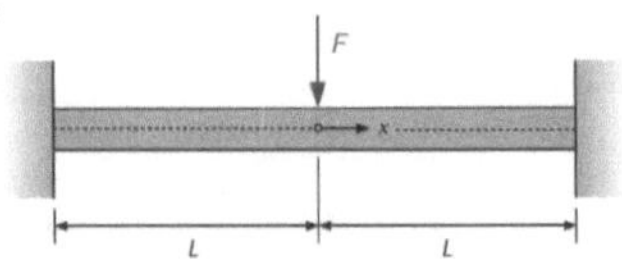	$l_M = \frac{L}{6}$	$l_M = \frac{s_f-1}{2s_f}L$
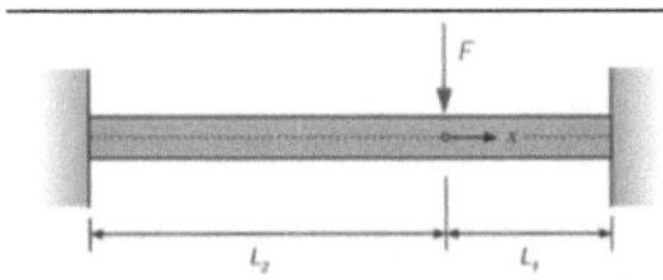	$l_{M_1} = \frac{L_1}{6}$ $l_{M_2} = \frac{L_2}{6}$	$l_{M_1} = \frac{s_f-1}{2s_f}L_1$ $l_{M_2} = \frac{s_f-1}{2s_f}L_2$
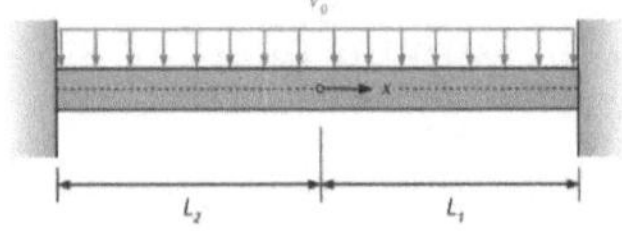	$l_M = \frac{L}{\sqrt{6}}$	$l_M = \frac{L}{\sqrt{2(1-s_f^{-1})}}$

The membrane strain, ε_N, on the right-hand side of a fully-clamped beam, due to a normal force N, becomes

$$\varepsilon_{N_1} = \Delta L_1 / l_{N_1} = \sqrt{1 + (W/L_1)^2} - 1,$$

which can be averaged with a similar expression for the left-hand side of the beam, leading to

$$\varepsilon_N = (1 + \xi^2)\,(hw)^2\,/4,$$

assumed to be distributed evenly along the beam span and throughout the cross-section.

The bending strains dominate the response up to $w \approx 1$. For larger displacements, the axial strain is mainly due to membrane effects. It follows that the equivalent strain at the impact point of a clamped beam is

$$\varepsilon_{eq} = \varepsilon_M = (3/2)(1 + \xi^2)h^2 w \qquad \text{for} \qquad w \leq 1$$

and

$$\varepsilon_{eq} = (1 + \xi^2)h^2(w^2 + 5)/4 \qquad \text{for} \qquad w > 1,$$

where the last equation was obtained by imposing the requirement $\varepsilon_M = \varepsilon_N$ at $w = 1$.

Note that the shear strain contribution to the equivalent strain was disregarded here so the equations above are expected to give poor results for beams hit too close to the support. Other expressions can be constructed based on the different boundary conditions and loading.

5.7.2 Strain rate

The strain rate is simply the time derivative of the equivalent strain. It can be shown that, for a clamped beam hit by a mass, that the strain rate during the bending phase is larger than during the membrane phase. The time derivative of the strain expression amounts to

$$\dot{\varepsilon}_{eq} = \frac{3}{2}h^2(1 + l^2)\dot{w}.$$

Now, simple models for unidimensional structures indicate that the maximum strain rate occurs when $\dot{W} = V_0/\sqrt{2}$ so that

$$\dot{\varepsilon}_{eq} = \frac{3h^2(1 + l^2)V_0}{2\sqrt{2}H},$$

which we simplify to

$$\dot{\varepsilon}_{eq} = \frac{h(1+l^2)V_0}{L_1}$$

and for central impact to

$$\dot{\varepsilon}_{eq} = \frac{2HV_0}{L^2}.$$

The next figure shows that the above equation gives a reasonable prediction of the strain rate in a beam when compared with finite element results.

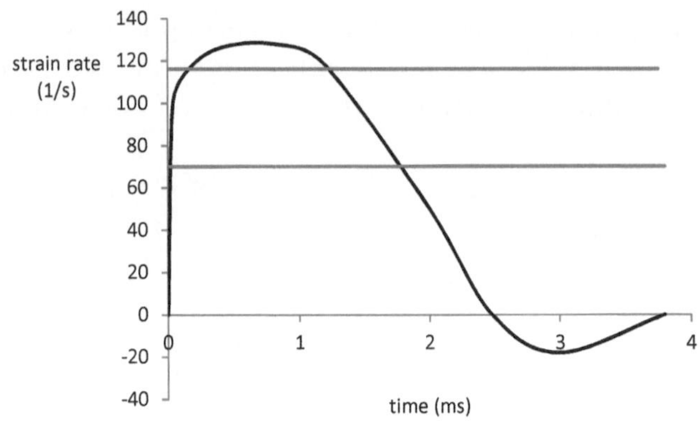

Figure 5.24 Finite element results for the strain rate versus time underneath an aluminium beam impacted on the top surface at the one–quarter span position (from J.L. Yu and N. Jones, Numerical simulation of a clamped beam under impact loading. Computer & Structures, 32, p. 281—293, 1989). $G = 5$ kg, $V_0 = 5.34$ m/s, $2L = 101.6$ mm, $H = 7.62$ mm. The horizontal lower line is the prediction according to $\dot{\varepsilon}_{eq} = h(1+l^2)V_0/L_1$, with the superior line considering also the cooperation of the shear strain rate.

Once we know the strain rate we can correct the flow stress by adopting a constitutive equation. This dynamic flow stress is then used in the calculation of M_0. Note that we opt here for simple expressions for the strain rate, which renders them constant throughout the beam motion. This is a gross simplification of what actually takes place in the beam response.

5.8 Circular plates impulsively loaded

Plates stand in the next level of complexity, after beams. They are bidimensional structures and have various shapes, although here we will deal only with circular plates, for the sake of simplicity. In fact, we will explore only two loading cases: a plate under an impulsive loading represented by an initial velocity and a plate under the transverse impact of a dropping mass. For the impulsive case, the plate is analysed under the assumption of small transverse displacements. It will be adopted a simple failure criterion that, when met, will let the plate to fly free in space, where both changing in the plate geometry and rigid body motion will be obtained. For the plate under the transverse impact of a mass, large transverse displacements will be considered.

5.8.1 Equilibrium equations

We now present the equilibrium equations for circular plates, the adopted yield surface and a simple failure criterion. When met, for the impulsive loading case, the plate will detached from the support, moving freely in the space, eventually changing its shape.

The governing equation for plates undergoing small transverse displacements is derived using the infinitesimal element of the next figure. Equilibrium of this element requires that

$$\frac{\partial}{\partial r} r M_r - M_\theta - r Q_r = 0,$$

and

$$\frac{\partial}{\partial r} r Q_r + r p = \mu r \ddot{w},$$

where r and θ are the cylindrical coordinates, M_r and M_θ are the unit radial and circumferential bending moments, Q_r the unit transverse shear force, p the distributed load, μ the plate mass per unit area and w the plate midplane transverse displacement, with the double dots representing the second derivative of time, t.

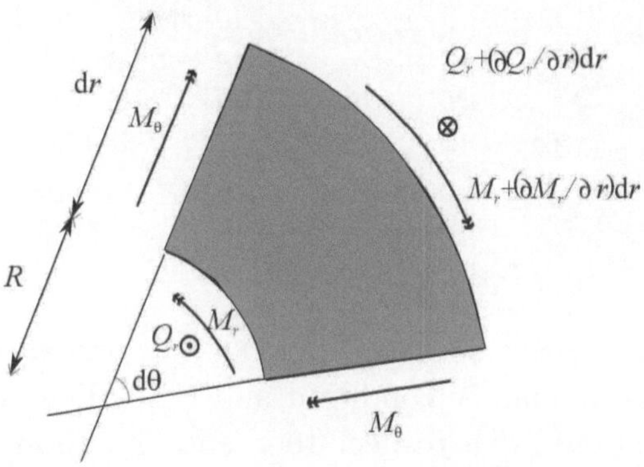

Figure 5.25 Circular plate infinitesimal element.

The kinematics of this infinitesimal element implies that the curvatures in the r and θ directions are

$$\kappa_r = -\frac{\partial^2 w}{\partial r^2}$$

and

$$\kappa_\theta = -\frac{1}{r}\frac{\partial w}{\partial r},$$

respectively, where membrane strains are disregarded since neither in-plane loads or finite displacements are considered in the present analysis.

The yield condition we adopt is shown in the next figure and it relates the radial and circumferential bending moments plus the transverse shear force. On grounds to be seen later, we assume that the velocity profile for the plate varies linearly with the radius, implying that

$$\dot{\kappa}_r = 0 \qquad \text{and} \qquad \dot{\kappa}_\theta \geq 0,$$

which results in $M_\theta = M_0$ but with M_r remaining to be determined from the analysis. Here M_0 is the plastic bending strength moment of the plate cross–section of unit width,

$$M_0 = \frac{\sigma_0 H^2}{4}.$$

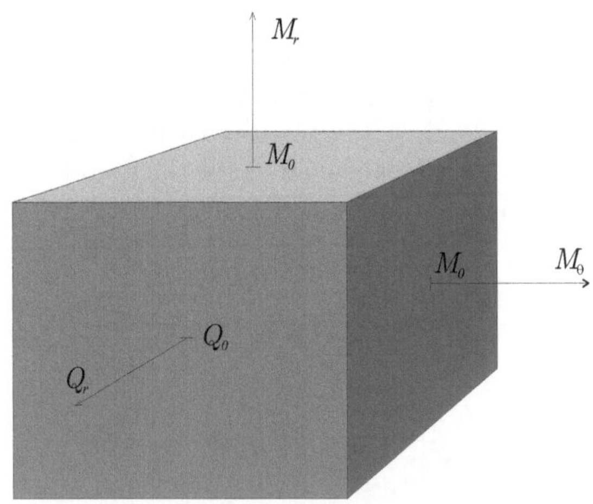

Figure 5.26 Yield surface used in the analysis.

Moreover, the transverse shear force, Q, may assume different values depending on the boundary condition and time. The derivative of the equilibrium equations with respect to r and the above considerations lead to the governing equation for circular plates as

$$\frac{\partial^2}{\partial r^2}(rM_r) = \mu r\ddot{w},$$

for $M_\theta = constant$ and $p = 0$.

We introduce the dimensionless parameter for a solid cross–section,

$$\nu = \frac{Q_0 R}{2M_0} = \frac{R}{H},$$

with $Q_0 = \sigma_0 H$ being the transverse shear force necessary for plastic flow of the plate cross–section per unit length. ν is a ratio between the shear and bending strengths and depending on its value leads to different solutions.

Here we deal only with plates with $\nu \geq 2$. This is a more complicate but realistic case.

This solution is explored in more details in N. Jones and M. Alves, Post–failure behaviour of impulsively loaded circular plates, International Journal of Mechanical Sciences, 52, p. 706–715, 2010.

As for the load applied to the plate, we here analyse the impulsive case, *ie* a pressure pulse of high amplitude and short duration and represented by an initial velocity V_0 applied throughout the plate top face. We also adopt for the first phase of motion the transverse velocity profile shown in the next figure and given by

$$\dot{w} = \dot{W}_c = V_0 \quad \text{for} \quad 0 \leq r \leq \xi_0$$

and

$$\dot{w} = \dot{W}_s + (V_0 - \dot{W}_s)\frac{R-r}{R-\xi_0} \quad \text{for} \quad \xi_0 \leq r \leq R,$$

with ξ_0 defining the position of the fixed plastic hinge.

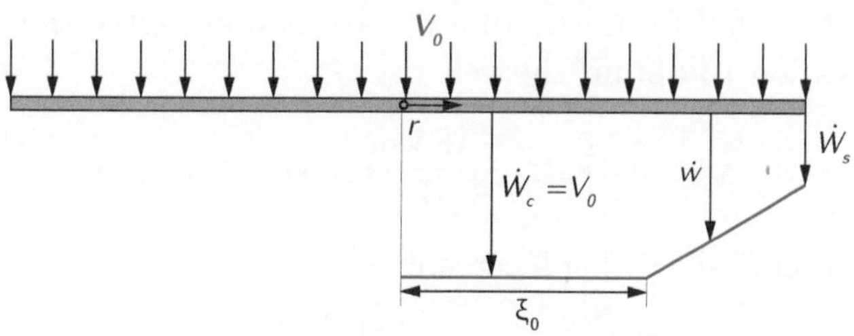

Figure 5.27 First phase of motion velocity profile for a circular plate with $\nu \geq 2$.

The equilibrium equations and the velocity profile leads to

$$M_r = M_0 + \frac{\mu \ddot{W}_s}{12(R-\xi_0)} \frac{(r+\xi_0)(r-\xi_0)^3}{r}$$

and

$$Q_r = \frac{\mu \ddot{W}_s}{6(R-\xi_0)} \frac{(2r+\xi_0)(r-\xi_0)^2}{r},$$

for $\xi_0 \leq r \leq R$, when applying the boundary conditions $M_r(\xi_0) = M_0$ and $Q_r(\xi_0) = 0$.

By knowing that $M_r(R) = 0$ for a simply supported plate, we can obtain the acceleration at the support as

$$\ddot{W}_s = \frac{-12M_0}{\mu\Gamma},$$

with

$$\Gamma = \frac{(R+\xi_0)(R-\xi_0)^2}{R}.$$

It follows that

$$\frac{M_r}{M_0} = 1 - \frac{(r+\xi_0)(r-\xi_0)^3}{r\Gamma(R-\xi_0)}$$

and

$$\frac{Q_r}{Q_0} = \frac{-R(2r+\xi_0)(r-\xi_0)^2}{\nu r\Gamma(R-\xi_0)},$$

which allows us to obtain the position of the stationary plastic hinge by using $Q_r(R) = -Q_0$ and $Q_0/M_0 = 2\nu/R$ as

$$\xi_0 = \frac{R}{2\nu}\left(\sqrt{4\nu^2 - 8\nu + 1} - 1\right).$$

5.8.2 Critical velocity

We now set ourselves the task to determine the critical velocity that leads the plate to fail from its support. We time integrate the transverse shear acceleration to obtain the velocity

$$\dot{W}_s = \frac{-12M_0 t}{\mu\Gamma} + V_0$$

and the transverse shear displacement

$$W_s = V_0 t - \frac{6M_0 t^2}{\mu\Gamma}.$$

Failure is assumed to take place at the support when $W_s = kH$, *ie* a fraction of the plate thickness, with k being a material parameter. Failure occurs at

This is a failure criterion, albeit very simple.

$$t_{fa} = \frac{\mu\Gamma\beta}{12M_0},$$

with

$$\beta = V_0\left(1 - \sqrt{1 - \frac{24kHM_0}{\mu\Gamma V_0^2}}\right).$$

At the failure time, the displacement of the plate in the region $0 \leq r \leq \xi_0$ is

$$W_{\bar{c}} = V_0 t_{fa} = \frac{\mu \Gamma \beta V_0}{12 M_0}$$

and the velocity at the support is

$$\dot{W}_{\bar{s}} = V_0 - \beta,$$

from which it follows the critical velocity

$$V_{cr} = \sqrt{\frac{24kHM_0}{\mu\Gamma}}.$$

Note that, if $V_0 < V_{cr}$, $W_s < kH$ at the support, when $\dot{W}_{\bar{s}} = 0$. Also, the failure criterion $W_s = W_{fa} = kH$ is very crude but it does give an indication of how the plate will behave after being removed from the supports. But before giving more details, we need to deal with the moving hinge phase.

5.8.3 Motion after severance — moving hinge phase

The transverse velocity is now

$$\dot{w} = \dot{W}_{c_m} = V_0 \quad \text{for} \quad 0 \leq r \leq \xi$$

and

$$\dot{w} = \dot{W}_{s_m} + (V_0 - \dot{W}_{s_m}) \frac{R - r}{R - \xi} \quad \text{for} \quad \xi \leq r \leq R,$$

as illustrated in the next figure.

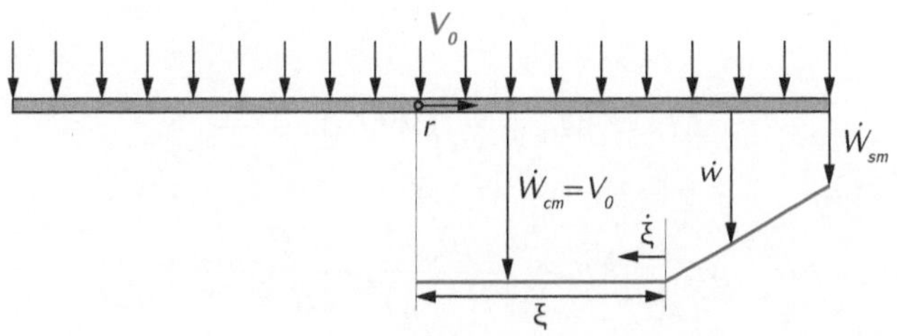

Figure 5.28 Post–severance velocity profile in the moving hinge phase for a circular plate with $\nu \geq 2$.

Note that, in contrast with the previous phase of motion, ξ is time dependent so the hinge moves towards the middle of the plate. This is, at first, an assumption for the kinematic field but it can be proven to be the correct one.

The acceleration now reads

$$\ddot{w} = \ddot{W}_{s_m}\frac{r-\xi}{R-\xi} + \frac{(V_0 - \dot{W}_{s_m})(R-r)\dot{\xi}}{(R-\xi)^2},$$

which can be integrated once and, using $M_\theta = M_0$, gives

$$Q_r = \frac{1}{r}\left(\mu\int r\ddot{w}\mathrm{d}r - M_0 + c_1\right),$$

or

$$Q_r = \frac{\mu\ddot{W}_{s_m}}{R-\xi}\left(\frac{r^2}{3} - \frac{\xi r}{2} + \frac{\xi^3}{6r}\right) + \frac{\mu(V_0 - \dot{W}_{s_m})\dot{\xi}}{(R-\xi)^2}\left(\frac{Rr}{2} - \frac{r^2}{3} - \frac{\xi^2 R}{2r} + \frac{\xi^3}{3r}\right),$$

since $Q_{r_{|r=\xi}} = 0$.

The transverse shear force, Q_r, is zero at the support and noting that $\frac{R^2/6-\xi^2/2+\xi^3/3R}{R^2/3-\xi R/2+\xi^3/6R} = \frac{R+2\xi}{2R+\xi}$, we obtain

$$\ddot{W}_{s_m} = -\frac{(V_0 - \dot{W}_{s_m})\dot{\xi}}{(R-\xi)}\frac{(R+2\xi)}{(2R+\xi)},$$

which can be written as $\ddot{W}_{s_m} = (\mathrm{d}\dot{W}_{s_m}/\mathrm{d}\xi)\dot{\xi}$ and integrated to yield

$$V_0 - \dot{W}_{s_m} = \frac{\beta\Xi}{(2R+\xi)(R-\xi)}$$

since $\dot{W}_{s_m} = \dot{W}_{\bar{s}}$ when the hinge is at $\xi = \xi_0$ and defining

$$\Xi = (2R+\xi_0)(R-\xi_0).$$

Now,

$$M_r = M_0 + \frac{1}{r}\int rQ_r\mathrm{d}r + \frac{c_2}{r},$$

or

$$M_r = M_0 - \frac{\mu\dot{\xi}\beta}{(R-\xi)^3}\frac{\Xi}{(2R+\xi)}\times \left[\frac{R+2\xi}{2R+\xi}\left(\frac{r^3}{12} - \frac{\xi r^2}{6} + \frac{\xi^3}{6} - \frac{\xi^4}{12r}\right) - \left(\frac{r^2R}{6} - \frac{r^3}{12} - \frac{\xi^2 R}{2} + \frac{\xi^3}{3} - \frac{\xi^4}{4r} + \frac{R\xi^3}{3r}\right)\right]$$

for $\xi \le r \le R$, when using $M_{r_{|r=\xi}} = M_0$ to obtain the constant c_2.

$M_{r|r=R} = 0$ predicts that the hinge velocity is

$$\dot{\xi} = \frac{-12M_0}{\beta\mu} \frac{R(2R+\xi)^2}{\Xi(R^2+4\xi R+\xi^2)},$$

giving a final expression for the radial bending moment as

$$\frac{M_r}{M_0} = 1 + \frac{12R(2R+\xi)}{(R-\xi)^3(R^2+4\xi R+\xi^2)}$$

$$\left[\frac{R+2\xi}{2R+\xi}\left(\frac{r^3}{12} - \frac{\xi r^2}{6} + \frac{\xi^3}{6} - \frac{\xi^4}{12r}\right) - \left(\frac{r^2 R}{6} - \frac{r^3}{12} - \frac{\xi^2 R}{2} + \frac{\xi^3}{3} - \frac{\xi^4}{4r} + \frac{R\xi^3}{3r}\right)\right]$$

and for the transverse shear force

$$\frac{Q_r}{Q_0} = \frac{-6R^2(2R+\xi)}{\nu(R-\xi)^3(R^2+4\xi R+\xi^2)}$$

$$\left[-\frac{R+2\xi}{2R+\xi}\left(\frac{r^2}{3} - \frac{\xi r}{2} + \frac{\xi^3}{6r}\right) + \frac{Rr}{2} - \frac{r^2}{3} - \frac{\xi^2 R}{2r} + \frac{\xi^3}{3r}\right]$$

Note that the acceleration at the support can be rewritten as

$$\ddot{W}_{s_m} = \frac{12M_0}{\mu} \frac{R(R+2\xi)}{(R-\xi)^2(R^2+4\xi R+\xi^2)}.$$

Now, the plastic hinge speed can be integrated to yield the moving hinge position with time

$$\xi(t) + \frac{3R^2}{2R+\xi} = \xi_0 + \frac{3R^2}{2R+\xi_0} - \frac{12M_0 R}{\beta\mu\Xi}(t - t_{fa}),$$

since $\xi = \xi_0$ when $t = t_{fa}$.

The displacements at the end of this phase of motion both at the support and at the center of the plate can be obtained by knowing that the end of this second phase of motion occurs at time t_m, when the moving hinge reaches the center of the plate. It follows that

$$t_m = \frac{\beta\mu\Xi}{24M_0}$$

and at this time the final displacement at the plate center is $W_{c_m} = V_0 t_m$ or

$$W_{c_m} = \frac{\beta\mu V_0 \Xi}{24M_0}.$$

The displacement at the support when $t = t_m$ is

$$W_{s_m} = kH + \int_{t_{fa}}^{t_m} \dot{W}_{s_m} \mathrm{d}t = kH + \int_{\xi_0}^{0} \dot{W}_{s_m} \frac{\mathrm{d}\xi}{\dot{\xi}},$$

or

$$W_{s_m} = kH + V_0(t_m - t_{fa}) + \frac{\mu\beta^2\Xi^2}{12M_0R} I_1,$$

with $I_1 = \int_{\xi_0}^{0} (R^2 + 4\xi R + \xi^2)/(2R+\xi)^3(R-\xi)\mathrm{d}\xi$, or

$$I_1 = \frac{2\ln 2}{9R} + \frac{7}{24R} - \frac{2}{9R}\ln\frac{2R+\xi_0}{R-\xi_0} - \frac{1}{2R+\xi_0}\left[\frac{1}{3} + \frac{R}{2(2R+\xi_0)}\right].$$

5.8.4 Motion after severance — hinge at the centre

The moving hinge now reaches the centre of the plate and the new transverse velocity is as depicted in the next figure, or

$$\dot{w} = \dot{W}_{s_f} + (\dot{W}_{c_f} - \dot{W}_{s_f})\left(\frac{R-r}{R}\right), \qquad \text{for} \qquad 0 \le r \le R.$$

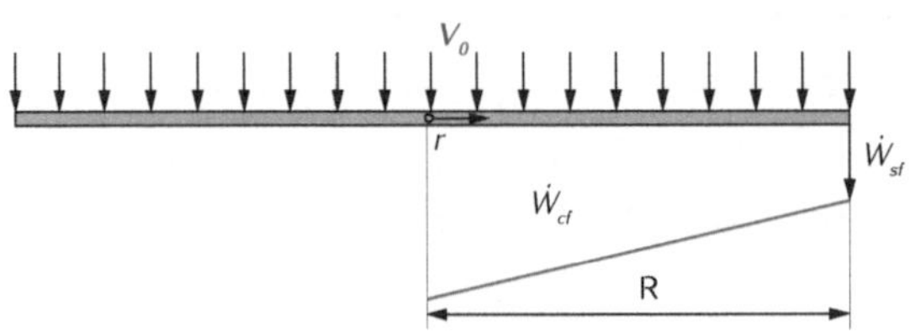

Figure 5.29 Post-severance velocity profile for a circular plate with $\nu \ge 2$ when the hinge is at the center.

The radial bending moment comes from a double integration, giving

$$\frac{M_r}{M_0} = 1 + \frac{\mu}{M_0}\left[\ddot{W}_{s_f}\frac{r^2}{6} + \left(\ddot{W}_{c_f} - \ddot{W}_{s_f}\right)\left(\frac{r^2}{6} - \frac{r^3}{12R}\right)\right],$$

where the integration constants were obtained from $M_r(0) = M_0$. We can also obtain the transverse shear force

$$Q_r = \mu r\left[\ddot{W}_{s_f}\frac{r}{3R} + \ddot{W}_{c_f}\left(\frac{1}{2} - \frac{r}{3R}\right)\right],$$

since $Q_r(0) = 0$.

The accelerations are

$$\ddot{W}_{s_f} + \ddot{W}_{c_f} = \frac{-12M_0}{\mu R^2}$$

and

$$2\ddot{W}_{s_f} + \ddot{W}_{c_f} = 0,$$

since both Q_r and M_r are zero at the support. This system of equations gives

$$\ddot{W}_{s_f} = \frac{12M_0}{\mu R^2},$$

and

$$\ddot{W}_{c_f} = -2\ddot{W}_{s_f} = \frac{-24M_0}{\mu R^2},$$

which can be integrated with the conditions $\dot{W}_{s_f} = \dot{W}_{s_m}$ and $\dot{W}_{c_f} = V_0$ at $t = t_m$ to give, respectively, the velocity at the plate support and at the center

$$\dot{W}_{s_f} = \frac{12M_0}{\mu R^2}(t - t_m) + V_0 - \frac{\beta\Xi}{2R^2}.$$

and

$$\dot{W}_{c_f} = \frac{-24M_0}{\mu R^2}(t - t_m) + V_0.$$

A further time integration of these equations yields the displacements at the centre and support,

$$W_{c_f} = V_0 t - \frac{12M_0}{\mu R^2}(t - t_m)^2$$

and

$$W_{s_f} = kH + V_0(t - t_{fa}) + \frac{6M_0}{\mu R^2}(t - t_m)^2 - \frac{\beta\Xi}{2R^2}(t - t_m) + \frac{\mu\beta^2\Xi^2}{12M_0 R}I_1,$$

when adopting as initial conditions the displacements given in the previous phase.

Note that these equations are valid until rigid body motion starts, which is the last phase of motion.

5.8.5 Rigid body motion

We can notice that there is a positive acceleration at the support, which implies that the support region moves faster than the center region of the plate. These two velocities are equal at a time t_r, $\dot{W}_{c_f} = \dot{W}_{s_f}$, or

$$t_r = t_m + \frac{\beta\mu\Xi}{72M_0} = \frac{\beta\mu\Xi}{18M_0}.$$

Thus, the duration of this phase of motion, from the time the plastic hinge reaches the center until the plate ceases to change its shape, is the difference

$$\Delta t = t_r - t_m = \frac{t_m}{3}.$$

The rigid body velocity, V_r, of the plate comes from the evaluation of either $\dot{W}_{c_f}$ or $\dot{W}_{s_f}$ at $t = t_r$, which gives

$$V_r = V_0 - \frac{\beta\Xi}{3R^2}.$$

It is interesting to note that the linear velocity profile for the plate during this phase of motion indicates that $\dot{\kappa}_r = 0$. Also, the circumferential curvature rate is

$$\dot{\kappa}_\theta = -\frac{1}{r}\frac{\partial\dot{w}}{\partial r} = \frac{\dot{W}_{c_f} - \dot{W}_{s_f}}{rR} = 0,$$

which means that no plastic energy is absorbed during this phase of motion, as expected.

When rigid body motion starts, the displacements at the plate center and support can be found as

$$W_{c_r} = V_0 t_r - \frac{12M_0\Delta t^2}{\mu R^2}$$

and

$$W_{s_r} = kH + V_0(t_r - t_{fa}) + \frac{6M_0\Delta t^2}{\mu R^2} - \frac{\beta\Xi\Delta t}{2R^2} + \frac{\mu\beta^2\Xi^2}{12M_0 R}I_1.$$

Note that W_{c_f} is less than $V_0 t_r$, because the plate center region decelerates during the last phase of motion.

This study on the severance of a circular plate due to a high impulsive loading discloses many aspects of the plate motion. Once a plate has been detached from its supports, it follows a free trajectory with a high residual kinetic energy. The solution has revealed a phase of motion in which a circular plastic hinge propagates towards the plate centre while the plate is travelling freely in space. During this phase of motion, the plastic hinges alter the displacement profile of the plate, when compared with the one at severance. This behaviour might have importance for forensic studies.

■ MOTION OF A PLATE

Describe in detail the motion of a simply supported circular plate with the properties listed below.

TABLE 5.2 Plate properties.

R(mm)	H(mm)	ρ	σ_0(MPa)	ν	k
50	2	2800	200	10	0.7

The critical velocity is the high value of $V_0 = 522.97$m/s for the assumed parameters and the failure time is 5.35μs. A stationary hinge is formed at $\xi_0 = 8.46$mm. Adopting an impulsive velocity of $V_0 = V_{cr}$, the time for the hinge to arrive at the plate center, once in motion, is $t_m = 26.8\mu$s. At $t = t_r = 35.7\mu$s, plastic deformation ceases and rigid body motion starts. Clearly, all these events occur very rapidly. It is interesting to note that the critical velocity when the plate has $\nu = 1$ is 63.25m/s.

The next figure shows the hinge speed with time. The hinge starts to move at a relatively slow speed, which increases to 800m/s for this plate when the plastic hinge reaches the plate centre. This maximum hinge speed is close to the speed of plastic waves in aluminium tubes.

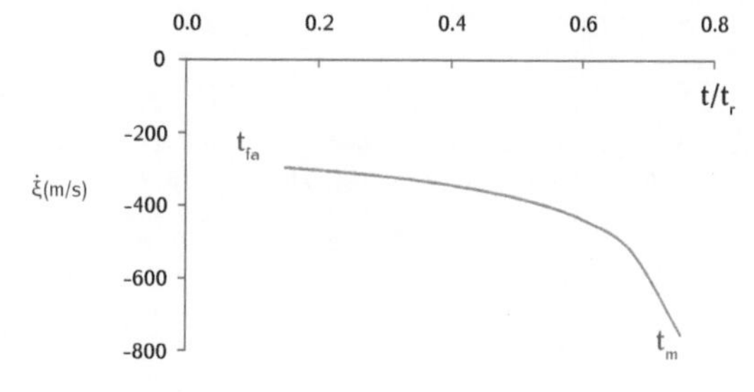

Figure 5.30 Hinge speed against dimensionless time.

The velocities at the plate support and center are next depicted against dimensionless time. The initial condition is the critical velocity. The velocity at the support decreases to zero at $t = t_{fa}$, and then increases quickly after the plate breaks free from the support. The velocity at the centre starts to decrease once the travelling hinge reaches the plate centre, at $t = t_m$, and rigid body motion starts once both velocities are equal.

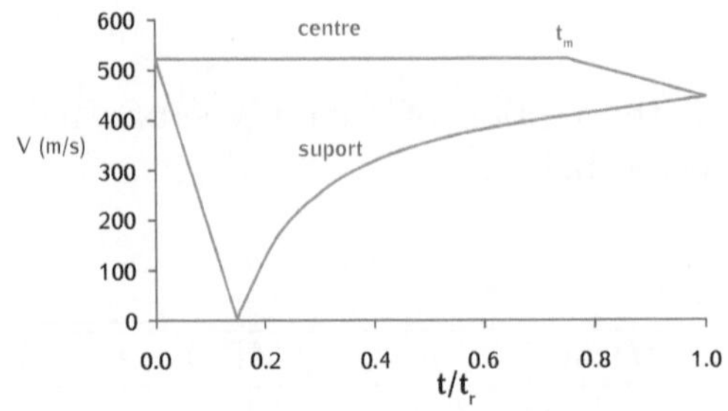

Figure 5.31 Velocities prior rigid body motion.

If the impulsive velocity were just below V_{cr} the plate would only deform plastically, eventually reaching a permanent deformed shape with some local partial tearing at the supports. For velocities just above V_{cr} the plate flies away and potentially can inflict severe damage on any object lying in its trajectory. This is so because the residual kinetic energy is 73% of the initial kinetic energy for the plate parameters adopted here. The displacements at the plate support and at the plate centre evolve with time as depicted below.

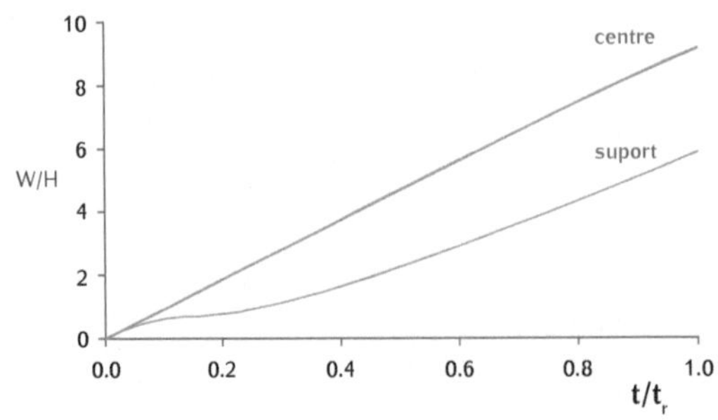

Figure 5.32 Plate displacement with time.

5.9 Circular plates under low-velocity impact of a mass

Let us now study the case of a clamped circular plate subjected to the central impact of a mass G. We aim to obtain the central displacement of a circular plate of thickness H and radius R, as illustrated in the next figure.

This section is based on N. Jones, On the mass impact loading of ductile plates, Defence Science Journal, 53, p. 15–24, 2003 and on R.C. Santiago, Impact performance of thermoplastic reinforced fibre–metal laminates (in Portuguese), Ph.D. Thesis, University of Sao Paulo, 2014.

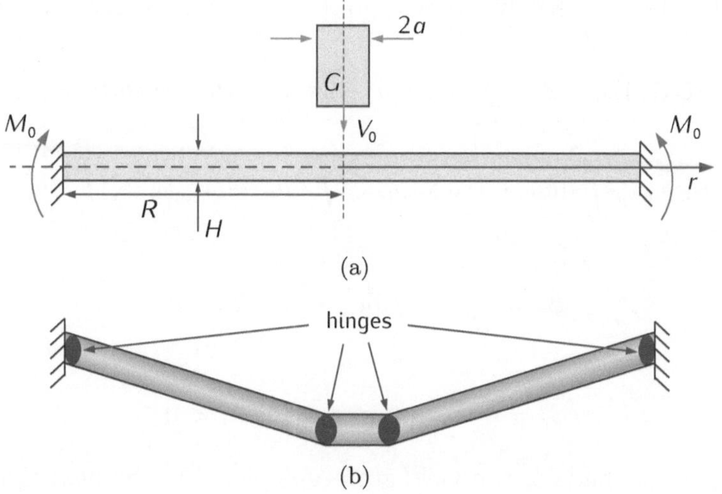

Figure 5.33 (a) A clamped circular plate under central impact of a cylinder and (b) a kinematic field representation.

We start the analysis by equating the external power (force × velocity) applied to a general structure to the internal power counter generated. We have then

$$-G\ddot{W}_0\dot{W}_0 = \int_A \mu\ddot{w}\dot{w}\mathrm{d}A = \int_A (M_r + wN_r)\dot{\kappa}_r\mathrm{d}A +$$

$$\int_A (M_\theta + wN_\theta)\dot{\kappa}_\theta\mathrm{d}A +$$

$$\sum_{m=1}^{n}\int_h (M_r + wN_r)\frac{\partial\dot{w}}{\partial r}|_m\mathrm{d}h_m + \sum_{p=1}^{q}\int_h Q_r\dot{w}|_p\mathrm{d}h_p,$$

with μ being the area density of the plate material.

The statement above, here specialized to the case of circular plates, is a very powerful one and it can be used to solve various problems. The generalized stresses M_r, M_θ, N_r and N_θ contribute to the power dissipated in the plate. The terms under summation are the power dissipated in the possible hinges to be formed during the deformation process. These localized hinges can be of bending moments, normal forces or shear nature and take place in various positions of the plate domain. Of course that their location depends on the adopted kinematic field, which is left to the analyst to suggest. The solution is then developed and in a final stage checked against the upper and lower theorems of plasticity.

Turning now to our problem, the impact mass is a cylinder of radius a and the plate has a flow stress σ_0 so that the expression above becomes

$$-G\ddot{W}_0\dot{W}_0 = 2\pi\mu\int_a^r \ddot{w}\dot{w}r\mathrm{d}r + 2\pi\int_a^R (M_\theta + wN_\theta)r\dot{\kappa}_\theta\mathrm{d}r+$$

$$M_0\dot{\phi}|_{r=a}\int_a \mathrm{d}h_m + W_0N_0\dot{\phi}|_{r=a}\int_a \mathrm{d}h_m - M_0\dot{\phi}|_{r=R}\int_R \mathrm{d}h_m,$$

with

$$\dot{\phi}_r = \frac{\dot{W}_0}{R-a} \quad \text{for} \quad r = a, \dot{\kappa}_r > 0,$$

and

$$\dot{\phi}_r = \frac{-\dot{W}_0}{R-a} \quad \text{for} \quad r = R, \dot{\kappa}_r < 0,$$

where use was made of the yield surface depicted in the next figure.

Now, adopting the dimensionless variables $\bar{r} = a/R$ and $\gamma = G/\mu\pi R^2$ and the following velocity field

$$\dot{w} = \dot{W}_0 \qquad \text{for} \qquad 0 \leq r \leq a$$

and

$$\dot{w} = \dot{W}_0\frac{R-r}{R-a} \qquad \text{for} \qquad a \leq r \leq R,$$

we have that

$$-G\ddot{W}_0\dot{W}_0 = \frac{2\pi\mu\ddot{W}_0\dot{W}_0}{(R-a)^2}\int_a^r (R-r)^2\mathrm{d}r$$

$$+2\pi M_0\int_a^R \frac{\dot{W}_0}{r(R-a)}r\mathrm{d}r + 2\pi N_0\int_a^R \frac{W_0(R-r)}{R-a}\frac{\dot{W}_0}{r(R-a)}r\mathrm{d}r$$

$$+2\pi a M_0\frac{\dot{W}_0}{R-a} + 2\pi a W_0 N_0\frac{\dot{W}_0}{R-a} + 2\pi R(-M_0)\frac{-\dot{W}_0}{R-a}.$$

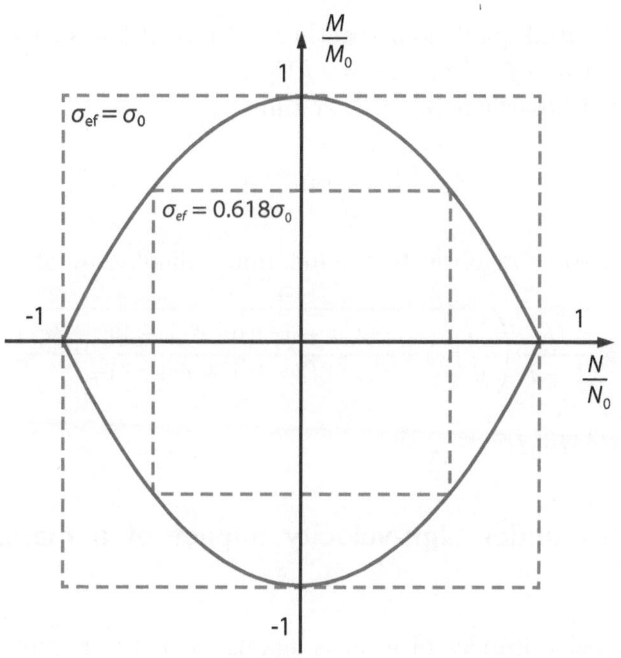

Figure 5.34 Exact and adopted upper and lower yield surfaces (dashed lines).

This equation, when it has its integrals solved, can be written as

$$\ddot{W}_0 + g^2 W_0 = f,$$

with

$$g^2 = \frac{\pi N_0 (R+a)/(R-a)}{G + \pi\mu(R^2 - 3a^2 + 2aR)/6},$$

and

$$f = \frac{-4\pi R M_0/(R-a)}{G + \pi\mu(R^2 - 3a^2 + 2aR)/6},$$

whose solution is

$$W_0 = c_1 \cos(gt) + c_2 \sin(gt) + \frac{f}{g^2}.$$

Now, equating the pre and post impact momentum we have

$$GV_0 = G\dot{W}_0 + \mu\pi a^2 \dot{W}_0 + \int_a^R 2\mu\pi r \dot{W}_0 \frac{R-r}{R-a} \mathrm{d}r,$$

from which it follows that

$$\dot{W}_0 = kV_0, \qquad k = \left(1 + \frac{1 + \bar{r} + \bar{r}^2}{3\gamma}\right)^{-1}$$

which is the initial condition, together with $W_0(0) = 0$, to be used in determining $c_1 = -f/g^2$ and $c_2 = -kV_0/g$.

Motion finishes when $\dot{W}_0 = 0$, at time

$$T = \frac{1}{g} \tan^{-1} \frac{c_2}{c_1},$$

leading, after some algebra, to the maximum displacement

$$W_{0_{max}} = \frac{H}{1+\bar{r}} \left(\sqrt{1 + \frac{3\gamma^2\lambda(1-\bar{r}^2)(6\gamma+1+2\bar{r}-3\bar{r}^2)}{8(3\gamma+1+\bar{r}+\bar{r}^2)^2}} - 1 \right),$$

with $\lambda = GV_0^2/2\sigma_0 H^3$.

5.10 Plates under high-velocity impact of a mass: ballistic limit

The high velocity impact of a mass against a plate or, more generally, against a flat surface, is of great importance. Suffices to say that military protecting structures should all be designed to avoid penetration of bullets fired against them.

When a projectile is launched against a structure, it can be refrained from penetrating it due to characteristics like material strength, projectile velocity, structural characteristics, etc. We refer to ballistic limit as the minimum projectile velocity, V_b, that causes full perforation of the armour, eventually just ejecting a slug or plug with a velocity, V_s. It is not so easy to have such limited experimental conditions, but the next figure shows an exception, where a sphere was just held in place by the plate. A little bit more of initial velocity, V_0, and the sphere would exit on the other side of the plate with velocity V_r.

It is of interest to evaluate the ballistic limit and to this end we first consider a simple model. The work done in fracturing the material so as to generate a plug, together with friction losses, sound and elastic energy, all given by W, should be added to the kinetic energy of the projectile of mass m_p and of the slug (plug) with mass m_s, both travelling with a residual velocity V_r, and equated to the initial kinetic energy, so that

$$\frac{1}{2} m_p V_0^2 = \frac{1}{2} m_p V_r^2 + \frac{1}{2} m_s V_r^2 + W.$$

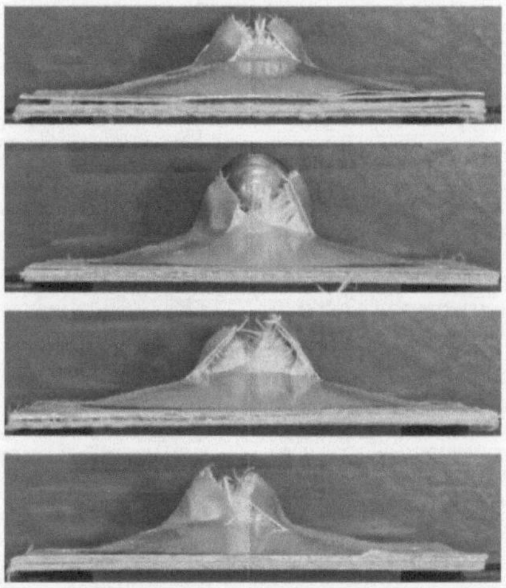

V_I/V_R (m/s)	P4B4
-98.81/7.14	
-100.70/0.00	
-101.6/-21.75	
-105.02/-3.93	

Figure 5.35 A series of experiments of a sphere impacting a composite circular plate. From top to bottom, the second figure shows a residual velocity, $V_r = 0$, and a ballistic limit of $V_b = V_0 = 100.7$ m/s.

At the limit, when the initial projectile velocity is just the ballistic limit, *ie* $V_0 = V_b$, the residual velocity is $V_r = 0$, and the above equation becomes

$$\frac{1}{2} m_p V_b^2 = W,$$

from which it follows that

$$V_r = \sqrt{\frac{V_0^2 - V_b^2}{1 + \frac{m_s}{m_p}}}.$$

This equation, known as the Recht–Ipson model, fits well various experiments involving the nose shapes indicated in the next figure. Note that the equation above does not actually predict the ballistic limit, V_b. This has been the topic of many researchers for decades and it is a difficult problem inasmuch as details such as projectile nose shape, material strength, plate geometry, wave speed and so on should be taken into account.

For more details see N.K. Gupta, R. Ansari and S.K. Gupta, Normal impact of ogive nosed projectiles on thin plates, International Journal of Impact Engineering 25 (2001) 641–660.

5.11 Experiments on beams and plates

When experimenting with the transverse impact of a mass on beams and plates, it is necessary to involve various equipment. It is imperative

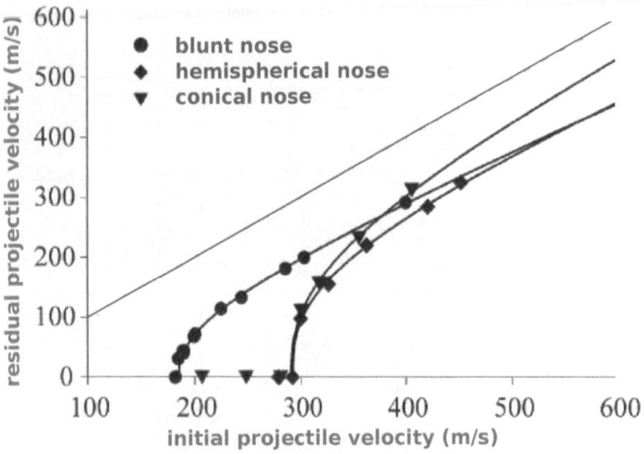

Figure 5.36 Adapted from Perforation of 12 mm thick steel plates by 20 mm diameter projectiles with flat, hemispherical and conical noses, Part I: Experimental study, T. Borvik, M. Langseth, O.S. Hopperstad, K.A. Malo, International Journal of Impact Engineering. 27 (2002) 19--35.

to have a system to accelerate the impact mass against the structure, together with a rig that holds the specimen in place, and importantly the instrumentation for monitoring the impact event.

There are two major systems for accelerating the impact mass. The most common one is called drop hammer rig. A drop hammer comprises guides, a mass lifting system and an anvil, where the specimen to be tested is located. A good design should also allow space for servicing the machine and setting up the tests. In particular, from experience and reasoning, and as seen in the theory in Chapter 2, it is evident that the anvil must have, comparatively, a very large mass, so as not to displace when impacted by the falling mass. Indeed, the drop hammer shown in Chapter 1 has an anvil mass of some 38 tons.

In impact events of the type considered here, we must also give full attention to the rig that holds the specimen. The tup geometry must also be defined according to the aims of the test. In general, sharp corner tups induce shear failure whereas round ones may lead to failure by tearing, although this is not a rule.

The next figure shows a beam under the impact of a round tup. It also shows typical laser velocimeter signals. We can note that the laser detects the free fall and the impact event. In the case on the left, the beam does not break and the tip bounces back and forth. In the case on the right, we have the beam broken by the impact and the laser signal registers the event. As already commented in Chapter 3, to be able to extract information from these signals one needs to acquire them at a high sample rate.

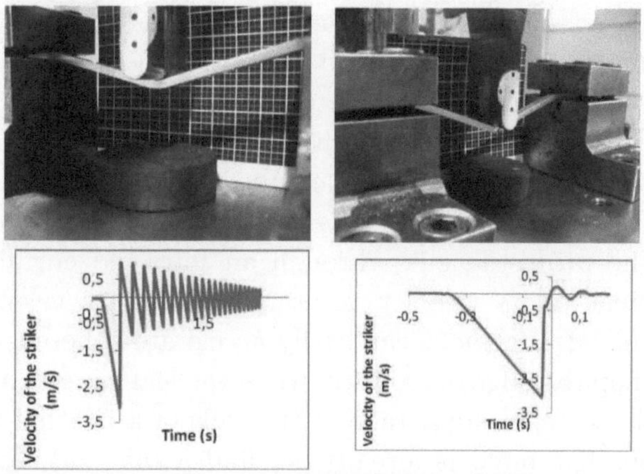

Figure 5.37 A round tip hitting a clamped beam and the associated impact mass velocity signals.

We collect in the next figure various beam profiles after the impact event. We highlight in this figure that the profiles are all linear, which is due to the fact that the impact speed is rather low so the hinge travelling phase is negligible.

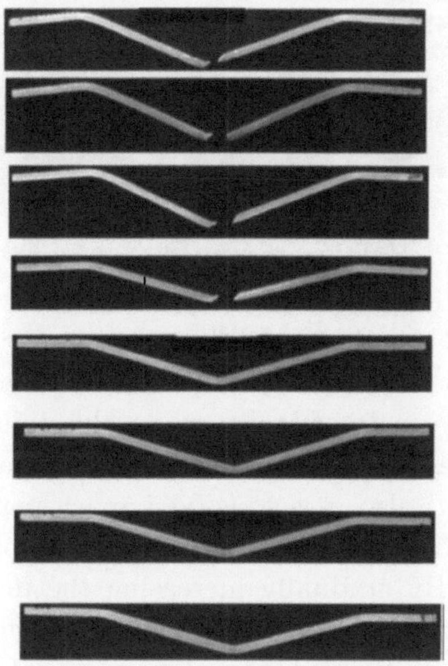

Figure 5.38 Various beam configurations for low impact velocity. Note the linear profile of the beams. Decreasing impact velocity from top (3.9 m/s) to bottom (2.8 m/s).

As for applications related to bullet penetration, the impact tests are carried out in a piece of equipment that is called the gas gun. The next figure shows a gas gun, which comprises a gas reservoir and a long tube. They are connected by a fast valve, *ie* the valve can open quickly and completely so as to allow the pressurized gas to flow through the tube. The gas meets what is called sabout, which is a cylinder of a polymeric material, say nylon, and pushes it along the tube. Lodged in the sabot is the projectile so both projectile and sabot travel in the tube with increasing velocity. At the tube exit, there is a system that holds the sabot but not the projectile, which then travels freely and hits the target. The projectile cuts through an infra-red curtain, allowing the measurement of its velocity prior to the impact velocity. Such a relatively simple arrangement can easily accelerate spheres of some 50 g to velocities of up to 250 m/s (900 km/h). A vertical gas gun arrangement is also possible. It takes advantage of the rails of a free fall tower to fix it on the tube. The anvil is already set and a safe cage can be easily placed.

(a) (b)

Figure 5.39 (a) A horizontal gas gun, the cage where impact takes place, a fast release valve and a detail of how to load the gas gun. (b) A vertical gas gun fit into a drop hammer.

One example of a test that can be done with a gas gun is the impact on FML plates. FML stands for Fibre–Metal Laminates and in the next figure we see a FML made of layers of an aluminium alloy plus a set of thermoplastic fibres bundled in a fabric. These materials exhibit high fatigue resistance and also a very good impact performance. A series of impacts were performed so as to determine the ballistic limit of this FML configuration. By gradually increasing the impact speed, it was possible to establish a range of velocities in which a small variation would lead the sphere to pass through the plate or not. Comparisons among the experimental ballistic limit with results from a numerical simulation based on the finite element model agree positively. The numerical model is not commented on here but the next figure does show that it is capable of predicting well even the shape of the left hole in the panel after the sphere has passed through.

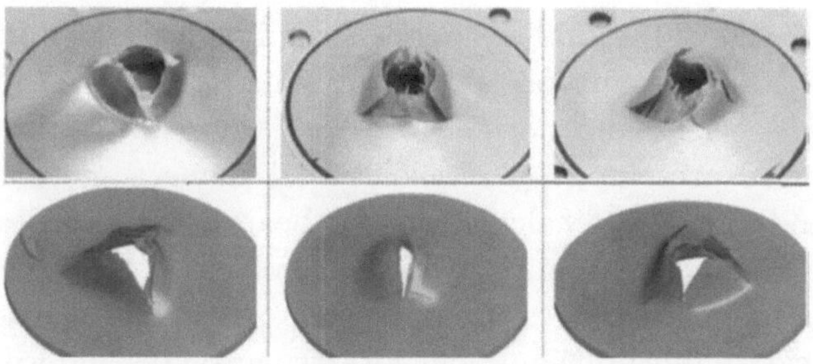

Figure 5.40 FML square plates clamped in a circular ring and submitted to high speed impact of a sphere. Top row: experiments, low row: numerical simulation.

We add further some comments on another experimental issue related to the measurement of the load and displacement in impact tests. We have already discussed this topic before but it is interesting to reassure that a laser Doppler, pointing towards the impact mass, can disclose the acceleration and displacement of the mass, giving the force–displacement curve. One example is shown in the next figure for the same FML experimental campaign previously quoted.

For more details see: The modelling of impact loading on thermoplastic fibre–metal laminates, R.C. Santiago, W.J. Cantwell, N. Jones and M. Alves, Composite Structures 189 (2018) 228--238.

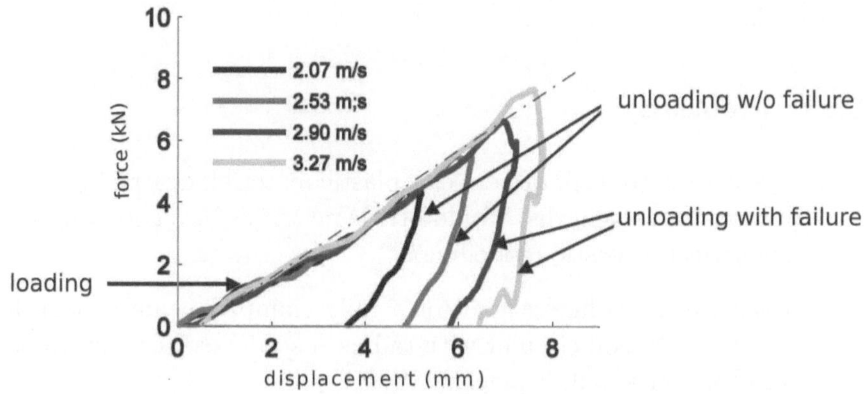

Figure 5.41 Load–displacement curve in circular FML clamped plates as measured by a laser system.

In closing this section, it is important to emphasise the safety aspects of operating equipment related to structural impact. Ideally, a trained person must always perform impact tests in a clear and clean environment. More than one operator must be at the site, although it is recommended that the same person who gives the yes command be also the one who actually operates the impact rig. It is also recommend one to write down the detailed procedure for the test, with each phase being checked step by step. The use of gloves and goggles is also recommended. In the test area there should be a first aid kit and ear protection devices.

5.12 Closure

The basic aspects of the analyses of beams and plates subjected to impact and blast loads were approached in this chapter. We considered large displacements, shear and normal forces on the yield surface, loads so high that were presented by a velocity, let alone some consideration on strain and strain rate calculations for beams. We also described the calculation of equivalent pulse, which allows us to treat any pressure loading as a rectangular pressure pulse, of easier solution. We presented here the basic equilibrium equations for circular plates and we solved two important problems of a plate under a blast load and under the impact of a mass, at low and high velocity.

In this chapter we commented on sophisticated experiments with plates, supported by high speed filming, image analysis, triggering system and sophisticate illumination.

We now advance further by studying a more complex structure, shells, which will be explored in the context of buckling and impact mitigation. We will also study the propagation of plastic waves in rods and shells.

5.13 Problems

1. Discuss the concept of perfectly plastic material, describing its limitations and strengths. Would brittle materials be a good candidate for perfectly plastic modelling?

2. Evaluate the collapse load for a fully clamped beam of length L and cross section circular with radius R subjected to a central load F. Consider small displacements.

3. Obtain the solution for a beam centrally loaded and simply supported.

4. Perform a literature review on the concept of moving hinge length. Are moving hinges real? Suggest a way to measure them in a centrally impacted clamped beam.

5. Obtain the final displacement of a simply supported square beam of length L subjected to an initial velocity V_0 applied on its top surface. Consider shear effects and small displacements.

6. Obtain the equilibrium equations for a circular plate as well as the kinematic relations.

7. In the evaluation of the ballistic limit of plates, sophisticated failure criteria are used side by side with finite element complex models, leading to a good prediction of the phenomenon. Bearing in mind this remark, discuss the importance of developing closed form solutions for the ballistic limit.

8. Suggest a design for a vertical gas gun giving details of how the impact of a sphere launched by it on a clamped plate can be filmed.

9. How to measure the velocity of the projectile on its way to impact a plate.

10. Suppose you are not aware that the performance of a guard rail depends on the strength of the bolts that connect the vertical and horizontal parts of the structure. A bolt supplier offers you bolts with a proof stress larger than the one specified in the design. Being stronger bolts you do not see any relevant problem in purchasing them only to learn later that people would be unprotected when travelling in roads covered by these guardrails when assembled with the wrong strength bolts. Shall you make public your decision of buying stronger bolts even considering that you could be prosecuted by a victim possibly not well protected by these guardrails?

CHAPTER 6

Axial impact in shells and plastic waves

with Dora Karagiozova

Figure 6.1 Buckling of cylindrical tubes.

We refer to shells as structures whose thickness is small compared to their other dimensions. A shell is capable of sustaining in–plane loads and transverse loads. Hence, we deal here with a bi–dimensional stress field. A shell also exhibits some curvature, in contrast to plates. In fact, the majority of natural and man–made structures can be classified as shells.

We present in this chapter various features of the axial impact in shells. But we start with the dynamic buckling of bars, establishing a

DOI: 10.1201/9781003485292-6

series solution for the dynamic governing equation. We briefly transfer these concepts to cylindrical shells, moving on to plastic waves propagation in bars.

We also study the fascinating subject of waves in shells, followed by the use of shells, here tubes of circular and square cross–sections, as kinetic energy absorbers. We explore further the topic of progressive buckling via experiments, including details of buckling transition.

6.1 Dynamic buckling of a bar

This section is based on N. Jones, Structural Impact, CUP, 1989

Let us consider the bar in the next figure is subjected to a suddenly applied constant axial compressive load of intensity F. Note that in the model there is an initial imperfection in the bar, due to reasons like manufacturing. These imperfections have an unknown transverse displacement magnitude and can be expressed by the sine series

$$w^i = \sum_{n=1}^{\infty} W_n^i \sin\left(\frac{n\pi x}{L}\right),$$

with

$$W_n^i = \frac{2}{L}\int_0^L w^i \sin\left(\frac{n\pi x}{L}\right) \mathrm{d}x,$$

being the Fourier coefficients already seen in Chapter 3, with W_n^i the imperfection amplitude for a given mode n.

Equilibrium in the load direction when disregarding axial inertia gives

$$\frac{\partial N}{\partial x} = 0 \quad \rightarrow \quad N = -F,$$

while equilibrium in the transverse direction results in

$$\frac{\partial Q}{\partial x} + N\frac{\partial^2 (w + w^i)}{\partial x^2} + \frac{\partial N}{\partial x}\frac{\partial (w + w^i)}{\partial x} = \rho A\frac{\partial^2 w}{\partial t^2},$$

or

$$EI\frac{\partial^4 w}{\partial x^4} + F\frac{\partial^2 (w + w^i)}{\partial x^2} = -\rho A\frac{\partial^2 w}{\partial t^2},$$

when using the fact that $Q = \partial M/\partial x$ and

$$M = -EI\frac{\partial^2 w}{\partial x^2},$$

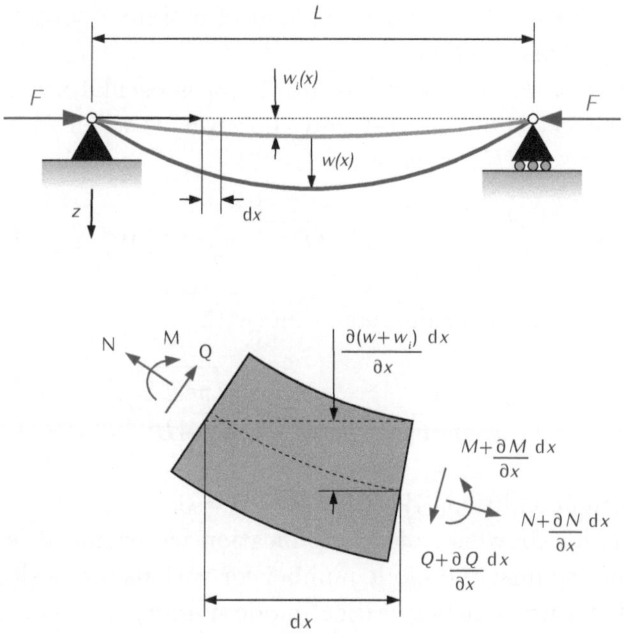

Figure 6.2 A rod under the action of an axial load and its infinitesimal element.

ie an elastic material behaviour is considered. Note that shortening of the bar and transverse shear deformation are neglected and that the variables above are the same as the ones used in Chapter 4 for beams.

It is expected that the solution for the above governing equation has a behaviour similar to the imperfection function, or

$$w = \sum_{n=1}^{\infty} W_n(t) \sin\left(\frac{n\pi x}{L}\right),$$

which satisfies the boundary conditions of zero displacement and moment at the extremes of the bar. If we now substitute this solution in the governing equation it results in

$$\rho A \frac{\mathrm{d}^2 W_n(t)}{\mathrm{d}t^2} + \left(\frac{n\pi}{L}\right)^2 \left[EI\left(\frac{n\pi}{L}\right)^2 - F\right] W_n(t) = F\left(\frac{n\pi}{L}\right)^2 W_n^i,$$

for each integer $n \geq 1$.

Note that if the load is static and if the column has no imperfection, the above equation gives

$$F = F_b = \frac{\pi^2 EI}{L^2},$$

which is the classical Euler buckling load of a simply supported elastic column loaded quasi–statically.

Now, for $F < F_b$, the solution for $W_n(t)$ is oscillatory in time and vibration is ensued. Our interest here, however, is when $F \geq F_b$, which implies that $n^2 \leq F/F_b$, yielding

$$W_n(\tau) = \frac{\bar{F}}{\bar{F}-1}\left[\cosh\left(n^2\sqrt{\bar{F}-1}\,\tau\right) - 1\right] W_n^i = A_n W_n^i$$

as a solution for the governing equation, with

$$\bar{F} = \frac{F}{n^2 F_b} \quad \text{and} \quad \tau = \sqrt{\frac{\pi^4 EI}{\rho A L^4}}\, t,$$

using the initial conditions $W_n(0) = \dot{W}_n(0) = 0$.

The term A_n above is called amplification factor and it is plotted in the next figure against the mode number for various dimensionless time. It is clear that there exists a critical mode number, n_c, associated with initial imperfections lying around it that would be more amplified than others. Indeed, it is possible to show that

Make $\partial A_n/\partial n = 0$ and assume that $\sinh(p_c\tau) \approx \cosh(p_c\tau) - 1$ and that $(p_c\tau - 1)/p_c\tau \approx 1$, with $p_c = \sqrt{n_c^2 F/F_b - n_c^4}$.

$$n_c \approx \sqrt{\frac{F}{2F_b}},$$

indicating that larger loads lead to a rod more wrinkled marked.

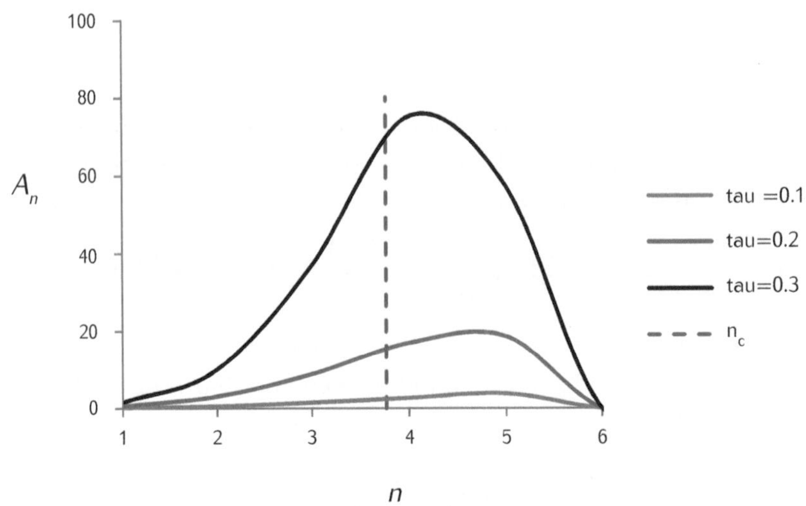

Figure 6.3 Amplification factor *versus* mode number.

Now, take m as the largest integer that satisfies

$$m^2 \leq F/F_b$$

and add the contributions to the displacement field of the rod of the $m+1 \leq n \leq \infty$ harmonics to obtain

$$w(x,t) = \sum_{n=1}^{m} A_n(\tau) W_n^i \sin\frac{n\pi x}{L} + \sum_{n=m+1}^{\infty} W_n(\tau) \sin\frac{n\pi x}{L}.$$

In the case of a material assumed to be rigid and with a slight linear strain hardening, the bending moment–curvature relation is given by

$$M = E_h I \kappa,$$

with E_h being the hardening modulus. All the equations developed in the previous section become the same, except that E must be substituted by E_h. This is then the case of dynamic plastic buckling of a bar. Note that such a simplification requires that the membrane force and bending moment lie on the plastic range, with no strain unloading.

A remark on dynamic plastic buckling is that the fact that an actual rod (not a rod model) displaced transversely may cause one side of it to unload elastically. This will not happen if there is a continuous increase in the magnitude of the load, which is not considered in our model. Incidentally, if there is no unloading, we refer to sustainable plastic flow. Note also that the displacement magnitudes here are assumed to be small, in line with the derivation of the equilibrium equations.

6.2 Dynamic buckling of cylindrical shells

The next figure shows a map of buckling of cylindrical shells, where it can be seen that only the narrow combination in region B of impact mass and impact energy gives rise exclusively to the dynamic buckling phenomenon. It also occurs in region C_1, but followed by a buckling mode called progressive, to be studied later.

See D. Karagiozova, M. Alves and N. Jones, Inertia effects in axisymmetrically deformed cylindrical shells under axial impact, International Journal of Impact Engineering, 24, p. 1083–1115, 2000.

Clearly, a cylindrical shell being axially impacted by a mass ensues various types of response. For certain combination of parameters, one such response is called dynamic plastic buckling. Dynamic buckling in the plastic regime is characterized by a series of permanent wrinkles that are spread along all the shell length, as depicted in the next figure. This phenomenon is the counter–part of dynamic plastic buckling for rods, seen in the previous section. It is generally associated with impact of light mass at high velocities, although light and high are relative to the shell geometry and material characteristics.

Details of the analysis can be found in N. Jones, Structural Impact, CUP, 2012.

The analysis of dynamic buckling requires one to develop the equilibrium equations for a shell, written in a configuration deformed by the shell imperfections. An elastic, linear strain hardening model is used to relate stress to strain, which are integrated along the shell thickness to obtain the generalized stress. They, in turn, are substituted in the equilibrium equations. The resulting governing equation is solved and a pattern of solution very similar to the case of elastic and plastic dynamic buckling of rods is obtained. The procedure discloses the preferable mode of deformation and the threshold velocity for dynamic buckling. We do not detail this solution here, to be found elsewhere.

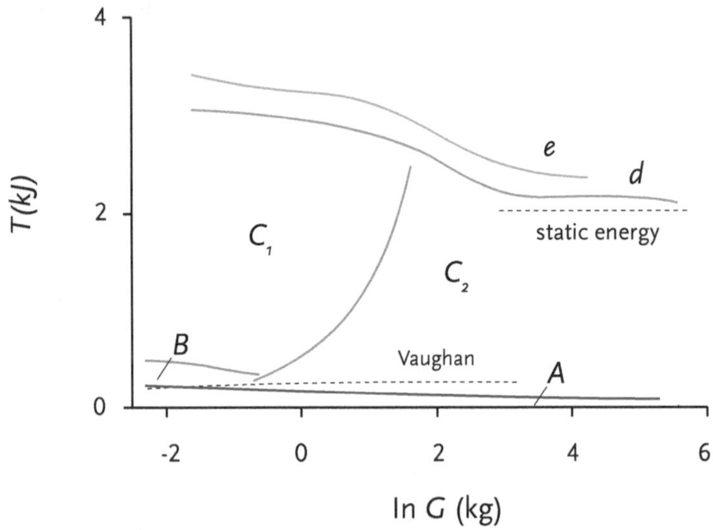

Figure 6.4 Map of buckling of an aluminium cylindrical shell with $L = 106.68$ mm, $D = 23.75$ mm and $H = 1.65$ mm where different buckling phenomena can occur: A – uniform axial compression; B – dynamic plastic buckling; C_1 – initial dynamic plastic buckling and a subsequent progressive buckling; C_2 – progressive buckling; d – maximum energy which can be absorbed by a shell made of strain-rate-insensitive material; e – maximum energy which can be absorbed by a shell made of strain-rate-sensitive material

6.3 Plastic waves in bars

Some buckling modes in rods and shells experiencing plastic deformation occur at high impact speed, within the range of dozens of micro–seconds. For this time span, plastic waves play a role in the overall response of the structure. Hence, it is important to consider disturbances whose amplitude exceeds the material elastic limit. These plastic waves can be generated in many situations. Remember from Chapter 3 that a wave, when reflected from a fixed end, doubles its intensity. Hence, an elastic wave can be transformed into a plastic one by just meeting a boundary.

Let us consider the propagation of a wave disturbance in a rod made from the material represented in the next figure. Force equilibrium against the axial inertia gives

$$\frac{\partial F}{\partial x} = \rho A \frac{\partial^2 u}{\partial t^2}.$$

Using $F = \sigma A$, it follows that

$$A \partial \sigma / \partial x + \sigma \partial A / \partial x = \rho A \partial^2 u / \partial t^2.$$

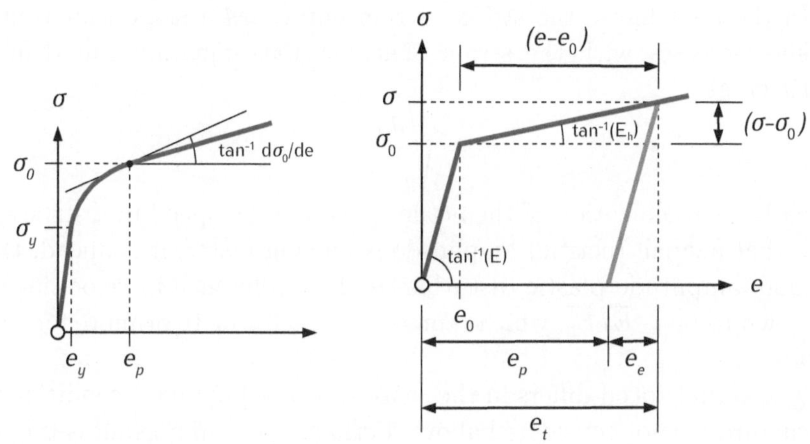

Figure 6.5 A nominal stress–strain curve and a front wave.

We shall see in Chapter 7 the concept of natural strain, which is displacement related by

$$\varepsilon = \ln(1 + \partial u / \partial x).$$

Now, assuming volume conservation, *ie* $\mathrm{d}A/\mathrm{d}\varepsilon = -A$, it follows that

$$\frac{\partial^2 u}{\partial x^2} = \frac{1}{c^2} \frac{\partial^2 u}{\partial t^2},$$

with

$$c = \sqrt{\frac{E_t - \sigma}{\rho} e^{-\varepsilon}},$$

use being made of $E_t = \partial \sigma / \partial x$.

This equation has the same format as the wave equations we encountered in Chapter 3. Indeed, for small strains, the above wave speed gives $c = \sqrt{E/\rho}$, for $\sigma = 0$ and $E_t = E$. Interesting, for a perfectly plastic material, $E_t = 0$ and $\sigma = \sigma_0$, such that

$$c = \sqrt{\frac{\sigma_0}{\rho}}$$

and for a rigid, linear hardening material, with $E_t >> \sigma_0$,

$$c_p = \sqrt{\frac{E_t}{\rho}},$$

which is the simplest expression for the speed of a plastic wave in a strain hardening rod.

In the next figure the stress—strain curve has a slope that continuously decreases with the strain. The previous equation can then be rewritten as

We follow here W. Johnson, Impact Strength of Materials, Edward Arnold, 1972.

$$c_p = \sqrt{\frac{d\sigma/d\varepsilon}{\rho}},$$

which is a simplification of the actual plastic wave speed in a structure. Note that a small (elastic) amplitude is moving faster, it is ahead, than the high amplitude plastic disturbance. The difference in velocities can be shown to be $\sqrt{E/E_p}$, which represents a factor of 10 or more for some metals.

The actual speed differs in the sense that the boundary conditions in a structure are not considered above. Perhaps more important is the fact that the passage of a wave quickly strains the material, which then responds with a different stress–strain curve, *ie* the plastic modulus used in the above equation is altered by the wave transit. It is also opportune to comment that a plastic wave packet disperses as it travels, since different strain levels lead to different propagation speed.

■ BREAKING OF A WIRE

What should be the velocity of the large impact mass in Figure 6.7 so that the wire breaks?

Knowing that $c^p = \mathrm{d}x/\mathrm{d}t$, we have

$$\mathrm{d}t = \frac{\mathrm{d}x}{\sqrt{\frac{\mathrm{d}\sigma/\mathrm{d}\varepsilon}{\rho}}},$$

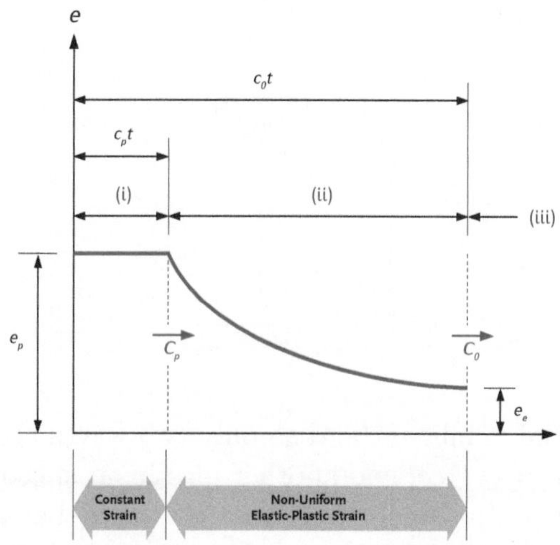

Figure 6.6 Strain distribution in a bar due to the transit of an elastic–plastic wave.

with no distinction between current and original configuration being made. This equation can be used in the momentum conservation of a bar element, $\mathrm{d}F\mathrm{d}t = m\mathrm{d}v$, to give

$$\mathrm{d}v = \frac{\mathrm{d}\sigma}{\rho\sqrt{\frac{\mathrm{d}\sigma/\mathrm{d}\varepsilon}{\rho}}} = \sqrt{\frac{\mathrm{d}\sigma/\mathrm{d}\varepsilon}{\rho}}\mathrm{d}\varepsilon \quad \rightarrow \quad v = \int_0^{\varepsilon}\sqrt{\frac{\mathrm{d}\sigma/\mathrm{d}\varepsilon}{\rho}}\mathrm{d}\varepsilon.$$

If the wire is 2 mm in diameter and made of an aluminium modelled as a bilinear material with $E = 70$ GPa, $E_p = 0.5$ GPa, $\rho = 2685$ kg/m^3 and $\sigma_y = 300$ MPa, it follows that

$$v = \varepsilon_y\sqrt{E/\rho} + (\varepsilon^p - \varepsilon_y)\sqrt{E_h/\rho} = \varepsilon_y c_0 + (\varepsilon^p - \varepsilon_y)c^p = 43.15 \quad \mathrm{m/s}$$

with $\varepsilon_y = \sigma_y/E$ and an assumed rupture strain of 10%.

Note that $c = 5105$ m/s and $c^p = 431$ m/s and that the impact mass is not relevant since we assumed that it is large enough to provide the necessary energy for the plastic deformation process.

■

Let us now examine the impact of a bar against a rigid wall, as depicted in Figure 6.8. Clearly, for sufficiently high impact velocities, part of the bar will be permanently deformed. This means that a plastic wave has traversed it. Of course elastic waves are also generated in this event and indeed they travel ahead of the plastic waves ones.

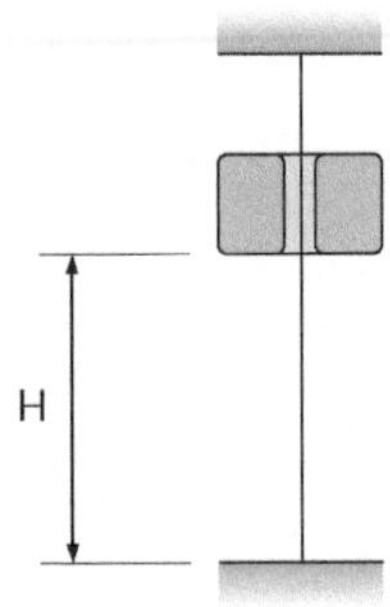

Figure 6.7 A wire being loaded by a falling mass (refered to as von Karman–Duwez experiment).

The bar material is bilinear and so only two waves are possible, one elastic, at speed $c_0 = \sqrt{E/\rho}$, and another plastic, at speed $c_1 = \sqrt{E_h/\rho}$. These waves start at the same time from the anvil–bar interface right after the impact, with the plastic wave lagging behind the elastic wave.

Now, the bar particles in region II have a speed of

$$v = V_0 - c_0\varepsilon_y,$$

which cause a compressive stress in region I of

$$\sigma = \sigma_y + \rho c_1(V_0 - c_0\varepsilon_y),$$

with the total compressive strain of

$$\varepsilon_I = \varepsilon_y + (\sigma - \sigma_y)/E_p = \varepsilon_y + (V_0 - c_0\varepsilon_y)/c_1.$$

If we remove the elastic strain contribution, the remaining plastic strain is

$$\varepsilon^p = \varepsilon_I - \varepsilon_e = (c_0^2 - c_1^2)(V_0 - c_0\varepsilon_y)/(c_0^2 c_1).$$

At time $t = L/c_0$, the elastic wave reaches the bar end and is reflected. Unloading of the bar starts and the so generated wave front decreases the bar velocity in this region to $V_0 - 2c_0\varepsilon_y$. This elastic wave front will meet the oncoming plastic wave at time

$$T_1 = \frac{x_1}{c_1} = \frac{2L - x_1}{c_0} = \frac{2L}{c_0 + c_1},$$

with $x_1/L = 2c_1/(c_0 + c_1)$.

We know that all region III of the bar has the speed $V_0 - 2c_0\varepsilon_y$. This region could be thought then as a bar of length $L - x_1$ that impacts the

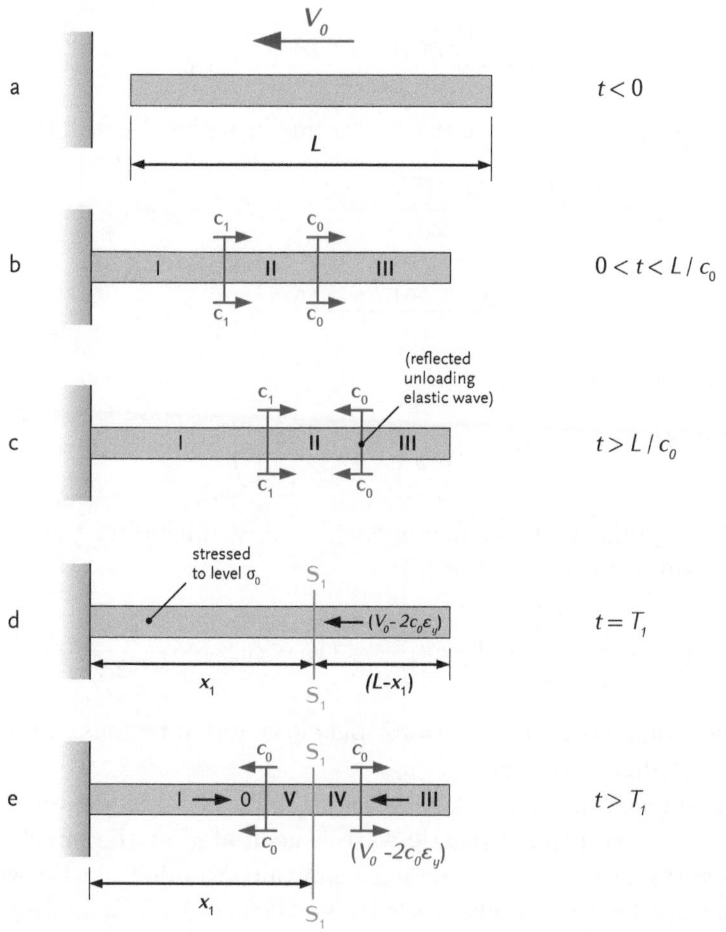

Figure 6.8 A bar impacting a wall and the various formed wave fronts.

bar of length x_1, now compressed to the level of σ_0. As we have seen in Chapter 3, following this impact event, waves will be reflected in both directions from this interface and the speed of the particle engulfed by these waves are now v, assuming that the waves are elastic.

The waves in region V in the figure have the particle speed in opposition to the wave direction, which characterizes a tensile wave propagating into a compressed zone which has zero speed. Hence, this reflected wave unloads elastically region V by $E_p\varepsilon^p + E\varepsilon_y - \rho cv$. Also, the wave reflected into region IV has amplitude $\rho c(V_0 - 2c_0\varepsilon_y + v)$ and force equilibrium in this interface gives

$$E_p\varepsilon^p + E\varepsilon_y - \rho cv = \rho c(V_0 - 2c\varepsilon_y + v),$$

or

$$v = \frac{(c_1 - 3c_0)(V_0 - c_0\varepsilon_y)}{2c_0} + V_0,$$

which leads to the maximum elastic strains in region IV of intensity

$$\varepsilon_y = \frac{\rho c}{E}\left[\frac{(c_1 - 3c_0)(V_0 - c_0\varepsilon_y)}{2c_0} + 2V_0 - 2c_0\varepsilon_y\right] =$$

$$\frac{(c_1 + c_0)(V_0 - c_0\varepsilon_y)}{2c_0^2},$$

from which it follows that

$$V_0 = c_0\varepsilon_y\left(1 + \frac{2c_0}{c_0 + c_1}\right).$$

The compressive strain in regions I and V, including the reflected elastic strain from the interface, is

$$\varepsilon_2 = \left(\varepsilon_y + \frac{V_0 - c_0\varepsilon_y}{c_1}\right) - \frac{\rho c_0 v}{E} = (V_0 - c_0\varepsilon_y)\frac{-c_1^2 + c_1c_0 + 2c_0^2}{2c_0^2c_1}.$$

Note that, unlike the elastic strains, plastic strain in regions I and V will no longer change with time.

If the impact velocity of the bar against the anvil is greater than V_0 above, then an additional plastic wave is generated at the interface and it propagates to the right, meeting again an incoming elastic wave. A new interface formation, also called a stationary front of second order, is once again formed. Interestingly, a plastic loading wave never reaches the bar free end. One way to explain it is that a free end is, of course, stress free and as such it cannot sustain plastic strains.

■ PLASTIC WAVE AND LOCALIZATION

A copper rod whose stress–strain relation is given by $\sigma = k\varepsilon^n$, k and n being material constants, is loaded with a velocity v. Determine the plastic wave velocity as a function of the strain. What should be v such that the bar experiences a strain that causes localization.

From the stress–strain law

$$\sigma = k\varepsilon^n,$$

the plastic wave speed is evaluated as

$$c = \sqrt{\frac{\mathrm{d}\sigma/\mathrm{d}\varepsilon}{\rho}} = \sqrt{\frac{kn\varepsilon^{n-1}}{\rho}}.$$

Localization, *ie* material necking, takes place under volume conservation and maximum load conditions, such that

$$\frac{\mathrm{d}}{\mathrm{d}\varepsilon}\sigma A = 0 \quad \text{and} \quad \frac{\mathrm{d}V}{\mathrm{d}\varepsilon} = 0,$$

from which

$$\frac{\mathrm{d}\sigma}{\mathrm{d}\varepsilon} = \sigma,$$

or

$$\varepsilon = n.$$

At this strain, the critical loading velocity should be

$$v = \int_0^{\varepsilon} \sqrt{\frac{\mathrm{d}\sigma/\mathrm{d}\varepsilon}{\rho}}\mathrm{d}\varepsilon = \sqrt{\frac{4nk}{\rho}}\frac{n^{(n+1)/2}}{n+1}.$$

The plastic wave speed as a function of the impact velocity is readily obtained as

$$c = \frac{n+1}{2}\frac{v}{\varepsilon}.$$

For $n = 0.3$, $k = 0.4$ GPa and ρ=8930 kg/m^3, the critical velocity for localization is 82 m/s, with the next figure showing the plastic wave velocity as a function of the strain.

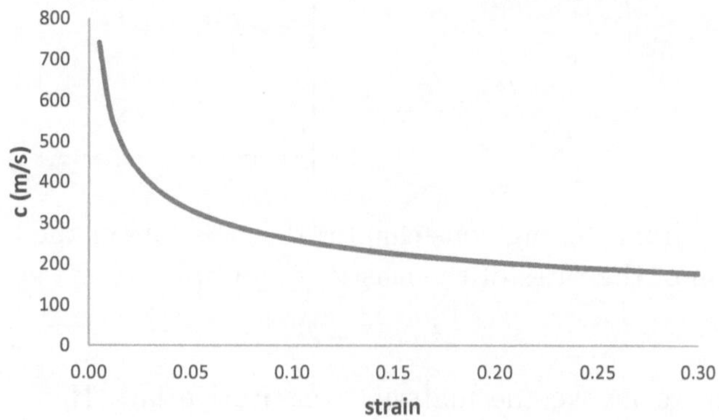

Figure 6.9 Plastic wave velocity in a copper bar as a function of the strain level, up to localization.

■

6.4 Plastic waves in cylindrical shells

Plastic waves also propagate in shells.

Elastic and plastic waves can be generated when tubes are subjected to an axial impact velocity. Features like peak force and buckling mode can be directly obtained from the intensity of plastic waves. The subject is new and to investigate it we need to adopt a yielding criterion, *ie* a relation between stresses that indicate whether the material deforms plastically. Details of yield criteria will be presented in the next chapter and their basics should be known to follow this section.

This section is based on D. Karagiozova, M. Alves and N. Jones, Inertia effects in axisymmetrically deformed cylindrical shells under axial impact, International Journal of Impact Engineering, 24, p. 1083–1115, 2000.

We note, first of all, that, to proceed with the study of plastic waves in shells, we work with local stresses, *ie* σ_{rr}, $\sigma_{\theta\theta}$ and σ_{xx}, instead of generalized stresses N_{ij}. Also, as our shell is axisymmetric, the stresses $\sigma_{\theta\theta}$ do not depend on the circumferential position.

Equilibrium in the axial and radial directions of the differential element shown in the next figure gives

$$\rho\frac{\partial v_x}{\partial t} - \frac{\partial \sigma_x}{\partial x} = 0$$

and

$$\rho R\frac{\partial v_r}{\partial t} - \sigma_\theta = 0$$

with ρ being the shell density and v standing for velocity.

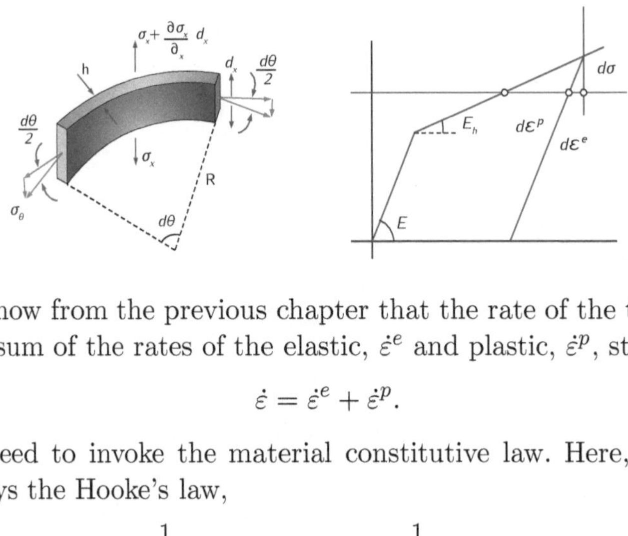

Figure 6.10 A cylindrical shell differential element and a bilinear stress–strain diagram.

We know from the previous chapter that the rate of the total strain, $\dot{\varepsilon}$, is the sum of the rates of the elastic, $\dot{\varepsilon}^e$ and plastic, $\dot{\varepsilon}^p$, strains

$$\dot{\varepsilon} = \dot{\varepsilon}^e + \dot{\varepsilon}^p.$$

We need to invoke the material constitutive law. Here, the elastic part obeys the Hooke's law,

$$\dot{\varepsilon}^e_x = \frac{1}{E}(\dot{\sigma}_x - \nu\dot{\sigma}_\theta) \quad \dot{\varepsilon}^e_\theta = \frac{1}{E}(\dot{\sigma}_\theta - \nu\dot{\sigma}_x),$$

whereas the rate of the plastic strains comes from the Prandt–Reuss relation (see next chapter). It states that

$$\mathrm{d}\varepsilon_x = \gamma s_x \qquad \mathrm{d}\varepsilon_\theta = \gamma s_\theta .$$

s is the deviator of the stress tensor and has the components

$$s_x = \frac{1}{3}(2\sigma_x - \sigma_\theta) \quad \text{and} \quad s_\theta = \frac{1}{3}(2\sigma_\theta - \sigma_x).$$

Substituting these expressions in the equivalent strain one given by

$$\mathrm{d}\varepsilon_{eq} = \frac{2}{3}\left[\mathrm{d}\varepsilon_x^2 + \mathrm{d}\varepsilon_\theta^2 - \mathrm{d}\varepsilon_x \mathrm{d}\varepsilon_\theta\right]^{1/2}$$

it follows that

$$\gamma = \frac{3}{2}\frac{\dot{\varepsilon}_{eq}}{\sigma_e},$$

with

$$\sigma_{eq} = \left[\sigma_x + \sigma_\theta - \sigma_x\sigma_\theta\right]^{1/2},$$

being the equivalent stress for a von Mises biaxial stress state.

The total strain rate components become

$$\dot{\varepsilon}_x = \frac{1}{E}(\dot{\sigma}_x - \nu\dot{\sigma}_\theta) + \frac{\dot{\varepsilon}_{eq}}{2\sigma_{eq}}(2\sigma_x - \sigma_\theta)$$

and

$$\dot{\varepsilon}_\theta = \frac{1}{E}(\dot{\sigma}_\theta - \nu\dot{\sigma}_\theta) + \frac{\dot{\varepsilon}_{eq}}{2\sigma_{eq}}(2\sigma_\theta - \sigma_x).$$

We now use these expressions in the cylindrical shell kinematic relations,

$$\dot{\varepsilon}_\theta = \frac{v_r}{R} \qquad \text{and} \qquad \dot{\varepsilon}_x = \frac{\partial \dot{u}}{\partial x},$$

to give, together with the previous equilibrium equations,

$$\begin{bmatrix} \rho & 0 & 0 & 0 \\ 0 & A_1 & 0 & A_2 \\ 0 & 0 & \rho R & 0 \\ 0 & A_2 & 0 & A_3 \end{bmatrix} \begin{Bmatrix} \dot{u}_{,t} \\ \sigma_{x,t} \\ \dot{w}_{,t} \\ \sigma_{\theta,t} \end{Bmatrix} +$$

$$\begin{bmatrix} 0 & -1 & 0 & 0 \\ -1 & 0 & 0 & 0 \\ 0 & 0 & 0 & 0 \\ 0 & 0 & 0 & 0 \end{bmatrix} \begin{Bmatrix} \dot{u}_{,x} \\ \sigma_{x,x} \\ \dot{w}_{,x} \\ \sigma_{\theta,x} \end{Bmatrix} + \begin{Bmatrix} 0 \\ 0 \\ -\sigma_\theta \\ -\dot{w}/R \end{Bmatrix} = \begin{Bmatrix} 0 \\ 0 \\ 0 \\ 0 \end{Bmatrix},$$

with

$$A_1 = \frac{1}{E}\left[1 + \frac{(1-\lambda)(2\sigma_x - \sigma_\theta)^2}{\lambda(2\sigma_e)^2}\right],$$

$$A_2 = \frac{1}{E}\left[-\nu + \frac{((1-\lambda)(2\sigma_x - \sigma_\theta)(2\sigma_\theta - \sigma_x)}{\lambda(2\sigma_e)^2}\right]$$

and

$$A_3 = \frac{1}{E}\left[1 + \frac{(1-\lambda)(2\sigma_\theta - \sigma_x)^2}{\lambda(2\sigma_e)^2}\right].$$

Note that use was made of the relation $\dot{\varepsilon}_e = g\dot{\sigma}_e$, being specialized to linear strain hardening material of hardening modulus E_h, as indicated in the figure above, so that

$$\lambda = \frac{E_h}{E}.$$

These equations can be written as

$$\boldsymbol{A}^t \boldsymbol{w}_{,t} + \boldsymbol{A}^x \boldsymbol{w}_{,x} + \boldsymbol{b} = 0$$

and the waves propagate at speeds c_1, roots of

$$\left| -c_1 \boldsymbol{A}^t + \boldsymbol{A}^x \right| = 0$$

or

$$-\rho^2 R\,(A_2^2 - A_1 A_3)\,c_1^4 - \rho R A_3 c_1^2 = 0$$

so that, for $c_1 \neq 0$, the wave speed is

$$c^p = \pm\sqrt{\frac{A_3}{\rho(A_1 A_3 - A_2^2)}}.$$

It is clear that the wave speed in a cylindrical shell depends on its stress state. It also depends on the geometry, material parameters and on the adopted yield criterion, in this case the von Mises.

Now, the next figure plots the wave locus for various combinations of stress states, with points 1, $\sigma_\theta/\sigma_x = \pm 1/2$, and points 2, $\sigma_\theta/\sigma_x = \sqrt{3}/8(1-\nu^2)$, being associated with minimum and maximum plastic wave speeds of

$$c^p_{min} = \pm\sqrt{\frac{E}{\rho}}\sqrt{\frac{4\lambda}{4\lambda(1-\nu^2) + 3(1-\lambda)}}$$

and

$$c^p_{max} = c^s_e = \pm\sqrt{\frac{E}{\rho(1-\nu^2)}},$$

with c^s_e standing for the elastic wave in a shell.

For a steel cylindrical shell, the minimum plastic wave speed is something in between 200 m/s and 400 m/s, depending on the plastic modulus, while the maximum plastic (and elastic) speed is some 5300 m/s.

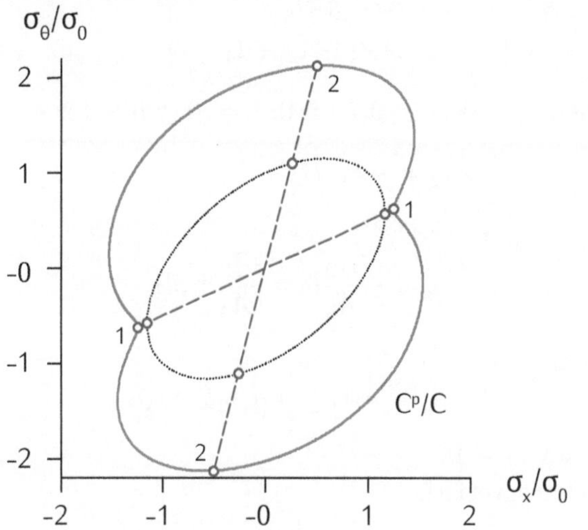

Figure 6.11 Plastic wave locus in von Mises cylindrical shell.

6.4.1 Peak load in a tube

The peak load experienced by a cylindrical shell impacted by a mass G with initial velocity V_0 is an important design parameter. When the axial impact velocity on a cylindrical shell is sufficiently high to initiate the propagation of a plastic wave, the relationship between the plastic wave speed and the stress at $t = 0$ can be obtained when using the method of characteristics applied to the governing equations. It can be shown that the following relationship holds along the characteristics

$$\mathbf{l}L[\mathbf{w}] = \mathbf{l}\mathbf{A}^t\left(\frac{\mathrm{d}\mathbf{w}}{\mathrm{d}t}\right) + \mathbf{l}\mathbf{b} = 0.$$

The vector $\mathbf{l} = [l_1 l_2 l_3 l_4]$ is obtained from $\mathbf{l}\left(-c_1\mathbf{A}^t + \mathbf{A}^x\right) = 0$ or

$$[l_1, l_2, l_3, l_4]\begin{bmatrix} -c_1\rho & -1 & 0 & 0 \\ -1 & -c_1A_1 & 0 & -c_1A_2 \\ 0 & 0 & -c_1\rho R & 0 \\ 0 & -c_1A_2 & 0 & -c_1A_3 \end{bmatrix}$$
$$= \begin{bmatrix} -c_1\rho l_1 - l_2 \\ -l_1 - c_1A_1l_2 - c_1A_2l_4 \\ -c_1\rho R l_3 \\ -c_1A_2l_2 - c_1A_3l_4 \end{bmatrix} = 0$$

If $c_1 = 0$ then $l_2 = 0, l_1 = 0, l_3 \neq 0, l_4 \neq 0$, while for $c_1 \neq 0$ we obtain

$$\begin{aligned} l_2 &= -c_1\rho l_1 \\ l_3 &= 0 \\ l_4 &= -\frac{A_2}{A_3}l_2 = \frac{A_2}{A_3}c_1\rho l_1 \end{aligned}$$

and the equation

$$-l_1 - aA_1l_2 - aA_2l_4 = 0$$

is satisfied for any $l_1 \neq 0$.

Along the characteristics

$$\boldsymbol{w}_{,t} = -c_1\boldsymbol{w}_{,x} + \frac{\mathrm{d}\boldsymbol{w}}{\mathrm{d}t}$$

and the equation

$$\mathbf{l}L[\mathbf{w}] = \mathbf{l}\left[\mathbf{A}^t\left(\mathbf{w}_{,t} + c_1\mathbf{w}_{,x}\right) + \mathbf{b}\right],$$

together with $l_1 \neq 0$ leads to the following relationship

$$\rho\frac{\mathrm{d}v_x}{\mathrm{d}t} - c_1\rho\left(A_1 - \frac{A_2^2}{A_3}\right)\frac{\mathrm{d}\sigma_x}{\mathrm{d}t} - \frac{v_r}{R}\frac{A_2}{A_3}c_1\rho = 0$$

along the characteristic $c_1 = \mathrm{d}x/\mathrm{d}t$, or

$$\frac{\mathrm{d}\sigma_x}{\mathrm{d}t} = \rho c_1\frac{\mathrm{d}v_x}{\mathrm{d}t} - \frac{v_r}{R}\frac{A_2}{A_3}c_1^2$$

Assuming that $v_r \approx 0$, the following relationship holds

$$\mathrm{d}\left(\sigma_x - \rho c_1 v_x\right) = 0.$$

Using the stress state at the beginning of the propagating of the plastic wave ($c_1 = c_{\min}$), we obtain

$$\mathrm{d}\left(\sigma_x - \rho c_{\min} v_x\right) = 0.$$

For an initial impact velocity

$$V_0 > \frac{2\sigma_0}{\sqrt{3}}\left(\frac{1-\nu^2}{\rho E}\right)^{1/2},$$

a precursor axial elastic wave propagates in an elastic–plastic shell. The velocity, $\bar{V}_0$, initiating the plastic wave due to an impact with initial velocity V_0, has been already deduced, so that

$$v_x = \overline{V_0} = V_0 - \frac{2\sigma_0}{\sqrt{3}}\left(\frac{1-\nu^2}{\rho E}\right)^{1/2}.$$

As a result, the axial stress σ_x, which is used to obtain the peak load at $t = 0$, is

$$\sigma_x = \frac{2\sigma_0}{\sqrt{3}} + \left[V_0 - \frac{2\sigma_0}{\sqrt{3}}\left(\frac{1-\nu^2}{\rho E}\right)^{1/2}\right]\sqrt{\rho E}\left[\frac{4\lambda}{4\lambda\,(1-\nu^2)+3(1-\lambda)}\right]^{1/2}.$$

It is not difficult to show that, for $\lambda << 1$, which is the case for most metals, the above equation, after being multiplied by the shell cross-section, yields

$$F_{peak} = \frac{4}{\sqrt{3}}\pi h R(\sigma_0 + V_0\sqrt{\rho E_h}).$$

This expression compares favourably with numerical simulations, as noted in the reference quoted in this section. It is valid for impact velocities which produce plastic deformations in the shell.

6.5 Progressive buckling of cylindrical shells

Buckling is an unstable response of a structure to loads. The term progressive buckling refers to the sequential formation of wrinkles in a tube under axial loading. This folding process may lead the material to the plastic regime and we refer to plastic progressive buckling. It has the useful characteristic of absorbing the loading energy in a somewhat smooth way and many devices for impact energy absorption have been proposed based on this phenomenon, as we shall see later in this chapter.

When a shell is axially loaded, depending on its geometry and material, various patterns of buckling wrinkles can be formed, as shown in the next figure.

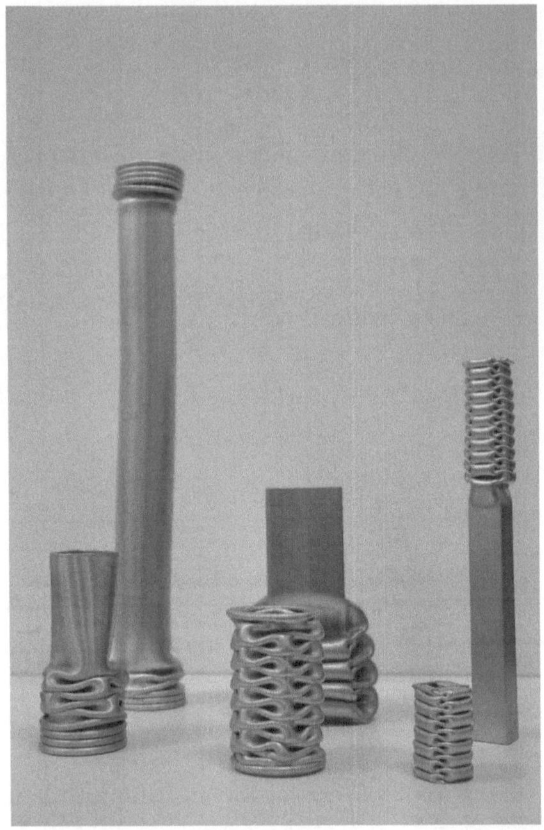

Figure 6.12 Various buckling patterns of tubes.

When considering an axisymmetric folding mechanism in a cylindrical shell, as in the next figures, it is possible to device a relative simple analysis of the phenomenon of progressive buckling. The next figure shows the load–displacement curve of a tube that is axially compressed. The load increases up to a certain limit and when a maximum load value is achieved, the first folding is formed. Clearly, it is much easier to form one fold than to compress the whole shell, as indicated by the decrease in the load magnitude during fold formation. This deformation mechanism is completely formed when the top part of the fold touches the bottom one. In the sequence, the load increases again, compressing once more the whole shell until the next fold is formed.

We remark that, although not considered in the model below, as the load increases, the shell experiments compression along all its length. This is to say that, if we were able to unfold a tube after it has been fully compressed, we would end up with it shorter than at the beginning of the test.

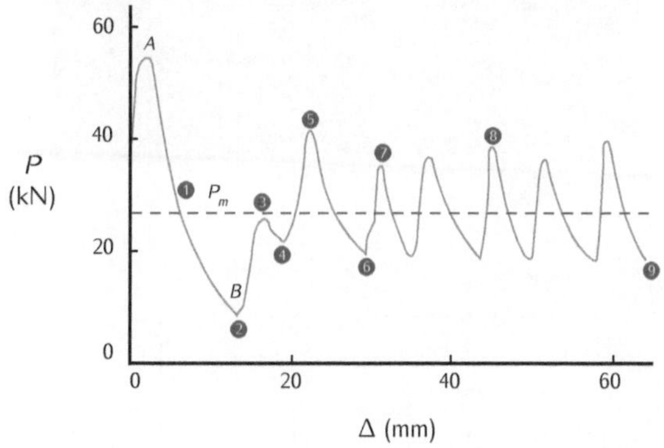

Figure 6.13 Load–displacement curve in the progressive plastic buckling of a cylindrical shell.

To model the axisymmetric folding, let us consider for now that the load is applied quasi–statically. We proceed by reasoning that the applied energy to form one fold, $P_m \times 2l$, is distributed in the various mechanisms of energy absorption of the tube model sketched in the next figure. We need to calculate the energy they will take from the outsource. From the figure, the external energy is used to form three circumferential hinges and to stretch circumferentially the material in between the hinges. If it were not for the tube integrity, the circumferential stretching energy would be nil and the whole deformation mechanisms would be similar to the axial compression of a bunch of bars circumferentially disposed, as illustrated next. In other words, upon compression, circumferential stresses arise so as to keep the shell integrity.

The energy to form the bottom hinge equals the one needed by the top hinge and amounts to

This solution is due to J.M. Alexander, An approximate analysis of the collapse of thin cylindrical shells under axial load, Quart. J. Mech. Appl. Math., 1960, 13, 10-15.

$$E_1 = 2\pi R M_0 \frac{\pi}{2},$$

with the collapse bending moment per unit width being

$$M_0 = \frac{2\sigma_0}{\sqrt{3}} \frac{H^2}{4}.$$

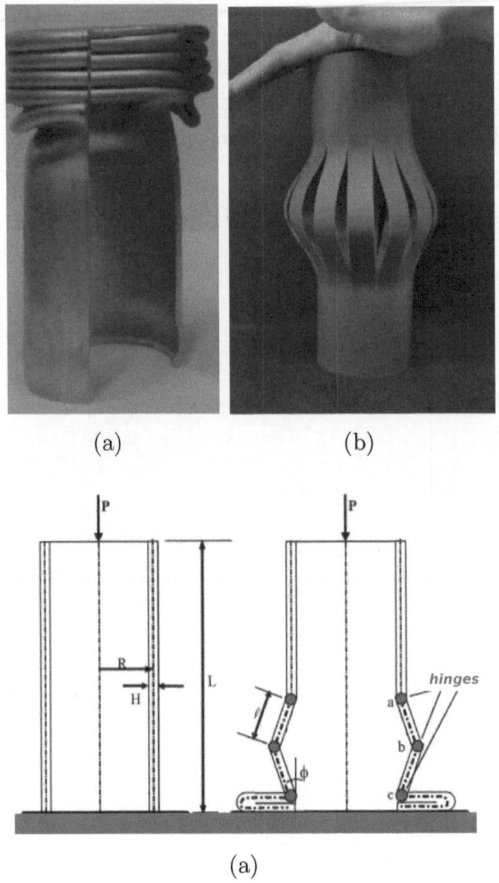

Figure 6.14 (a) Progressive buckling of a cylindrical shell. (b) Demonstration that shell strips are kept together by circumferential stresses. (c) Basic dissipation mechanism kinematics of a cylindrical shell under progressive buckling.

This last equation is obtained assuming a plane strain state with $\sigma_r = 0$ and $\varepsilon_\theta = (\sigma_\theta - \nu\sigma_z)/E{=}0$. The von Mises equivalent stress leads to $\sigma_z = 2\sigma_0/\sqrt{3}$, whose integration along the shell thickness gives M_0.

The absorbed energy in hinge b is

$$\mathrm{d}E_2 = 2\pi(R + l\sin\phi)M_0(2\mathrm{d}\phi) \rightarrow E_2 = 4\pi M_0\left(R\frac{\pi}{2} + l\right).$$

The energy to stretch circumferentially the material in between the hinges is

$$\mathrm{d}E_3 = \sigma_0 \mathrm{d}\varepsilon_\theta 2lH2\pi R,$$

with $\mathrm{d}E_3$ being the average increment in the strain given by

$$\mathrm{d}\varepsilon_\theta = \frac{2\pi\frac{l}{2}\sin(\phi + \mathrm{d}\phi) - 2\pi\frac{l}{2}\sin(\phi)}{2\pi R} = \frac{l\cos\phi\mathrm{d}\phi}{2R},$$

so that

$$E_3 = \int_0^{\frac{\pi}{2}} \sigma_0 l \cos\phi \mathrm{d}\phi 2lH\pi = 2\sigma_0 l^2 H\pi.$$

Hence, the total energy absorbed by the model is

$$E_T = 2\pi\sigma_0 H^2(\pi R + l)/\sqrt{3} + 2\pi\sigma_0 l^2 H,$$

to be equated to the external applied energy, giving

$$P_m = \sigma_0 \pi H \left[H\left(\pi R/l + 1\right)/\sqrt{3} + l\right].$$

The average load cannot yet be calculated since we do not know the value of the fold length. But we can assume that l is such that the load is a minimum, *ie* $\partial P_m/\partial l = 0$, yielding

$$l = \sqrt{\frac{\pi R H}{\sqrt{3}}},$$

so that

$$P_m = \sigma_0\left(6H\sqrt{DH} + 1.8H^2\right).$$

Experimentally, it is seen that the inwards and outwards portion of the tube wall is not the same, which is captured by the so called eccentricity factor, introduced by T. Wierzbicki, S.U. Bath, W. Abramowicz and D. Brodkin, Alexander revisited: a two fold folding elements model for progressive crushing of tubes, International Journal of Solids and Structures, 29, 3269–3288, 1992.

A closer look at the previous figure shows that the folding mechanism also happens inwardly so the factor $1.8H^2$ becomes negative, averaging with the above expression and cancelling out, finally giving

$$P_m = 6\sigma_0\sqrt{DH^3}.$$

It is interesting to note that in our model the fold length, l, does not depend on the material property, σ_0.

So far, our model was formulated for a quasi–static load. If we apply the load suddenly then there will be two additional effects that we need to consider, inertia and material strain rate sensitivity, with the latter being more important for moderate impact velocities.

The increase in the material strength from σ_0 to σ_d due to the loading rate is captured by a constitutive law, as the one used in the previous chapter, *ie*

$$\sigma_d = \sigma_0\left(\frac{\dot{\varepsilon}}{\dot{\varepsilon}_0}\right)^m,$$

with $\dot{\varepsilon}_0$ being a reference strain rate for which σ_0 was measured and m is a material constant. This new corrected stress will affect the bending moment and it is necessary to determine the strain rate of the folding mechanisms. We can do so by assuming an average circumferential strain of $\varepsilon = l/D$ and a total time to form one fold of $t = 2l/V_0$, giving

$$\dot{\varepsilon} = \frac{V_0}{2D},$$

so that the average dynamic load, P_{m_d}, is

$$P_{m_d} = 6\sigma_0 \left(\frac{V_0}{2D\dot{\varepsilon}_0}\right)^m \sqrt{DH^3}.$$

We plot in the next figure the dynamic to static average load for tubes of various diameters and made of steel, with $m = 0.077$ and $\dot{\varepsilon}_0 = 0.001/\text{s}$, subjected to an impact velocity of $V_0 = 30$ km/h. It is clear from the figure the importance of considering strain rate effects on this phenomenon.

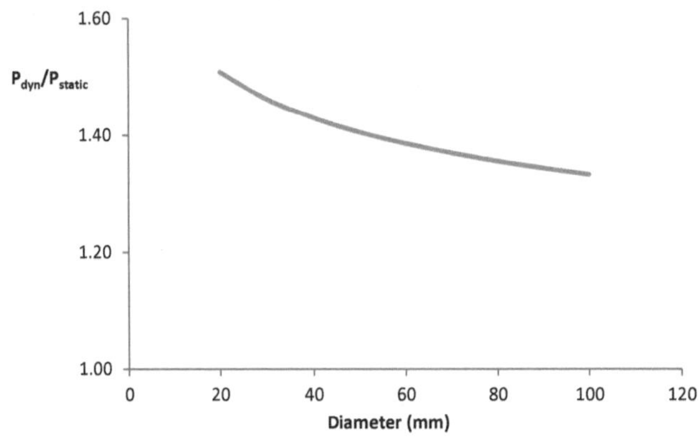

Figure 6.15 Dynamic to quasi–static average load for various diameters of a steel cylindrical shell under the impact of a mass at 30 km/h.

■

IMPACT OF A TUBE

A mild steel tube 300 mm long, $D = 60$ mm, $H = 1.2$ mm, flow stress of 300 MPa, is used to absorb the energy of a mass of 100 kg travelling at 10 m/s. Evaluate the shortening of the tube and the deceleration of the mass. How long is the crushing event? Compare it with the time for an elastic wave to traverse the tube length. Perform the calculations with and without strain rate considerations on the material response. Consider that the material strain rate sensitivity is given by the so called Cowper—Symmonds equation, $\sigma_d = \sigma_0[1 + (\dot{\varepsilon}/C)^{1/q}]$, with $C = 40/\text{s}$, $q = 5$.

The average crushing load is

$$P_m = \pi D H \sigma_0 \sqrt{\frac{2\pi H}{\sqrt{3}D}} \left\{ 1 + \left(\frac{V_0}{2DC} \right)^{1/q} \right\}$$

and using the given parameters leads to:

with strain rate effects	no strain rate effects
$P_m = 39.4$kN	$P_m = 20.4$kN
$T = 25.4$ms	$T = 49.1$ms
$\Delta = 126.9$mm	$\Delta = 245.3$mm

The time for an elastic wave to propagate in the tube is around 60μs, about 440 times shorter than the crushing time. This supports analyses whose transient aspects are disregarded.

The next figure shows the force in this tube as a function of the impact speed for the mild steel and also for a titanium alloy ($C = 120$/s, $q = 9$) and a less strain rate sensitive aluminium alloy ($C = 6500$/s, $q = 4$).

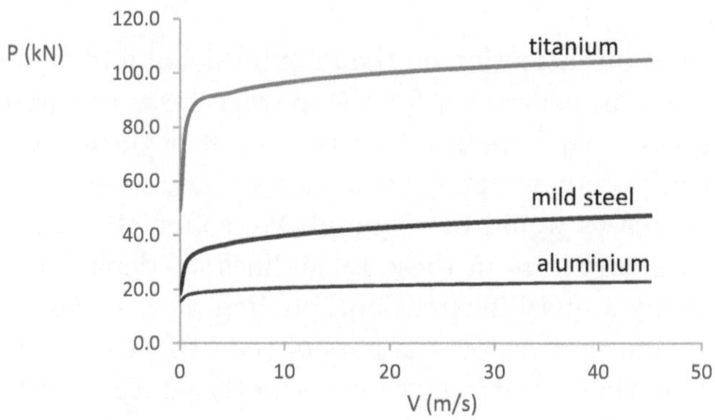

Figure 6.16 Force in different alloys tubes against the impact speed.

■

6.6 Progressive buckling of rectangular tubes

Inextensional and extensional modes occurs, approximately, for $C > 40.8H$ and $C < 7.5H$, respectively. In between one has an asymmetric mode

The analysis of plastic buckling of rectangular tubes is somewhat involved. The geometrical relations are more difficult to visualise and the kinematics bear some peculiar characteristics.

When a square cross section tube is axially loaded, it responds by an initial full length compression which soon gives away to localized buckling. The general patterns that develop in the buckling process can be sketched as shown in the next figure. There we see only the corner of the tube, with the pattern (a) represented by a mechanism whose hinges marked in the figure move one outwards and the other inwards so that the circumferential length is kept nearly the same. Accordingly, we call it in-extensional folding mechanism. Conversely, as shown in figure (b), the extensional folding mechanism occurs when both horizontal hinges move outwards, modifying the initial circumferential length.

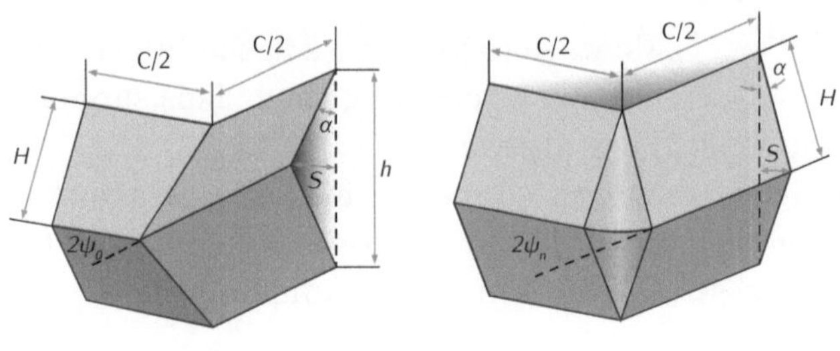

Figure 6.17 Corner of a square tube showing the (a) inextensional and (b) extensional buckling modes.

Focusing now our attention on the in-extensional folding mechanism, we aim to obtain an expression for the average force required to attain this buckling mode. We note first that the hinges of the idealized mechanism are somewhat non realistic given that an infinity amount of energy to bend these creases would be required. We soften then this model by allowing a finite curvature in these hinge lines, as depicted in the next figure, so creating a quasi-inextensional folding mechanism.

We notice here that region 1 is part of the surface of a circular tube bounded by lines that move away from each other during the folding. We associate the name "moving hinge" to this characteristic but "spreading hinge" could equally be used. Region 2 is part of a conical surface whose bottom moves with Region 3, a toroidal curve surface bent in two directions, towards regions 1 and 2, therefore having a double curvature. It is in this region that stretching, *ie* in-plane strain, occurs.

Similar to the analysis of the circular tubes, the procedure here is to evaluate the energies associated with each of these regions and equate them to the external energy. We shall see that the tube thickness, its side dimension and its material flow stress will be presented in the final equation.

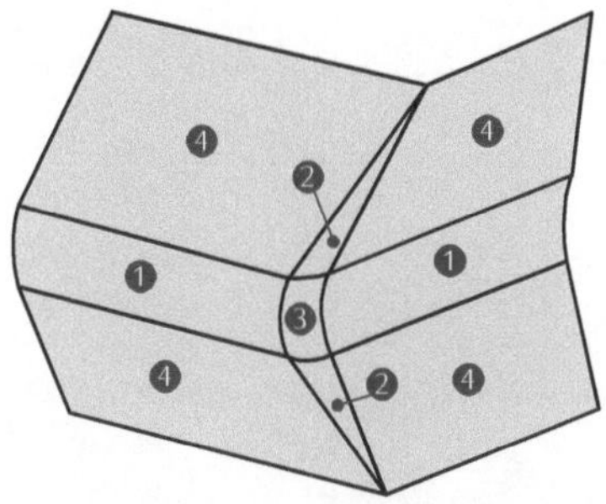

Figure 6.18 Corner of a buckled square tube and the associated main parts that dissipate energy.

This solution is due to T. Wierzbicki and W. Abramowicz, On the crushing mechanics of thin walled structures, Journal of Applied Mechanics, 50, p. 727–734, 1983. For square tubes see X. Zhang and H. Zhang, Crush resistance of square tubes with various thickness configurations, International Journal of Mechanical Sciences, 2016, p 58-68. This section is based on G. Lu and T. Yu, Energy Absorption of Structures and Materials, CRC Press, 2003 and on Dai-heng Chen, Crush Mechanics of Thin-Walled Tubes, CRC Press, 2016.

We start by studying the next figure, which reveals that

$$\sin\alpha = \frac{Bb}{Bu} = \frac{Uu}{Ub} = \frac{Pb}{Ub}.$$

We use this relation in

$$\tan\gamma = \frac{Bu}{Uu}$$

to obtain

$$\tan\gamma = \frac{\tan\psi}{\sin\alpha}.$$

Also,

$$\tan\beta = \frac{BP}{PU} = \frac{TU}{BT} = \frac{\frac{Tu}{BU\sin\psi}}{\frac{Bu\sin\alpha}{BU}} = \frac{Tu}{Bu\sin\psi\sin\alpha}$$

or

$$\tan\beta = \frac{TU\sin\psi}{TU\cos\psi}\frac{1}{\sin\psi\sin\alpha} = \frac{\tan\psi}{\sin\psi\sin\alpha} = \frac{\tan\gamma}{\sin\psi}.$$

The crushing distance, δ, is related to α according to

$$\delta = 2H(1-\cos\alpha)$$

and we also have that

$$S = H\sin\alpha,$$

so time differentiation of these equations yield

$$\dot{\delta} = 2H\sin\alpha\dot{\alpha} \qquad \text{and} \qquad V = \dot{S} = H\cos\alpha\dot{\alpha}.$$

The internal energy has been disclosed to comprise of three basic mechanisms, namely

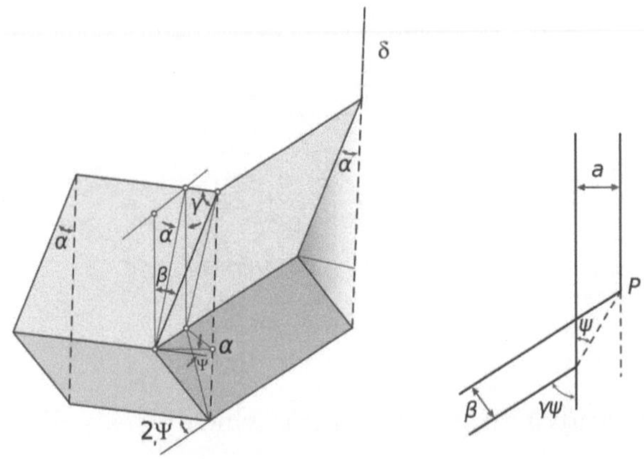

Figure 6.19 Kinematics for the in-extensional folding mechanisms.

1. Energy dissipation in horizontal plastic hinge lines, W_1
2. Energy dissipation in the toroidal surface, W_2
3. Energy dissipation in inclined plastic hinge lines, W_3

In the dissipation mechanism 1, energy is consumed by the hinge lines AB and BC to an amount of

$$\dot{W}_1 \propto M_0\dot{\alpha} \rightarrow \dot{W}_2 = 2M_0C\dot{\alpha},$$

or

$$W_1 = 2\int_0^{\pi/2} M_0 C \mathrm{d}\alpha = \pi M_0 C.$$

The zone of the toroidal surface, mechanism 2, is part of a circular tube with radius b, which is bent with radius a. To obtain an expression for the energy necessary to this mechanism we note that the yield surface should involve the bending moment and the normal force via

$$\frac{M_\phi}{M_0} + \left(\frac{N_\phi}{N_0}\right)^2 = 1.$$

It can be shown that N_ϕ dominates the flow so that the energy for this mechanism is

$$\dot{W}_2 = \int N_\phi \dot{\varepsilon}_\phi \mathrm{d}S = \int_{-\beta}^{\beta}\int_{\pi/2-\psi}^{\pi/2+\psi} N_\phi \dot{\varepsilon}_\phi r \mathrm{d}\phi b \mathrm{d}\theta.$$

To evaluate $\dot{\varepsilon}_\phi$ we notice that

$$\dot{\varepsilon}_\phi = \frac{\dot{w}}{r},$$

or

$$\dot{\varepsilon}_\phi = V_t \frac{\sin\theta}{a + b\cos\theta} = \frac{\dot{S}}{\tan\psi_0}\frac{\sin\theta}{a + b\cos\theta} = \frac{H\cos\alpha\dot{\alpha}}{\tan\psi_0}\frac{\sin\theta}{a + b\cos\theta}$$

with the help of the next figure.

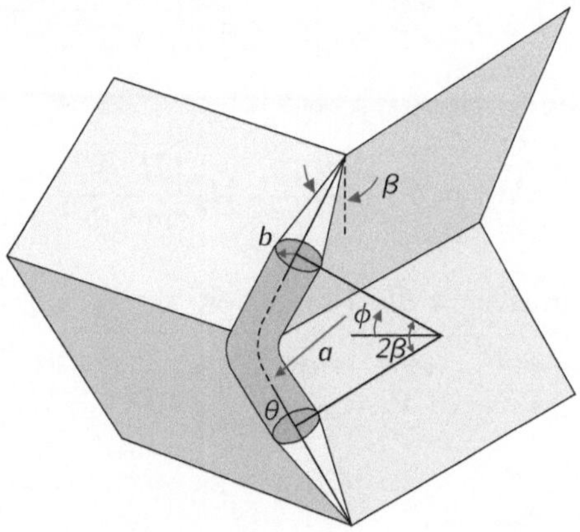

Figure 6.20 Details of the kinematics of a rectangular tube.

Substituting in the energy rate expression it will follow that

$$\dot{W}_2 = \frac{2N_0 bH\cos\alpha\dot{\alpha}}{\tan\psi_0}\int_0^\beta \int_{\pi/2-\psi}^{\pi/2+\psi} \sin\theta \mathrm{d}\theta \mathrm{d}\psi = \frac{4N_0 bH\cos\alpha\dot{\alpha}}{\tan\psi_0}\int_0^\beta \sin\psi \mathrm{d}\phi,$$

which results in

$$\dot{W}_2 = 4N_0 bHI_1$$

when assuming a linear relation between ψ and ϕ, *ie*

$$\psi = \psi_0 + \frac{\pi - 2\psi_0}{\pi}\phi,$$

with

$$I_1 = \frac{\pi}{(\pi - 2\psi_0)\tan\psi_0}\int_0^{\pi/2} \cos\alpha \left[\cos\psi_0 - \cos\left(\psi_0 + \frac{\pi - 2\psi_0}{\pi}\beta\right)\right] \mathrm{d}\alpha.$$

For a square tube, $\psi_0 = \pi/4$ and using the relation $\tan\beta = \tan\alpha/\sin\psi_0$, the integral is numerically evaluated at $I_1 = 0.577$.

To obtain the energy W_3, associated with the travelling inclined hinges, consider that, when a strip of material is pushed through an anvil, see next figure, energy is required for a segment to bend and unbend it, for segments CD and DE for bending, giving

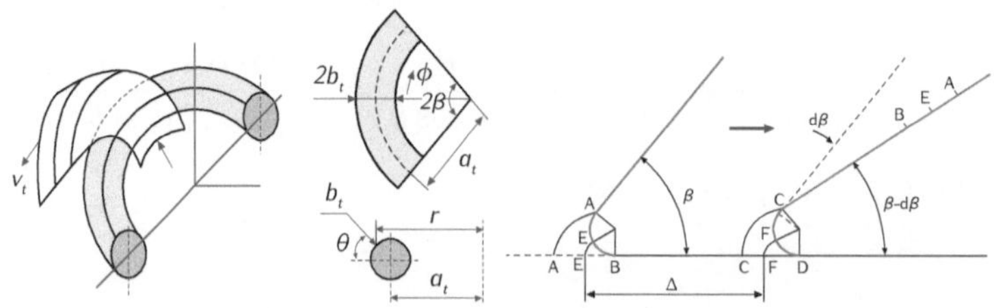

Figure 6.21 Bending of a tube conical surface.

$$W_3 = 4M_0L\dot{\theta} = 4M_0\frac{H}{\sin\gamma}\frac{V_t}{b},$$

since $\sin\gamma = H/L$ and $V_t = \dot{\theta}b$. Using now the fact that

$$V_t = \frac{H\cos\alpha\dot{\alpha}}{\tan\psi_0},$$

we can arrive at

$$\dot{W}_3 = \frac{4M_0H^2}{b\tan\psi_0}\int_0^{\pi/2}\frac{\cos\alpha\sin\alpha}{\tan\psi_0\cos\gamma}\mathrm{d}\alpha,$$

or

$$\dot{W}_3 = \frac{4M_0H^2}{b}I_3,$$

with

$$I_3 = \frac{1}{\tan\psi_0}\int_0^{\pi/2}\frac{\cos\alpha}{\sin\gamma}\mathrm{d}\alpha.$$

$I_3 = 1.148$ for a square corner.

We now equate the external work, $2P_mH$, to the internal dissipation, $W_1 + W_2 + W_3$, to obtain

$$P_m = M_0\left(A_1\frac{b}{h} + A_2\frac{C}{H} + A_3\frac{H}{b}\right),$$

where $A_1 = 8I_1$, $A_2 = \pi/2$ and $A_3 = 2I_3$. The parameters b, the radius of the knee folding, and H, the height of the fold, are not known but they can be obtained from $\partial P_m/\partial H = 0$ and $\partial P_m/\partial b = 0$, such that

$$b = \sqrt[3]{\frac{A_2 A_3}{A_1^2} C h^2} \quad \text{and} \quad H = \sqrt[3]{\frac{A_2^2}{A_1 A_3} C^2 h},$$

which are material independent for our model. We finally obtain,

$$P_m = 3M_0 \sqrt[3]{A_1 A_2 A_3 \frac{C}{h}},$$

with the three basic plastic dissipation mechanisms sharing equal contributions. The above equation becomes

$$P_m = 9.56\sigma_0 \sqrt[3]{h^5 C}$$

for a square tube of thickness h and side C.

The present analysis could be carried on for the case of extensional mode but we leave this case now, remarking that the subject is a rich one, specially if one pauses to wonder about the influence of strain hardening and strain rate hardening, which would require some measures of strain and strain rate.

It is of notice also that, based on mixed mode collapse mechanisms, it has been proposed that the average forces for square and hexagonal cross-section tubes are given by

$$P_m = 48.64M_0 \left(\frac{C}{h}\right)^{0.37}$$

and

$$P_m = 80.92M_0 \left(\frac{C}{h}\right)^{0.40}$$

respectively, which represents various experimental programmes on the crushing of tubes.

See W. Abramowicz and T. Wierzbicki, Axial crushing of multiconer sheet metal columns. Journal of Applied Mechanics, 56:113–120, 1989.

Tubes like the ones we have been analysing are used to absorb kinetic energy of cars, trains, package goods, etc. Hence the importance of analyses like these. As illustrated, the next table presents some recent studies and ingenious principles to absorb energy.

These energy absorbers are scrutinized in terms of efficiency and some parameters that are used are as follows:

- Structural effectiveness
$$\eta = \frac{P_m}{A\sigma_0}$$
- Solidity ratio
$$\phi = \frac{A}{A_c},$$
A_c is the cross section area enclosed by the tube cross section. For a thin circular tube, $A_c = \pi R^2$ and $A = 2\pi RH$, so $\phi = 2H/R$.
- Specific energy
$$S_e = \frac{D_a}{m},$$
where D_a is the total energy absorbed and m is the mass of the device.

6.7 Experiments with shells: buckling transition

See D. Karagiozova and M. Alves, Dynamic elastic–plastic buckling of structural elements: a review, Applied Mechanics Review, 61(4), 2008.

There is a vast coverage concerning experiments made with shell–like structures subjected to axial impact in the literature.

The major variables to be measured in a shell axial impact are the load and the displacement along time. They allow the calculation of the absorbed impact energy by the shell, which is useful in terms of crashworthiness. A smart way to obtain these variables is by using a laser Doppler system, which was already described in Chapter 3. This device allows us to measure the velocity of the impact mass during its whole descending motion on an impact rig. During the impact event, the laser is apt to measure the change in velocity of the mass. Hence, the whole velocity variation of such an impact event can be captured by the laser.

Another way to measure the load–displacement curve in the impact event of a mass colliding with a tube is by image analysis. There, the impact phenomenon is filmed and software traces a point or a set of points chosen by the experimentalist. These points along time reveal the displacement behaviour of the mass, from which immediately follows, by derivative, the velocity, acceleration and force, the latter from impact mass times acceleration. The great advantage of this system is that it measures right away both the displacement of the top of the tube and

Various principles using tubes to absorb kinetic energy.

Principle	Advantages	References
Windowing tubes 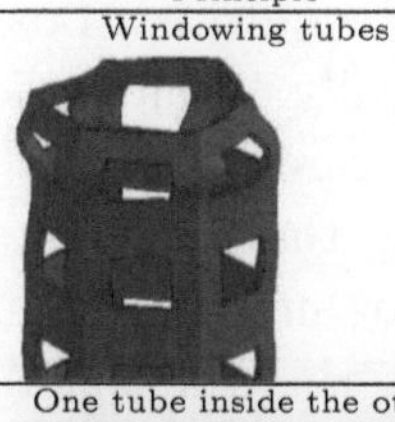	Low peak load	author
One tube inside the other 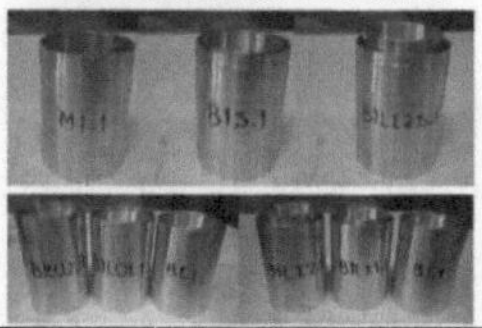	low peak load, smoother mean load	Experimental investigation of bitubal circular energy absorbers under quasi-static axial load, S. Sharifi, M.Shakeri, H. Ebrahimi Fakhari, M.Bodaghi, Thin-Walled Structures, 89 (2015) 42–53
Origami–like tubes 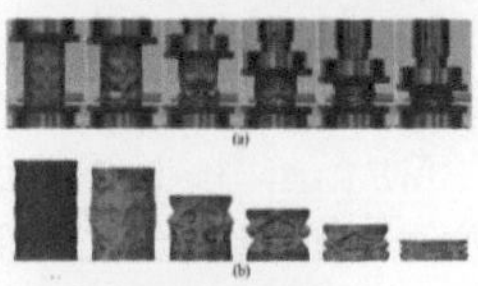	reduces peak force and increases specific energy absorption	Energy absorption of thin-walled tubes with pre-folded origami patterns: Numerical simulation and experimental verification, K. Yang, S. Xu, J. Shen, S. Zhou, Y.M. Xie, Thin-Walled Structures, 103 (2016) 33–44
Composite corrugated plates 	absorbed deformation energy of elements made from filament wound composite tubes depends on the laminate layup	Development of composite energy absorber, R. Milana, K. Viktora, B. Sergiia, S. Vít, Procedia Engineering. 96 (2014) 392 - 399
Expanded tubes 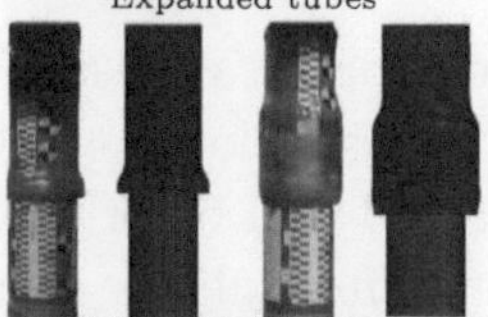	expansion angle and friction were main factors influencing energy absorption	Theoretical prediction and numerical studies of expanding circular, J. Yan, S. Yao, P. Xu, Y. Peng, H. Shao, S. Zhao, International Journal of Mechanical Sciences, 105 (2016) 206–214
Splitting of tubes 	expansion angle and friction were main factors influencing energy absorption — bending energy surpasses splitting and friction energies	Study on the energy absorption of the expanding–splitting circular tube by experimental investigations and numerical simulations, J. Li, G. Gao, H. Dong, S. Xie, W. Guan, Thin-Walled Structures 103 (2016) 105–114

the force, with the integral of the force—displacement curve giving the absorbed impact energy .

Recall that the anvil mass should be quite large as to allow us to assume that the whole kinetic energy of the impact mass is transferred to the shell.

Both the image and laser give, in a single measurement, the basic information of an axial impact event in a relative simple way. Note that, for the laser signal, care should be exercised with the acquisition board, which should have a good amplitude resolution, typically 12 bits, and a good sample rate, typically 1 Ms/s. Even taking such care, we note that the signal is prone to noise, one of the reasons being that the contact between the impact mass and the top of the tube is not smooth. A quick solution to this problem is to remove the noise by filtering the signal.

D. Karagiozova and M. Alves, Transition from progressive buckling to global bending of circular shells under axial impact. Part I: Experimental and numerical observations, International Journal of Solids and Structures, 41:1565–1580, 2004.

An extensive experimental programme using aluminium tubes is now partially described so to highlight some interesting features of the buckling phenomena in shells. Quasi–static and impact tests were performed on circular aluminium tubes with external diameter 50.8 mm and thickness 2.0 mm.

Consider first the quasi–static tests, where tubes of different lengths were crushed in a screw driven machine. As indicated in the next figure by the set of experiments 1, tubes of length 250 mm and 320 mm buckled in a progressive and in global (Euler) mode, respectively. Tubes of length longer than 250 mm and shorter then 320 mm were tested in the same way, experiment set 2, and such a procedure was carried out on other tubes lengths, with experiment set 3 revealing a length transition of around 300 mm. This means that, for tubes longer than 300 mm, the buckling mode is global and for shorter ones it is progressive. This transition zone defines a more (progressive) or less (global) energy absorption efficient device.

The same procedure can be set out using the drop of a mass. The transition length, L_{cr}, strongly depends on the impact velocity. In the experimental programme, tubes as long as $2L_{cr}$ could still deform progressively. Clearly, the impact velocity plays a fundamental role in the selection of the collapse mode, as seen in Figure 6.23. We emphasize that the phenomenon of buckling transition depends less on the material strain rate sensitivity and more on the early choice of the buckling mode by the shell, a choice function of lateral inertia and imperfections.

It has been found that tubes made from low strain hardening modulus materials present a lower impact velocity transition. The transition length decreases considerably as the yield stress decreases.

To further explore these findings, the support of finite element analysis is most important. Indeed, we see in the study of buckling transition

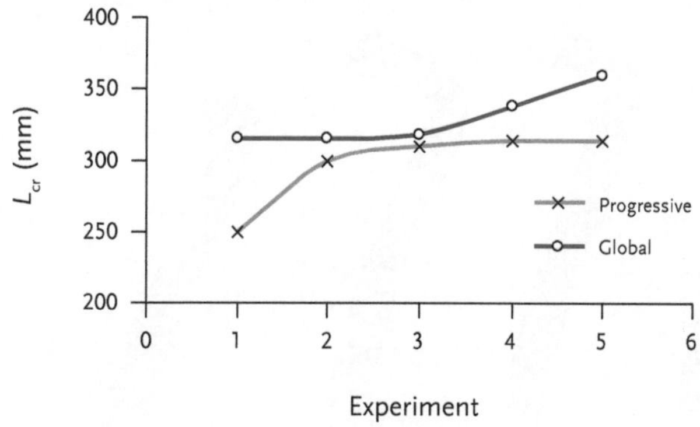

Figure 6.22 Quasi-static buckling transition length for aluminium alloy tubes.

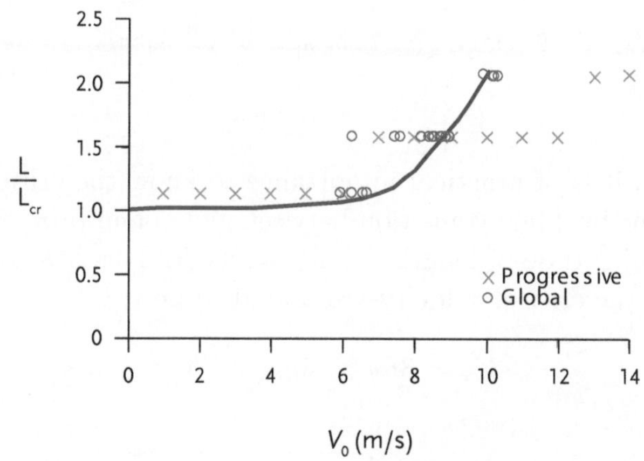

Figure 6.23 Impact velocity influence on buckling transition.

a useful interplay among theory, experiments and numerical analysis. Finite element results shown in the next figure indicate an anomalous response for an aluminium tube. As the impact velocity increases, the tube moves from global to progressive buckling. However, if the velocity is increased further, in this case to 7.5 ms^{-1}, the tube returns to the global mode. A further increase to the impact velocity leads the tube again to the progressive mode but with the wrinkles formed at the bottom.

Another interesting aspect is the influence of the strain rate on the tube response. As seen in Figure 6.25, for an impact velocity of 9.0 ms^{-1}, the tube deforms progressively. If the material model, however, is strain rate sensitive, at that impact velocity the tube would collapse progressively. An increase in the drop height from 4.13 m to 4.85 would make the strain rate sensitive tube to buckle again progressively.

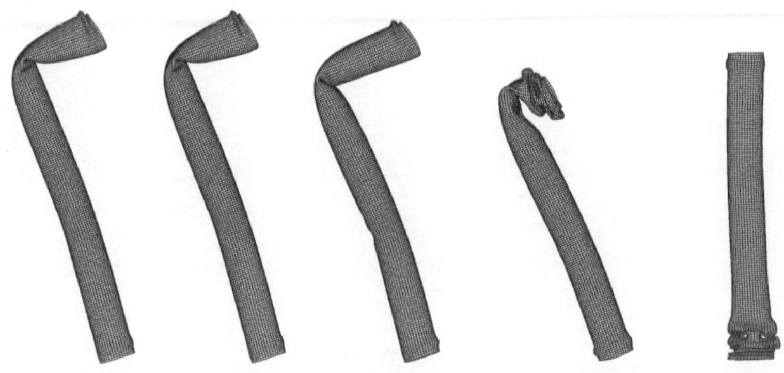

(a) 6.5m/s (b) 6.75m/s (c) 7.5m/s (d) 8.0m/s (e) 8.5m/s

Figure 6.24 Anomalous response of an aluminium tube with $L = 570$ mm.

Of course, it is of practical importance to know the critical velocity that causes the buckling transition between global and progressive buckling of a circular tube of length L and radius R. This has been studied in detail and the critical velocity was found to be

See D. Karagiozova and N. Jones, On the mechanics of the global bending collapse of circular tubes under dynamic axial load: Dynamic buckling transition, International Journal of Impact Engineering, v. 35, p. 397–424, 2008.

$$V_{cr} = 3\gamma\psi_0^2 L \sinh(2\gamma t^*),$$

with

$$\gamma = \frac{3}{L}\sqrt{\frac{6\sigma_0}{\rho}}\sqrt{1 - \frac{30E_h R^2}{\sigma_0 L^2(1+\sqrt{\lambda})^2}},$$

E_h being the hardening modulus of the tube material with yield stress σ_0 and density ρ, $\lambda = E_h/E$, ψ_0 the initial imperfection for the global bending and t^* the duration of the compression phase, which, unfortunately, could not be obtained analytically.

6.8 Closure

We have seen in this chapter quite a few new aspects of structural impact and wave propagation. Buckling was explored leading to the important concepts of plastic and progressive buckling. Circular and rectangular cross section shells were examined and closed form expressions for the crushing load were obtained. This led us to explore kinetic energy absorbers and buckling transition.

(a) 9.0m/s (b) 9.0m/s (c) 9.75m/s

Figure 6.25 Buckling transition for an aluminium alloy tube of $L = 630$ mm modelled as a (a) no strain rate sensitive material. (b) Strain rate effect is taken into account. (c) New transition velocity when strain rate effect is considered.

Another important area in the field of structural impact is how the material behaves under dynamic loads. As we now know, it is this behaviour that bridges the load and displacement fields. Given this importance, we dedicate the next chapter to study material behaviour and the means to probe their properties.

6.9 Problems

1. Perform a literature review and get acquainted with the concept of shock waves. What are the material conditions for it to occur?

2. Obtain the equilibrium equations for an axisymmetric shell.

3. For the square cross-section tube studied in the context of buckling show that

$$b = \sqrt[3]{\frac{A_2 A_3}{A_1^2} C h^2} \quad \text{and} \quad H = \sqrt[3]{\frac{A_2^2}{A_1 A_3} C^2 h}.$$

4. Prove that volume conservation implies that $\mathrm{d}A/\mathrm{d}\varepsilon = -A$. Use this to show that localization occurs at a strain of $\varepsilon = n$ in a material obeying the law $\sigma = k\varepsilon^n$.

5. Calculate the plastic wave velocities in a tube made of stainless steel and of titanium. How do they compare with the elastic wave velocities?

6. Reverse engineering can be used to discover principles of a mechanical application through analysis of its structure, function and operation. Comment on some possible ethical issues in reverse engineering a patented product. Would it be fair to support reverse engineering on the basis that it allows products to interoperate or to be checked against potential harmful or unethical use of it?

CHAPTER 7

Material behaviour and failure

Figure 7.1 A camera–light system filming a material test.

We have seen in Chapter 3 that the material behaviour bridges the kinematic and equilibrium equations of a structure. These differential equilibrium equations and their associate kinematic fields have long been established in the $19^{\text{th.}}$ and $20^{\text{th.}}$ centuries for many basic structures. But, as new materials are being conceived daily, it is always necessary to develop accurate mathematical models to describe their behaviour. With this in mind, the so called constitutive equations need constant improvement to reflect some odd material behaviour, as the ones depicted in the next figure.

Perhaps here is the best Chapter to bear in mind the importance of the material behaviour as the one bridging equilibrium and kinematics.

DOI: 10.1201/9781003485292-7

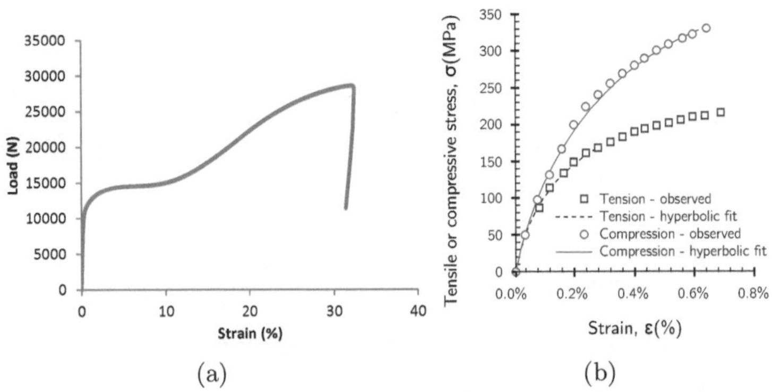

Figure 7.2 (a) A uncommon behaviour of a stainless steel alloy at -30°, as given by a force engineering strain curve. (b) A nonlinear stress—strain engineering curve at small strains for cast iron (from B. Rajani, Nonlinear stress—strain characterization of cast iron used to manufacture pipes for water supply, Journal of Engineering Materials and Technology October 2012, Vol. 134).

To further highlight the importance of knowing the material behaviour, the next figure indicates a typical numerical procedure for a structural, component or process analysis. Here is shown the material response to mechanical, thermal, radioactive, etc. loads should be well established so the analysis can be carried out. Indeed, poor material behaviour knowledge leads to a faulty analysis, regardless of the high quality of the numerical model in terms of meshing and accurate specification of the boundary and initial conditions.

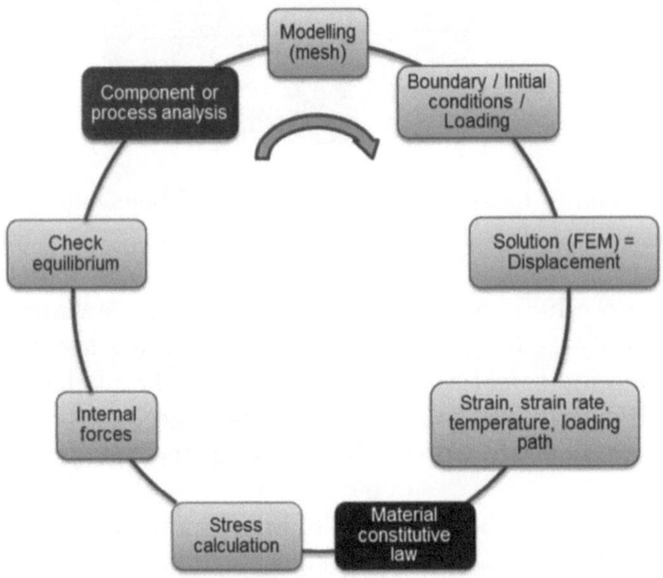

Figure 7.3 Spiral analysis highlighting the importance of knowing the material behaviour.

As new materials are developed so are the experimental techniques to gauge their behaviour. This is an important feature to bear in mind. All the mathematical material laws used in previous chapters require constants. These constants are obtained by best fitting experimental data, which in turn are obtained from some thermo–mechanical loading procedures associated with the measurement of the material kinematic response. We have together complex experimental techniques, mathematical material models and curve fitting procedures, sometimes based on optimization and finite element analysis. These features are covered in this chapter.

Traditionally, the tensile test is the most common procedure to establish a constitutive law. The simplicity of the test however, is only apparent and its analysis and data interpretation can be quite sophisticated. In fact, a full description of what happens when loading a tensile specimen up to rupture is not yet feasible. To interpret data recorded in a tensile test, we start this chapter by defining true stress and strain measurements as well as their equivalent counterparts. We also describe compression tests and introduce dynamic tests, both at mid and high strain rates, commenting on the machines used to perform them.

We next explore various material constitutive laws, moving on to the topic of inverse modelling and image analysis. A section is dedicated to the von Mises yield criterion, with hints of computational plasticity. We close this chapter with an overview on material failure, summarizing some failure criteria.

7.1 Stress–strain definitions

Let us now consider a bar pulled by a force, F, as depicted in the next figure. This causes the bar to increase its size from to l_0 to l.

We define the stretch, λ, as

$$\lambda = \frac{l}{l_0}$$

which measures how much the bar was stretched, *ie* it describes its kinematics. We seek to obtain a relation between cause, the load, and the effect, the stretching. By invoking static equilibrium we find out that

$$\frac{\partial F}{\partial z} = 0,$$

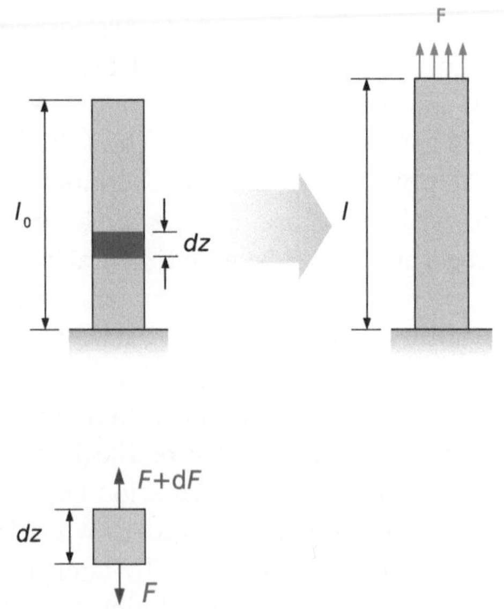

Figure 7.4 A pulled bar.

where no information regarding the kinematics is presented. On reflection, we may agree that there is no other way of relating the static field above, F, and the kinematic field, λ, but by measuring both the stretching and the load that causes it. This has to be done for many load levels and their corresponding stretching. If these various points fall in a straight line then we refer to a linear material and we have then established that

$$F_0 \propto l_0.$$

The knowledge gained from the thought experiment above will not be restricted to this bar. Indeed, other bars made of the same material can also be tested in a similar way and we will find that

$$F_1 = \frac{F_0}{l_0} l_1.$$

Of course it is possible to apply a force $F_1 > F_0$ and have $l_1 < l_0$ if $A_1 > A_0$. If we, on the other hand, divide the force in the bar by its cross-section, we will always have for our linear material that, if $F_1/A_1 > F_0/A_0$ then $l_1 > l_0$, which shows the convenience of working with the stress definition

$$\sigma = \frac{F}{A},$$

called true stress, where A is the bar cross–section area, which changes as the load F is applied. We can also define the engineering stress

$$s = \frac{F}{A_0},$$

which is a stress based on the initial cross–section, A_0.

Let us now consider two distinct cases where bars of unity length, $l = 1.00$, are stretched to final lengths of $l = 1.01$ and $l = 2.00$, giving $\lambda = 1.01$ and 2.00, respectively. The first stretching may well represent a metal bar while the second a rubber. If we define the engineering strain as

$$e = \lambda - 1,$$

applying it to the above bars gives $e = 0.01 = 1\%$ and $e = 1 = 100\%$. Besides, if the bar are not stretched at all, $\lambda = 1$ and $e = 0$. This points to the fact that it is better to work with the above equation to quantify the deformation of the bar rather than simply with the stretching definition.

On the other hand, consider again a bar of unit length which is so compressed that its size tends to zero, *ie* $l \to 0$. This gives and engineering strain $e = -100\%$, which is the minimum possible engineering strain value. Hence, the engineering strain definition is bound by $[-100\%, \infty]$ but it seems that the bounds $[-\infty, \infty]$ would be more appropriate. This can be obtained with the strain definition

$$\varepsilon = \ln \lambda = \ln(1 + e),$$

called natural, or true or, of course, log strain. This strain definition has important properties. The first, just seem, is the infinity bounds for a fully compressed and fully stretched rod. Other properties, addition of the log strains and symmetry are best understood with the following example.

■ Engineering and natural strains

For a tensile specimen of length 50 mm, evaluate the engineering and natural strains when its length is doubled and halved in steps of 25 mm.

In the tensile case, the specimen increases in size in two steps: 1- from 50 mm to 75 mm and 2- from 75 mm to 100 mm. For step 1, the engineering, e, and natural, ε, strains are

$$e_1 = \frac{75 - 50}{50} = 0.500 \qquad \varepsilon_1 = \ln(1 + 0.500) = 0.176.$$

In step 2, these strains are

$$e_2 = \frac{100-75}{75} = 0.333 \qquad \varepsilon_2 = \ln(1+0.333) = 0.125.$$

If we now consider that this stretching was performed in a single step 3, we have

$$e_3 = \frac{100-50}{50} = 1.000 \qquad \varepsilon_3 = \ln(1+1.000) = 0.301,$$

being clear that $e_3 \neq e_1 + e_2$ but $\varepsilon_3 = \varepsilon_1 + \varepsilon_2$.

For the compression case, these strains are

$$e_4 = \frac{25-50}{50} = -0.500 \qquad \varepsilon_1 = \ln(1-0.500) = -0.301,$$

which indicates the advantage of the log strain definition as far it is symmetric and additive.

■

Now, as a function of the stretching, λ, it is possible to define other strain measures, which can be arranged in the following strain family

$$\varepsilon_m = \begin{cases} \frac{1}{m}(\lambda^m - 1) & \text{if } m \neq 0, \\ \ln\lambda & \text{if } m = 0 \end{cases}$$

and in the next table.

TABLE 7.1 Deformation family and stress conjugated measures.

	m	symbol	strain	stress
Almansi	-2	ε_A	$\frac{1}{2}\left(1-\lambda^{-2}\right)$	$\sigma_A = \sigma_N\lambda^3$
Hyperbolic	-1	ε_H	$1-\lambda^{-1}$	$\sigma_H = \sigma_N\lambda^2$
Logarithmic	0	ε_L	$\ln\lambda$	$\sigma_L = \sigma_N\lambda$
Linear	1	ε_N	$\lambda - 1$	$\sigma_N = s = N/A_0$
Green	2	ε_G	$\frac{1}{2}\left(\lambda^2-1\right)$	$\sigma_G = \sigma_N\lambda^{-2}$

Note that all deformation measures tend to coincide for small stretching, but as the latter grows, the various deformation definitions diverge significantly. This is illustrated in the next figure.

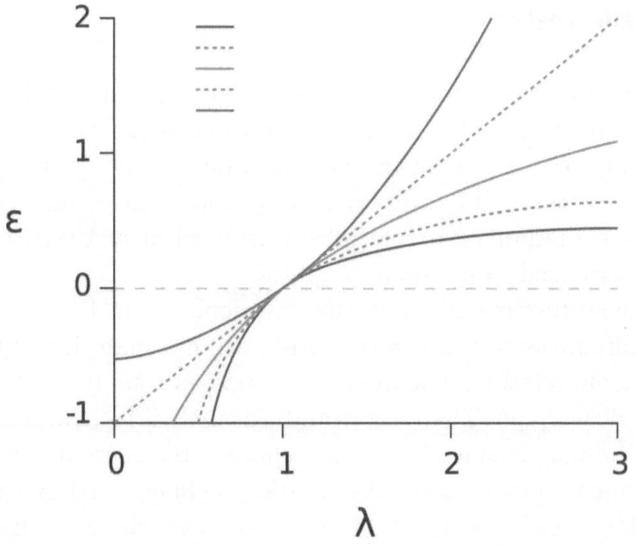

Figure 7.5 Different strain measures as a function of the stretching.

In passing, we can obtain new strain rates definition from the previous equation as

$$\dot{\varepsilon}_m = \lambda^{m-1}\dot{\lambda} = \lambda^m d,$$

with

$$d = \lambda^{-1}\dot{\lambda},$$

which is called instantaneous strain rate.

In structural analysis, one can choose any of the strain measures defined in the table above in order to describe the kinematics. To each strain measure we associate a work conjugated stress measure,

$$\sigma_m = \sigma_N \lambda^{1-m}.$$

where σ_N is the nominal or engineering stress (see table above). The stress–strain conjugate pairs indicated in the table above form a family of stress—strain measures called consistent.

In Chapter 9 we will introduce the concept of the gradient of deformation and use it to express strain and stress measures.

It is also interesting to note that the time derivative of the above equation gives the general stress rate,

$$\dot{\sigma}_m = \dot{\sigma}_N \lambda^{1-m} + (1-m)\lambda^{-m}\sigma_N\dot{\lambda},$$

which does not converge to a common value, even for $\lambda \approx 1$.

7.2 Tensile tests

From the previous section, basic knowledge of a material mechanical behaviour can be gained by loading a rod and measuring the force and the stretching. Once these variables are known, one could express the mechanical behaviour of the material by the stress–strain curve. This curve plays a fundamental role in any structural analysis, as we have already experienced in previous chapters.

The tensile test machine, as the one depicted in the next figure, is the basic apparatus to measure the strength of a material and to obtain its stress–strain relation. The material sample is usually machined flat or on a cylindrical shape. There are standards describing the geometry and the test, although given research may require alternative dimensions. The figure also shows a close–up of the tensile specimen fixed by the machine jaws. In the straight segment of the specimen, there is clip gauge to measure the displacement between the two knives of the gauge as the specimen is pulled by the machine. Also, on the top of the superior jaw, there is a load cell so it is possible to obtain the load and the stretching during the test.

We can use strain gauges to measure the strains in this test. One arrangement is two strain gauges glued on opposite sides of the specimen and connected in half–bridge configuration.

Figure 7.6 A tensile test machine and a composite material sample fixed in the jaws.

A observation is that tensile test machines also have an accurate sensor to measure the jaw displacement. However, this information needs to be handled with care since the machine transverse beam, which holds the jaws, deforms as the stretching load increases. This adds to the actual strain value experimented by the specimen. Hence, a local displacement measure, as the one yielded by strain or clip gauges, is far better to use for the calculation of the strains.

Currently, image processing techniques are being more and more used, allowing the mapping of the stretching of a sample only by tracking the motion of image pixels. We will explore this technique later in this chapter.

From the measured load, the initial specimen cross–section area and the jaw head displacement, we can obtain the nominal, or engineering, and the true stress–strain curves. From the previous sections, it is possible to change from the nominal to the true stress–strain definitions according to

$$\sigma = s(1+e) \qquad \text{and} \qquad \varepsilon = \ln(1+e),$$

when assuming that there is no volume change in the plastic regime, *ie* $AL = A_0L_0$.

Note that the difference between engineering and true stress–strain values can be quite large for ductile materials and the next figure presents the case of a steel alloy. In this particular case, the true stress–strain curve is in the equivalent space, a concept that we will explore next.

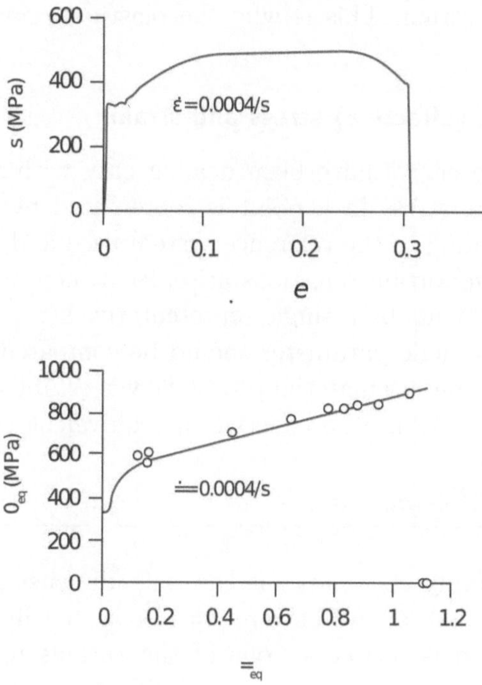

Figure 7.7 Stress–strain curves in engineering and true equivalent spaces for the same ductile steel alloy.

We see from the results in the figure that the material in play presents a nearly linear behaviour for small strains. For many materials, when this linear stress–strain behaviour ceases, the material enters into the plastic regime, *ie* the removal of the load leaves the material permanently strained. This transition, not clear for some metals, is the yield stress, an important design parameter. There are many cases though, that even metals do not behave linearly in small strain regimes, like the cast iron seen in the opening of this chapter.

After yielding, a typical metal continues to increase its strength as it enters in the strain hardening regime. Observe that, as the strain grows, the discrepancy between the nominal and true stress curves become more evident. The true stress grows quicker because the sample cross–section area decreases.

It is important to realize that the fact that the material behaviour ceases to be linear does not mean that it has entered the plastic regime. A macroscopic way to gauge plastic strains is by unloading the sample and measuring the left deformation. In other words, plastic strains are revealed only after load removal, given that what one measures during a test is the total strain. This is why the plastic strains are called an internal variable.

7.2.1 Equivalent (effective) stress and strain

So far in this chapter, we have been dealing only with uniaxial stress. More generally, the stress in a point is represented by a set of stress components, according to the reference system used and to the assumptions made. Two important questions arise. First, how to represent this set of stress components by a single representative stress value. Second, to what value this single parameter should be compared to in order to allow us to access whether material plastic flow is established. As for the first question let us define the effective, or equivalent (true) stress, or von Mises stress as

$$\sigma_{eq}^2 = \frac{(\sigma_x - \sigma_y)^2 + (\sigma_y - \sigma_z)^2 + (\sigma_z - \sigma_x)^2 + 6(\sigma_{xy}^2 + \sigma_{yz}^2 + \sigma_{zx}^2)}{2},$$

which becomes simply $\sigma_{eq} = \sigma_z$ for the uni—dimensional stress case. For the case of pure shear, we have $\sigma_{eq} = \sqrt{3}\sigma_{xy}$. It will be shown later that this expression is related to one of the various invariants of the stress tensor. Clearly, this stress state quantification takes into account all components of the stress tensor.

We can define a similar expression for the equivalent (true) plastic strain. We expect that this variable should always increase, for both positive and negative values of the various strain components, so as to go side by side with the equivalent stress, which is always positive. Accordingly, we define the equivalent plastic strain rate as

In finite element analysis, the material behaviour is usually informed by equivalent stress–strain pairs.

$$(\dot{\varepsilon}^p_{eq})^2 = \frac{2}{9}\left\{(\dot{\varepsilon}^p_x - \dot{\varepsilon}^p_y)^2 + (\dot{\varepsilon}^p_y - \dot{\varepsilon}^p_z)^2 + (\dot{\varepsilon}^p_z - \dot{\varepsilon}^p_x)^2 + \right\} +$$

$$+\frac{2}{9}\left\{6\left[(\dot{\varepsilon}^p_{xy})^2 + (\dot{\varepsilon}^p_{yz})^2 + (\dot{\varepsilon}^p_{zx})^2\right]\right\}$$

and from there the equivalent plastic strain

$$\varepsilon^p_{eq} = \int \dot{\varepsilon}^p_{eq} \mathrm{d}t.$$

For a circular bar being pulled in tension in the z direction, before necking takes place, the equivalent stress reduces to the (true) unidimensional case, *ie*

$$\sigma_{eq} = \sigma_z = \frac{F}{A}.$$

To obtain an expression for the equivalent strain in such a cylindrical bar, let us evaluate the variation of volume of the unity cube shown in the next figure.

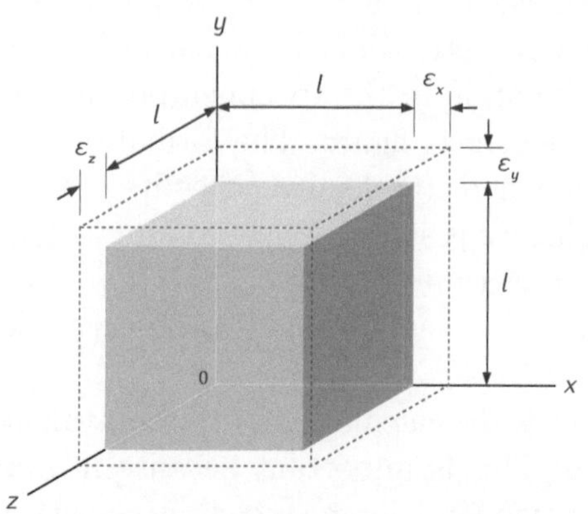

Figure 7.8 A unity cube experimenting a change in its dimensions.

The variation in volume of the unity cube is $\Delta V = (1+\varepsilon_x)(1+\varepsilon_y)(1+\varepsilon_z) - 1$, which equals to $\Delta V = \varepsilon_x + \varepsilon_y + \varepsilon_z = \varepsilon_x(1-2\nu)$ when high order cross products are disregarded and use is made of the Poisson ratio definition, $\nu = -\varepsilon_y/\varepsilon_x = -\varepsilon_z/\varepsilon_x$. Now, various experiments in metals have shown that $\nu \approx 1/2$ as plastic flow grows, which indicates from the ΔV expression above that volume is conserved in the plastic regime.

For the cylinder under tension, from the expression for the equivalent strain, in the proportional loading regime, we arrive at

$$\varepsilon_{eq} = \ln \frac{l}{l_0} = 2 \ln \left(\frac{d_0}{d} \right).$$

Note that it is only necessary to trace the specimen diameter to obtain the equivalent strain (in the probing area). As for the equivalent stress, this same diameter can be used to obtain the current area. One should be aware though of the phenomenon of necking, to be described next. In the presence of necking, the equivalent stress is no longer given by

$$\frac{F}{\pi d^2/4},$$

as we now discuss.

7.2.2 Necking

We have established that the true values of axial stress and natural axial strain that occur in a tensile test coincide with their equivalent counterparts. This is true up to a certain point in a test. If we continue to pull a tensile specimen, there will be a moment when material instability occurs and the deformation in the specimen becomes concentrated on a small necking zone, see next figure. The formation of the necking zone is roughly the point where the load reaches its maximum value, which is the ultimate engineering stress. Once necking starts, to measure the strain with the clip gauges and using $\varepsilon_{eq} = \ln(l/l_0)$ is meaningless since this measure will be a too vague value for the actual localized straining process in the neck.

The question now is the calculation of the equivalent stress and strain in the necking zone. This is important because it gives us information on the material behaviour at large strains, necessary when dealing, for instance, with crushing of a device. After necking, the stress field ceases to be unidimensional. An analysis of the stress field in the necking zone was developed by Bridgman. He found, and modern numerical models

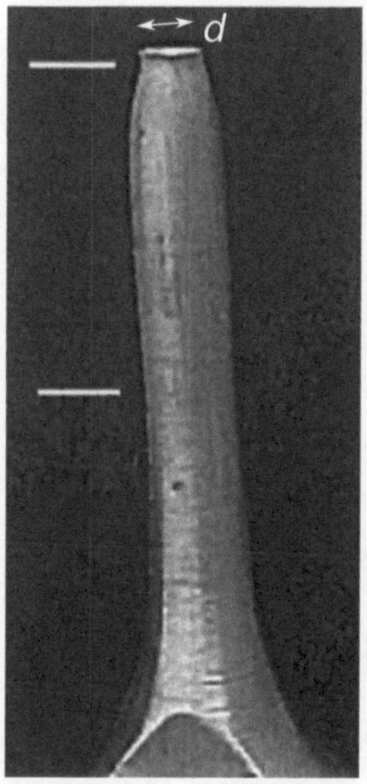

Figure 7.9 Necking zone in a tensile specimen.

corroborate with it, that the equivalent stress in this region is given by

$$\sigma_{eq} = f_c \frac{F}{(\pi d^2/4)},$$

with

$$f_c = \frac{1}{\left(1 + \frac{4R}{d}\right) \ln\left(1 + \frac{d}{4R}\right)}.$$

The term $4F/\pi d^2$ is clearly a true stress acting, not homogeneously, in the axial direction and the correction factor, f_c, takes into account the three dimensionality of the stress field. Here, R is the necking radius, not to be confused with $d/2$, which needs to be measured as the test progresses, a somewhat easy task with image analysis systems.

See M. Alves and N. Jones, Influence of the hydrostatic stress on failure of axisymmetric notched specimens. Journal of the Mechanics and Physics of Solids, 47(3):643–667, 1999.

The next figure plots the correction factor for a ductile steel alloy. We can see that the difference between the equivalent stress and the true unidimensional stress (F/A) can be significantly large and may lead to important errors in a non–linear finite element analysis if no distinction between true axial and true equivalent stresses are made.

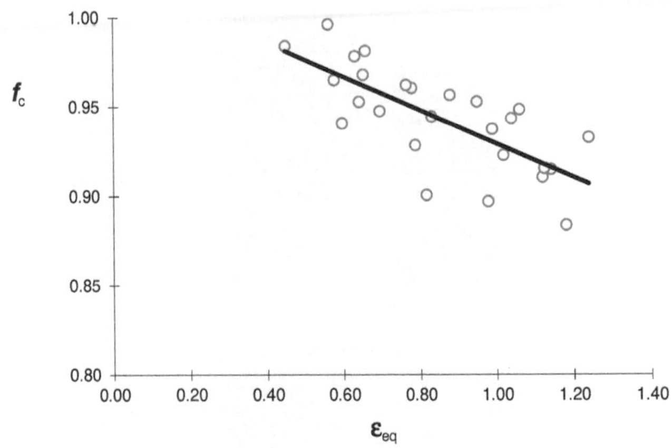

Figure 7.10 The factor f_c for a ductile steel alloy.

7.3 Compression tests

Many manufacturing procedures, as well as components, are subject to a compressive stress state. Hence the necessity of knowing better how the material behaves under this condition. The compression test serves this purpose as well as to investigate, together with the tensile test, material anisotropy.

The compression test is more complicated to perform than the tensile test. First, the loads involved are higher because, usually, the material cross–section in a disc form is larger than the one from the tensile specimen. Such a high load causes unwanted deflection of the machine due to its compliance (flexibility), so leading to errors in the measurement of the displacement, if done by the machine cross head sensor. Ideally, the disk compression should be measured by gauging the supporting plates movement. But, due to the possible high loads, these plates where the sample rests can tilt since it is nearly impossible to assure a proper alignment of the load, due for instance to the uneven thickness dimensions of the sample. To overcome the above problem, one possibility is to gauge the plates approach using three accurate and precise displacement gauges.

In compression tests, friction is also an important issue. When compressed, the material expands radially and this gives rise to friction forces between the plate and the sample. Hence, an unknown part of the measured compressive force is used to overcome friction. Lubrication is a must in these tests but it only partially compensates the high friction forces. Due to friction, at some stage of the loading process, the disks start to bulge, breaking down the uniformity of the strain and stress fields. From this point onwards, load over current area no longer represents the sought equivalent stress of the material.

In a compression test of a uniform cylinder, the axial stress coincides with the equivalent stress.

One of the reasons to go through the difficulties indicated above is that the material compressive curve is used as a reference to the compressive material response at higher strain rates . Also, by loading the material in tension, followed by compression, it is possible to obtain information on its kinematic parameters.

7.4 Medium strain rate tests

We have seen in the previous chapter that an average value of strain rate in a circular tube of diameter D impacted by an object with a velocity V_0 is given by $\dot{\varepsilon} = V_0/4R$. Applying this equation to the frame of a car of diameter 100 mm, impacting a rigid barrier at 50km/h, gives a rough strain rate value of 70/s. The example shows that many daily impact situations occur at intermediate strain rate levels, say between 1/s and 100/s. Hence, it is important to perform mechanical material characterization at these strain rate levels, since one can expect that many materials exhibit strain rate dependent behaviour. This is indeed the case for most of the steels, so the importance of having machines that can quickly pull a material sample.

Tensile test machines capable of pulling coupons rapidly are mainly based on a hydraulic system comprising a power unity, a piston, an oil pressure accumulator, sensors and a servo–valve driven by sophisticated software. All this amounts to increasing costs, which explains the paucity of material data in this range of strain rates. Moreover, medium strain rate machines may present, during their operation, oscillations that are captured by the instrumentation, disturbing the actual readings of a test. This is shown in the next figure for the load signal and indicates that data processing is a critical issue when one opts for tests with this machine.

A simpler alternative to these expensive tensile test machines consists of a frame where the material sample, in the shape of the so called dog–bone specimen, is pulled in tension by the movement of the lower rig. This motion is set by the drop of a mass, which impacts a shaft supporting one sample end. The machine uses an accurate pre–compressed load cell to gauge the test force and a laser sensor to measure the support

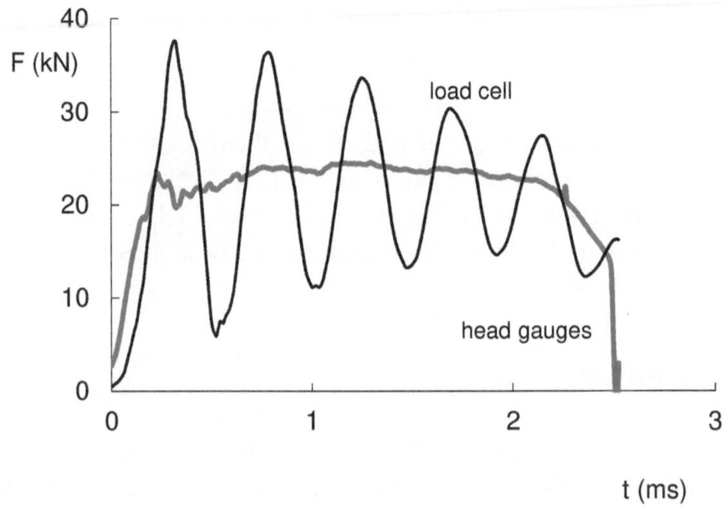

Figure 7.11 Load cell reading of a servo–hydraulic machine. The correct load readings were obtained by sticking strain gauges to the specimen shoulders and assuming a strain rate independent elastic modulus. The actual load cell machine readings indicates a poor performance of the instrumentation.

displacement. The next figure shows some a design solution together with one based on linear electrical motors, also capable of midrange strain rate tests provided one uses small gauge lengths.

7.5 High strain rate tests

We have worked out in the previous section a strain rate of 70/s when a 100 mm diameter tube is hit at 50 km/h. At this speed, an average strain rate of (50 km/h)/(10 mm)=1388/s is easily achieved in spot weld zones, in a rivet or in a bolt of some 10 mm in size. Hence, it is also important to cover the material behaviour at such high strain rates, which are clearly difficult to achieve using servo–hydraulic machines.

This machine is explored in details in W.W. Chen and B. Song, Split Hopinkson (Kolsky) Bar: Design, Testing and Applications, Springer Verlag, 2011.

High strain rate compression, tension, torsion and biaxial tests can be performed in the so called split Hopkinson pressure bar, SHPB. It was devised by H. Kolsky (1917-1992) in 1949 based on the various wave propagation studies by John Hopkinson (1849-1898) and his son, Bertram Hopkinson (1874-1918). Since then, this machine has been extensively used in many research laboratories due to its low cost, simple operation and good results. Yet, there are quite a few problems with this apparatus, which have been overcome along the years.

In fact, it seems more appropriate to call this machine an Elastic Wave Machine, EWM, and the name reveals its operating principle,

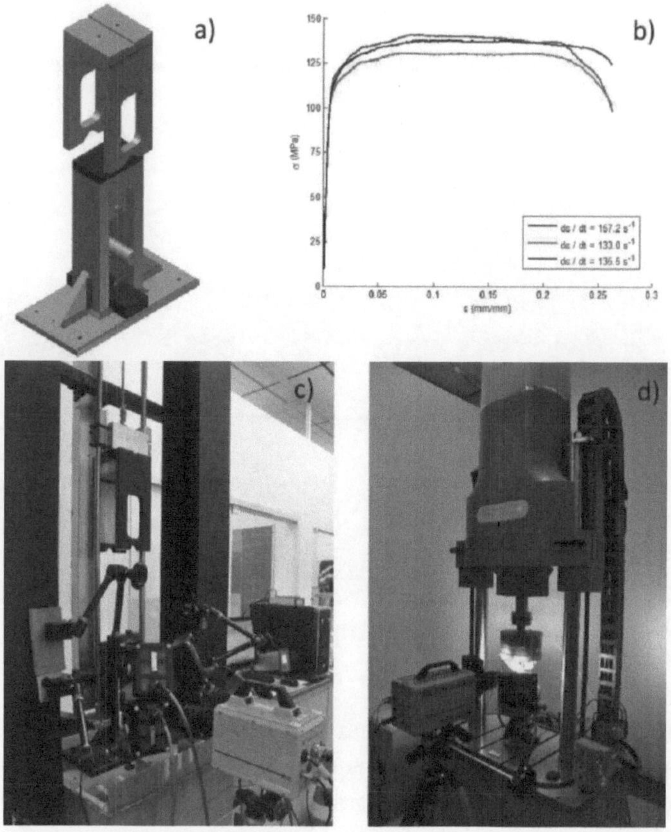

Figure 7.12 a) An overview of how a mid-strain-rate machine works, b) a typical stress–strain curve at mid strain rates of a fiber–metal laminate dog–bone shape tensile specimen, c) the full setup and d) an electrical driven midrate machine for tensile, torsion and fatigue tests.

depicted in the next figure for the case of a compressive test. The figure also shows the design stage of such a piece of equipment together with its realization.

A striker of length L_{st} impacts against a long input bar. As we have seen in Chapter 3, this impact event generates two compressive waves, one in the striker and another in the input bar. The wave in the striker is reflected at its left end in the figure as a tensile wave which, when reaching the striker–input bar interface becomes trapped in the striker.

Bars are typically between 1 and 6 m long. At ISPRA, Italy, there is an EWM of a total length of 202 m.

The wave in the input bar, encompassing a length $2L_{st}$, propagates up to its end, when it then meets the material sample being tested, usually in a disc shape. The amplitude of this incident wave, e_i, is recorded by strain gauges placed in the mid–length of the input bar. Due to the impedance mismatch in the input bar–sample material, part of the wave is reflected in the incident bar with amplitude e_r. The wave pulse generated by the

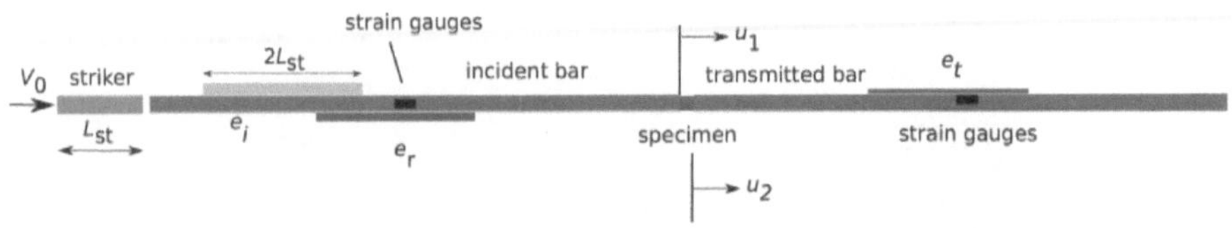

(a)

(b)

Figure 7.13 (a) Elastic wave machine: operating principle. (b) An EWM capable of tensile and compressive tests.

striker, by reflecting at the bar–material interface, displaces this end by u_1 with a velocity v_1, which causes compression of the sample material.

Part of the input wave is also transmitted to the output bar, causing the material–out put bar interface to displace by an amount of u_2, with a velocity v_2. The magnitude of the transmitted pulse, e_t, is also measured, usually, by strain gauges. This transmitted wave reaches the output bar end and is reflected back to the material–output bar interface, if no trap bar is presented. In any case, by this time, the apparatus has fulfilled its purpose of compressing the sample material at a high strain rate.

The input and reflected measured signals can be used to obtain the material stress–strain curve and the test strain rate. To this end, consider that the axial strain in a bar is given by $e = \mathrm{d}u/\mathrm{d}x$, so that the displacement in the bar–material interface is

$$u_1 = \int e_1 \mathrm{d}x = \int e_1 \frac{\mathrm{d}x}{\mathrm{d}t}\mathrm{d}t = \int_0^t c e_1 \mathrm{d}t$$

with a similar expression for u_2, with $c = \sqrt{E/\rho}$.

Considering that a compressive (tensile) wave displaces the material in the same (opposite) direction of its propagation, one deduces that

$e_1 = e_i - e_r$ and $e_2 = e_t$. Also, the engineering strain experienced by the sample material of thickness L_s is $e_s = (u_1 - u_2)/L_s$, so that

$$e_s = \frac{c}{L_s} \int_0^t (e_i - e_r - e_t)\, \mathrm{d}t.$$

The previous figure also helps us to establish the forces caused by the particles motion and they read $F_1 = E_b A_b (e_i + e_r)$ and $F_2 = E_b A_b e_t$, where E_b and A_b stand for the elastic modulus and cross–sectional area of the bars. Assuming that, during the test, equilibrium within the specimen is achieved, the engineering stress experienced by the sample material of transverse cross–section area A_s is

Equilibrium in such a dynamic test is not a trivial matter, as discussed by L.M. Yang and V.P.W. Shim, An analysis of stress uniformity in split Hopkinson bar test specimens, International Journal of Impact Engineering, 31, p. 129-150, 2005.

$$s = \frac{E_b A_b}{A_s}(e_i + e_r) = \frac{E_b A_b}{A_s} e_t,$$

with the total engineering strain deformation becoming

$$e_s = \frac{-2c}{L_s} \int_0^t e_r \mathrm{d}t.$$

The engineering strain rate is simply the derivative of the above equation and reads

$$\dot{e}_s = \frac{-2c}{L_s} e_r,$$

with the respective true value being

$$\dot{\varepsilon}_s = \frac{\dot{e}_s}{1 + e_s}.$$

The use of the elastic wave machine leads to the full characterization of the material in a somewhat simple way. However, the results are valid only when equilibrium is achieved, *ie* $F_1 = F_2$. This occurs after a few back and forward propagation of the waves within the tested material. This accounts for the fact that the initial part of the so measured stress–strain curve is not correct because of lack of equilibrium in this phase. Therefore, it is not correct to extract the material elastic modulus from this test, unless some additional data processing is performed.

The next figure shows various phases of the loading process of a tensile specimen by the bars. Note that the bar ends translate at different speeds so that the sample can be stretched.

Figure 7.14 Tensile test sequence in an EWM.

The previous equation indicates that the use of a thin disc cooperates with high strain rates since L_s is rather small. Indeed, the elastic wave machine shown in Figure 7.13 allows one to achieve strain rates in the range of 500/s to 10000/s. It should also be noted from the expressions above that it is necessary one measurement station only, at the input bar, to obtain the material response, *ie* one needs only to measure e_i and e_r. However, it is important in a test one to assure that the data are validated against equilibrium, which can be determined by the measurement of the transmitted pulse.

The simplicity of the EWM does not come at little cost. We can enumerate problems like friction between the bar ends and the specimen, inertia effects due to the rapid radial and axial motion of the disk sample and dispersion of the waves in the bars. These problems will now be commented on.

7.5.1 Dispersion

The measurement stations in an EWM are usually located in the middle of the bars. For the output bar signal, the strain gauges can be glued close to the material–bar interface. The readings of this transmitted strain immediately give the material strength. In the input bar, however, it is necessary to decouple the input and reflected pulses. Hence, the strain gauge station should be located at least one striker length from the material–bar interface in order to avoid superposition of these pulses. The problem with the gauges being far from the bar–material interface is that, as the signal travels from the interface it becomes distorted due to dispersion. This means that what is measured by the gauges is not what happened at the interface, so affecting the calculation of the strain and the strain rate in the sample.

Inertia is one of the causes of dispersion and we refer to geometrical dispersion. Because the bar is a bounded media, as a compression (tension) pulse travels in the bar, its cross-section expands (shrinks), which requires energy, so affecting the wave amplitude. The other source of dispersion is the bar material. As a pulse travels in the bar, its frequency components are damped differently. We can think that the higher the frequency the higher the bar material internal friction. This causes the high frequency components being attenuated in its amplitude, with the overall result that the pulse changes its shape.

One practical idea to avoid some of these problems is to eliminate the high frequency contents of the pulse generated in the impact between the striker and the input bar. This can be done by placing a soft material, like cardboard or a copper thin plate, in between them. Even grease or sprayed oil helps to get rid of high frequency components.

These dispersion effects can be all eliminated if one measures the strain locally and not via the bar gauges signals. This is possible by glueing strain gauges in the material sample, but this is rather troublesome. A better way is to measure the strains in the material via film recording, as illustrated in the next figure. One needs here to be careful with the filming rate. Remember that a typical EWM tests lasts some 100μs. If one targets to obtain 100 pairs of stress—strain data, it is necessary to record one image at every 1μs, *ie* the camera should operate at one million frames per second. As for the stress, one alternative is to measure the velocity of the free end transmitted bar with a laser velocity meter. Here, dispersion can be worked out and the force at the transmitted bar—material sample interface is obtained, allowing the calculation of the material strength.

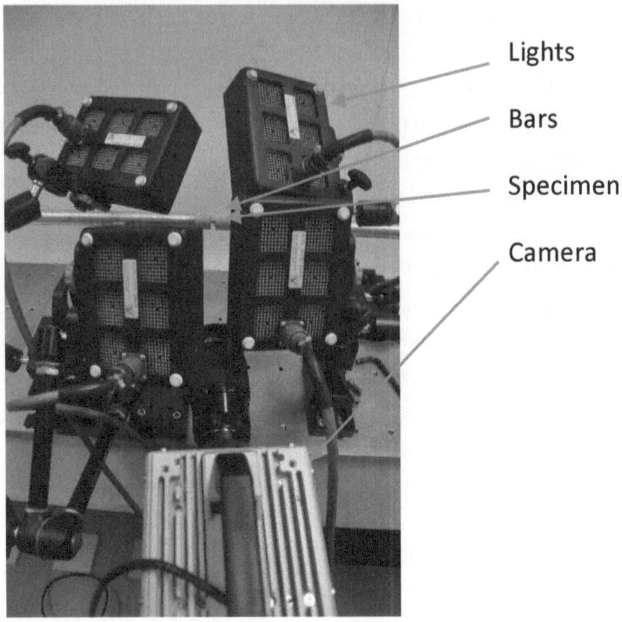

Figure 7.15 An EWM camera set up.

7.5.2 Friction correction

See T. Jankowiak, A.Rusinek and T.Lodygowski, Validation of the Klepaczko–Malinowski model for friction correction and recommendations on SHPB, Finite Elements in Analysis and Design, 47(2011), 1191–1208.

Perhaps the most important problem in the dynamic compression bar tests is friction. Friction arises from the radial expansion of the disk specimen. The classical theory of the Hopkinson bar does not take friction into account but there are some models which do so. However, these models require knowing the friction coefficient value, which arguably is a difficult parameter to measure under dynamic conditions.

The next figure shows how the friction coefficient affects the material stress–strain curve of an aluminium sample. The graph is the result of a finite element analysis of an EWM test, with the geometry of the sample being in the format of a disk. The input curve in the FE analysis is the actual material curve and the various other curves are obtained when applying the EWM formulae, with the strain readings yielded by the finite element results at the middle of the bars.

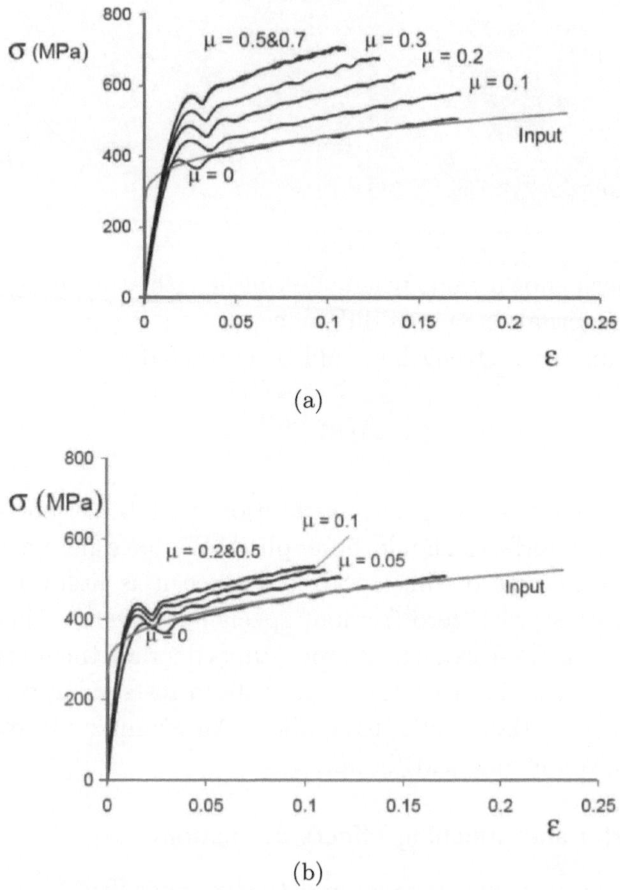

Figure 7.16 Friction influence for (a) disk and (b) ring on dynamic stress–strain curve of an aluminium alloy.

It is interesting to note that, by adopting a coupon ring instead of a disk, one obtains the curves (b) in the figure above. There we see that the error between the input curve and the various other numerical curves are bounded to a limit, regardless of the friction coefficient. The reason for this behaviour can be appreciated in the next figure, where a washer was submitted to a fast compression. We see that the material moves inwards and outwards and that some portion, indicated approximately by the white rings in the figure, hardly moves radially. For this portion, friction is not relevant, suggesting that the use of rings can counterfeit the problems caused by friction.

Figure 7.17 A ring (washer) under different stages of compression. White circles reveal a region that does not move much in the radial direction.

It has been shown that, when the ring length (thickness) equals the external to internal diameters difference, the traditional Kolsky expression for the material strength should be corrected to

$$s = 0.87\frac{E_b A_b}{A_s} e_t.$$

Such a correction takes into account friction; no lubrication between the bars–material interfaces should be applied. We have here an important advantage of the use of ring specimens since it is literally impossible to perform a test with zero friction, specially in tests at high and low temperatures, as these extreme temperatures degrade the lubricant properties. We also mention that the strain rate in tests using rings tends to be more constant then when using disks. An example of how the rings perform is given in the next figure.

See M. Alves, D. Karagiozova, G.B. Micheli and M.A.G. Calle, Limiting the influence of friction on the split Hopkinson pressure bar tests by using a ring specimen, International Journal of Impact Engineering, 49, p.130–141, 2012.

7.5.3 Inertia and punching effects correction

In the same way energy is required to overcome friction, it is also a consequence of the EWM test that the sample will be accelerated in the thickness and radial direction. This energy causes an inertia effect, which is marginal in some cases. For instance, the measured stress level is affected by some 1% for copper.

See Displacement correction for punching at a dynamically loaded bar end, K. Safa and G. Gary, International Journal of Impact Engineering, 37, p.371–384, 2010.

The so called punching effect is also quite small. It arises due to the fact that a disk shape specimen tends to deform elastically the bars in the specimen–bar contact zone. This has been quantified, being found that it is more important when brittle materials are tested, where attention to small strains are necessary. For ductile materials, only at strains of less than say 1% this would be important.

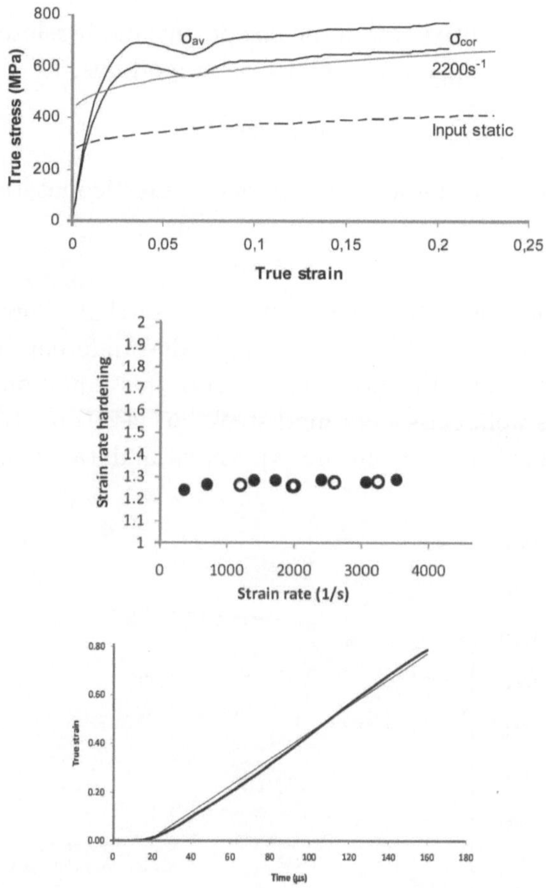

Figure 7.18 Performance of rings in EWM tests. (a) Simulation results for a ring and a disk made of steel. (b) Strain rate hardening, defined as the strength increase in relation to the quasi–static flow stress, versus strain rate, for a steel alloy. Open and bold symbols are for rings and discs, respectively. (c) Strain—time evolution for a typical experiment using a steel ring.

In closing this section, one may wonder why to pay such detailed attention to the EWM methodology. We recall that finite element program benchmarks calculate quite accurately the final configuration of a structure. But such an analysis hinges on a good knowledge of the material behaviour. Hence, it does not make much sense to use a powerful numerical tool for analysis if one inputs inaccurate material parameters.

7.6 Material constitutive laws

The topic of constitutive laws is a broad one. Basically, a constitutive law is a mathematical expression that aims to describe the material behaviour. In structural analysis, one inputs mainly the strain, strain rate and temperature to obtain the stress levels.

A simple constitutive law is the elastic one and it relates to the stress and strain via the single parameter elastic modulus, E,

$$\sigma = E\varepsilon.$$

We have also seen in Chapter 5 the perfectly plastic material constitutive law,

$$\sigma = \sigma_0$$

where σ_0 is the flow stress. Note in this case that there is no unique stress–strain relation, *ie* for any strain value different from zero, the stress level is always the flow stress. Yet, this equation can describe some materials well, classically mild steel, but also some titanium alloys, as seen in the next figure from the experimental data ($\times$) after the strain rate jump.

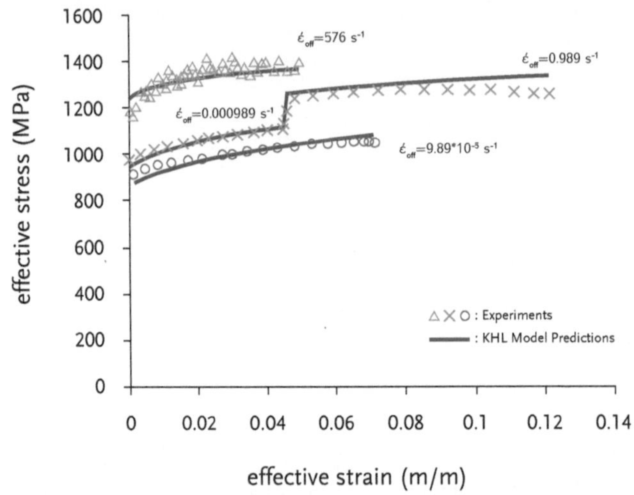

Figure 7.19 Titanium alloy Ti-6Al-4V compressive stress–strain behaviour at various strain rates. Adapted from Experimental and numerical analysis of high strain rate response of Ti-6Al-4V titanium alloy A.Bragov, A. Konstantinov, A. Lomunov, I.Sergeichev and B. Fedulov, DYMAT 2009, p. 1465–1470, 2009.

The perfectly plastic material law is a specific case of

$$\sigma = \frac{\sigma_y}{(\sigma_y/E)^n}\varepsilon^n$$

with $n = 0$. The constants σ_y and n are given in the next table for some steel alloys, while the next figure presents a titanium alloy compressed at a strain rate of 1000/s indicating again that the rigid–plastic model is a reasonable assumption for this material at such a rate.

As we have seen on other occasions, some materials increase their flow stress levels when loaded dynamically. If we plot this increase of the

TABLE 7.2 Constitutive parameters for various steel alloys with $E = 210$ GPa. Valid for $\sigma = \frac{\sigma_y}{(\sigma_y/E)^n}\varepsilon^n$ and adapted from A. Nayebi, R. El Abdi, O. Bartier and G. Mauvoisin, New procedure to determine steel mechanical parameters from the spherical indentation technique, Mechanics of Materials, 34, p. 243–254, 2002.

Material	σ_y (MPa)	n
16NC6	345	0.17
100C6	430	0.14
35CD4	473	0.09
35NC15	790	0.08
Z38CDV5	390	0.17
Z15CN17-03	804	0.10
Z2CN18-10	200	0.21
42CD4	515	0.10
XC18	350	0.10
XC38	576	0.11
XC48	419	0.16
XC65	345	0.24
XC80	303	0.17

flow stress, say $\sigma_d/\sigma_s - 1$, against strain rate, $\dot{\varepsilon}$, we can conjecture, and experimental results for many metals have confirmed, that the resulting relation is linear when using a log scale for the strain rate. It follows then that

$$\sigma_d = \sigma_s \left[1 + \left(\frac{\dot{\varepsilon}}{C}\right)^{1/p}\right],$$

with C and p being material constants. This is the well known Cowper–Symonds equation, implemented in most explicit finite element codes. It can also be written as

$$\sigma_d = \sigma_s\left(1 + m\dot{\varepsilon}^n\right), \qquad m = \frac{1}{C^{1/p}} \qquad n = \frac{1}{p}$$

Observe that the equation above can be adapted to strain hardening materials by writing the static flow stress as a function of the strain, *eg* $\sigma_s = \sigma_y + A\varepsilon^p$. The constants m and n are determined at a given plastic strain, say at the yielding strain ε_y. In this case, only the dynamic yielding stress can be well predicted. Using these constants to predict,

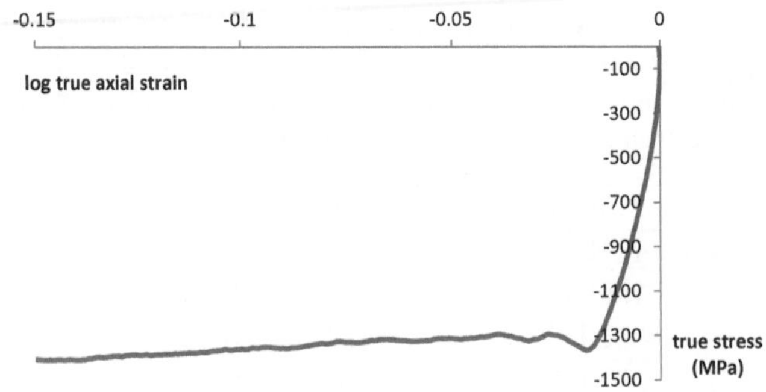

Figure 7.20 The behaviour of a titanium alloy at a strain rate of 1000/s.

for instance, the ultimate stress, σ_u, or the failure stress, σ_{fa}, would lead to an error, as shown in the next figure.

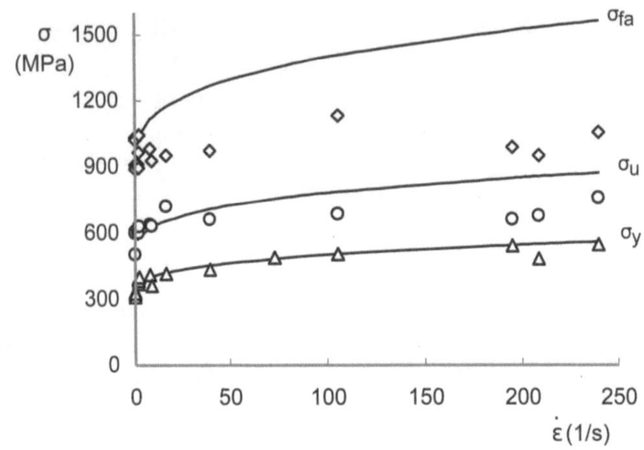

Figure 7.21 Various stress predictions for the steel alloy BS EN 10025 FE430A according to the Cowper–Symonds equation, with its constants determined from the yielding stress.

A remedy to this situation is to use the equation

$$\sigma_{eq_d} = \sigma_{eq_s} + \bar{m}\dot{\varepsilon}^{\bar{n}},$$

with

$$\bar{m} = \frac{\sigma}{C^{1/q}} \quad \text{and} \quad \bar{n} = \frac{1}{q},$$

valid for materials whose strain hardening is not affected by the strain rate. Here, σ_{eq_d} and σ_{eq_s} are the dynamic and static equivalent stresses, respectively. The coefficients C and q are determined in the same way as in Cowper–Symonds'. The parameters σ and $\dot{\varepsilon}$ can be freely chosen based on the available experimental data or application. Upon this choice, the coefficients $\bar{m}$ and $\bar{n}$ will assume specific values. This equation is just a

See M. Alves, Material constitutive law for large strains and strain rates, Journal of Engineering Mechanics, 216–218, 2000.

slight modification of the Cowper–Symonds equation, yet it avoids the problem of dealing with strain dependent coefficients. The next figure shows the predictions of these equations together with experimental data for a steel alloy.

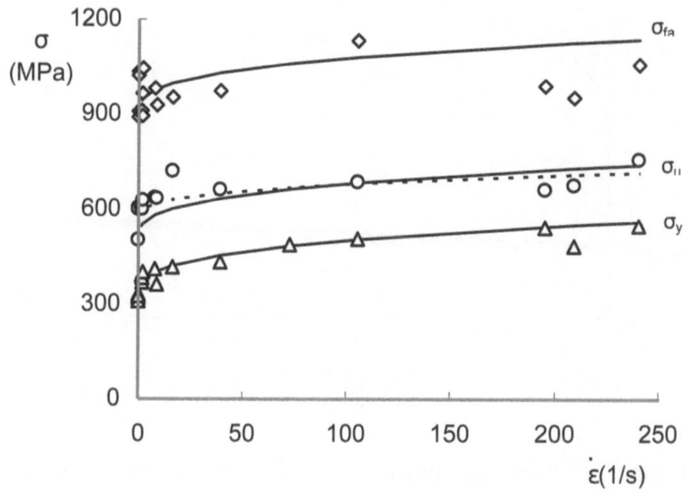

Figure 7.22 Prediction of the constitutive law $\sigma_{eq_d} = \sigma_{eq_s} + \bar{m}\dot{\varepsilon}^{\bar{n}}$ for steel alloy BS EN 10025 FE430A together with experimental results for $\triangle$: lower yield, σ_y, $\circ$: ultimate, σ_u, and $\diamond$: failure, σ_{fa}, stresses *versus* strain rate.

Another common constitutive equation, referred to as Norton equation, simply reads

$$\sigma_d = \sigma_0 \left(\frac{\dot{\varepsilon}}{\dot{\varepsilon}_0} \right)^q ,$$

where q is a constant assigned to minimize the error between experimental and predicted data. $\dot{\varepsilon}_0$ is a reference strain rate, usually very small, associated with a quasi–static tensile or compressive flow stress value, σ_0. This equation has the great advantage of facilitating the development of some theoretical models, as we have seen in Chapter 5 and we shall see in Chapter 10. Despite being simple, this equation compares well with the predictions given by the Cowper–Symonds equation, as illustrated in the next figure for mild steel.

In impact engineering, a much used constitutive equation is the one due to Johnson and Cook and it reads

$$\sigma_d = (A + B\varepsilon^n) \left(1 + C \ln \frac{\dot{\varepsilon}}{\dot{\varepsilon}_0} \right) \left[1 - \left(\frac{T - T_r}{T_m - T_r} \right)^m \right],$$

with the constants A, B, C, n, m being determined by best fitting a series of stress–strain curve measured at different strain rates and temperatures. Here, T is the temperature and its subscripts refer to melting

$A + B\varepsilon^n$ is the quasi-static material behaviour at $\dot{\varepsilon}_0$.

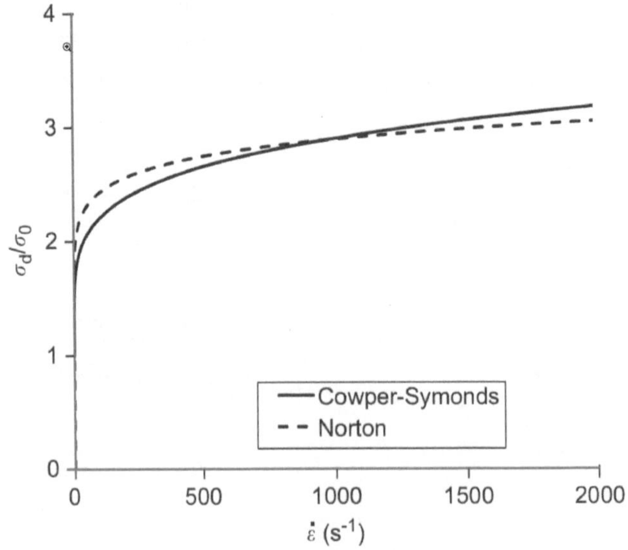

Figure 7.23 Predicted dynamic flow stress for a typical mild steel. Continuous line, Cowper–Symonds equation with $p = 5$ and $D = 40/s$. Dashed line, Norton equation with $\dot{\varepsilon}_0 = 0.001/s$ and $q = 0.077$. In both predictions $\sigma_0 = 235$ MPa.

and room (or reference) temperature. $\dot{\varepsilon}_0$ is a reference strain rate, usually the unity in the Johnson–Cook work but it can well be associated with the quasi–static tensile test, say 0.001/s.

Once the five constants of a material are known, we can enter any temperature, strain rate and plastic strain to obtain the stress level. It is possible to obtain a true equivalent stress if we input the equivalent plastic strain value and in this case, of course, the constants have to be obtained from true equivalent stress–strain pairs. It has been found that this equation performs relatively well for many materials but poor predictions are equally possible.

The next table lists the values of these constants for various materials as measured in the original work of Johnson and Cook.

Complex material behaviour requires an equally complex constitutive equation. This is the case in a polymer represented in the next figure by the stress–strain curve. The constitutive equation relating engineering stress and strain reads

See Mechanical models of cellular solids, M. Avalle, G. Belingardi and A. Ibba, International Journal of Impact Engineering, 2007, 34, p. 3–27.

$$\sigma = A\left[1 - exp^{(-E\varepsilon/A)(1-\varepsilon)^m}\right] + B\left(\frac{\varepsilon}{1-\varepsilon}\right)^n,$$

with A, B, E, m and n being fitting constants. As shown in the figure it can predict this polymer behaviour well.

TABLE 7.3 Johnson&Cook constitutive parameters for some metals.

material	ρ kg.m^{-3}	E GPa	c m/s	T_{melt} °K	A	B	n	C	m
copper	8960	124	3720	1356	90	292	0.31	0.025	1.09
brass	8520	100	3425	1189	112	505	0.42	0.009	1.68
nickel	8900	170	4370	1726	163	648	0.33	0.006	1.44
1006 Steel	7890	200	5034	1811	350	275	0.36	0.022	1.00
2024 Al.	2770	68	4954	775	337	343	0.41	0.010	1.00
4340 steel	7830	200	5054	1793	792	510	0.26	0.014	1.03
tungsten	17000	400	4850	1723	1506	177	0.12	0.016	1.00

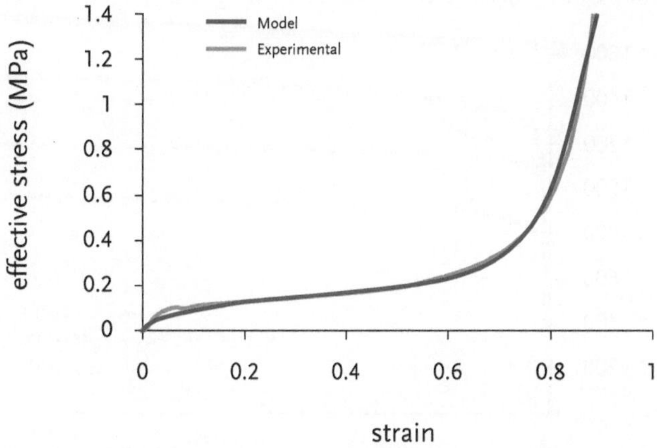

Figure 7.24 A polymer stress–strain behaviour and the corresponding model prediction.

In connection with the polymers behaviour, a well known constitutive equation is the Mooney–Rivlin, which can be used for stretches as large as 100%, as in rubber sheets. Because it is obtained from a strain energy function, we refer to it as a hyperelastic model. The relation between true stress and stretching reads

$$\sigma = \left(2C_1 + \frac{2C_2}{\lambda}\right)\left(\lambda^2 - \frac{1}{\lambda}\right),$$

with C_1 and C_2 being material constants that are determined from curve fitting to the experimental test. In terms of engineering stress, it follows that

$$s = \left(2C_1 + \frac{2C_2}{\lambda}\right)\left(\lambda - \frac{1}{\lambda^2}\right),$$

which is useful for obtaining the C_i parameters. We will have the chance to apply this equation in Chapter 11, in connection with the impact of tyres.

A material can also be pulled at different speeds during loading. This is exemplified in the next figure, where a titanium alloy is pulled at different strain rates. The various experimental curves are well fitted by the constitutive equation

See A. S. Khan, R. Kazmi and B. Farrokh, Multiaxial and non–proportional loading responses, anisotropy and modelling of Ti-6Al-4V titanium alloy over wide ranges of strain rates and temperatures, International Journal of Plasticity, 2007, 23, p. 931–950

$$\sigma = \left[A + B\left(1 - \frac{\ln \dot{\varepsilon}}{\ln D}\right)^a \varepsilon^b\right]\left(\frac{\dot{\varepsilon}}{\dot{\varepsilon}^*}\right)^c\left(\frac{T_m - T}{T_m - T_{ref}}\right)^d,$$

with A, B, D, a, b, c, d being material constants. Here, T is the current temperature and T_m is the material melting temperature. T_{ref} and $\dot{\varepsilon}^*$ are temperature and strain rate references.

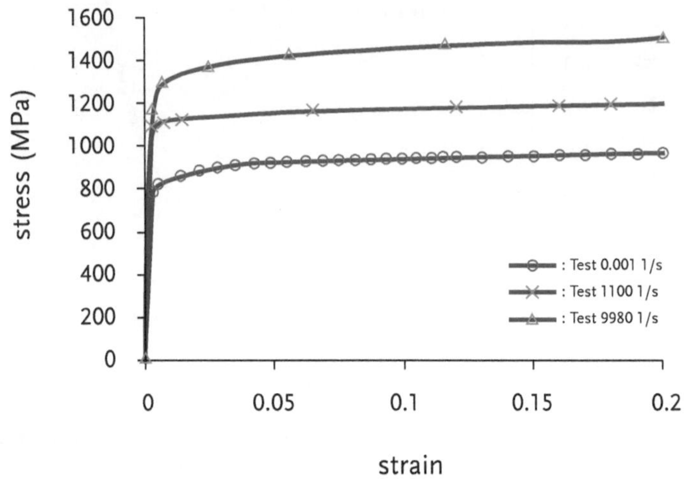

Figure 7.25 A titanium alloy mechanical behaviour at various strain rates.

Another constitutive equation, based on physics rather than empirical, is the Zerilli–Armstrong one, described by

$$\sigma = C_0 + C_1 e^{-C_3 T + C_4 T \ln \dot{\varepsilon}} + C_5 \varepsilon^n.$$

This equation has been used to predict a variety of material behaviour at a wide range of strains, strain rates and temperatures.

7.7 Inverse modelling and image analysis

Another way to obtain the true equivalent stress–strain curve of a material is by trial and error. This can be efficiently done using an optimization algorithm, as available in commercial finite element packages.

The idea is to experimentally load a specimen and record force and displacement. The specimen under the same load and boundary condition configuration is then analysed by the finite element method, a procedure which requires the material stress–strain curve. A set of stress–strain pairs is given, or else some constitutive parameters, and the analysis is run, giving as output a numerical load–displacement curve. This curve is then compared to the experimental one and the discrepancies are minimized by changing the stress–strain curve or the constitutive parameters, running the analysis again. This optimization procedure is automatically repeated until some acceptable discrepancy between the input and output material stress–strain curves is achieved.

As an application of this technique, consider the material stress–strain curve in the next figure. It was obtained using various techniques, including this optimization procedure. Note that the data come from a steel sheet, where certainly necking was developed. Even so there was no need to apply some sort of correction to obtain the equivalent stress, as we did before for cylindrical specimens using the Bridgman correction.

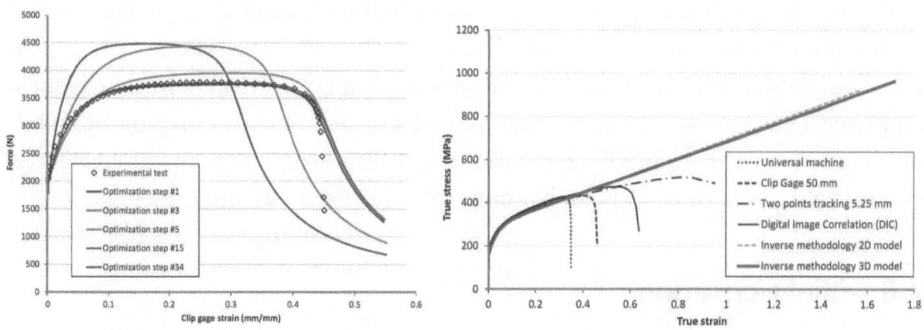

Figure 7.26 Various attempts in the optimization procedure and the equivalent stress–strain curve so obtained and compared with other techniques. The material is a thin sheet of steel.

An example of the necessity of using this optimization technique is when it becomes difficult to manufacture a sample of the material to be tested. Take for instance the small polymeric support in the next figure. It is difficult to machine a sample out of such a small piece, being easier just to squash it, measuring load and displacement. A FE model of this test procedure requires the mechanical properties. After 5 interactions, the Ramberd–Osgood coefficients in

$$\sigma = \sigma_0 + K\varepsilon^n,$$

of $K = 26.27$ MPa, $n = 0.96$ and $\sigma_0 = 29.59$ MPa were obtained.

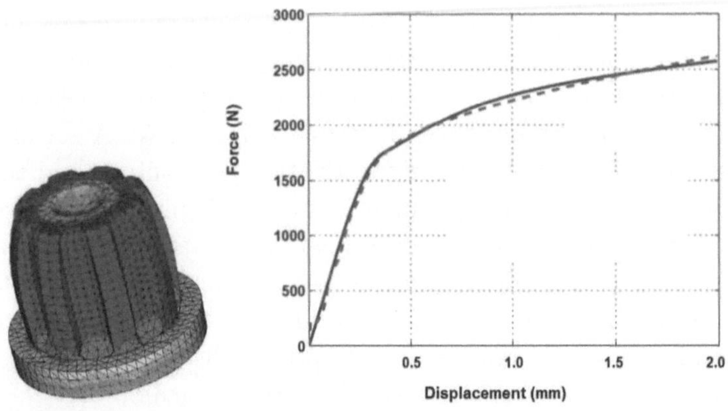

Figure 7.27 A polymeric part of height some 10 mm and the experimental (continuous) and numerical optimized (dashed) load–displacement curve from a compression test.

As for the image analysis, there has been a great development in such techniques. The advantages of measuring a strain field via image are various. This is a non contact technique so that the measurement procedure does not interfere with the phenomenon. Extreme temperatures of the sample can be well handled by the contact-less procedure and the accuracy of the technique is ever increasing due to the use of high image resolution and sophisticated software. Nowadays, it is possible to measure elastic strains in metals with a somewhat simple camera, meaning that a resolution of some 50—100 microstrain can be obtained, depending on the field of view. An example of the measurement of the strain field on a metal sheet is shown in the next figure.

7.8 Yield criterion

We have previously presented expressions for the equivalent stress and strain and we now investigate them further. The problem we need to address is what stress state a material, component or structure should be submitted to in order to trigger the basic micromechanical aspects of plastic flow. Macromechanically, plastic flow is said to occur when some permanent deformation is left in a material initially stress free after all the load is removed. In a uniaxial tensile test, the stress state associated with initial plastic flow is, by definition, the stress σ_y.

Suppose now that we apply a uniaxial stress of magnitude $\sigma < \sigma_y$ in a rod together with a torsion load that causes a shear stress τ. Depending on the magnitude of τ, it is possible to trigger the plastic flow

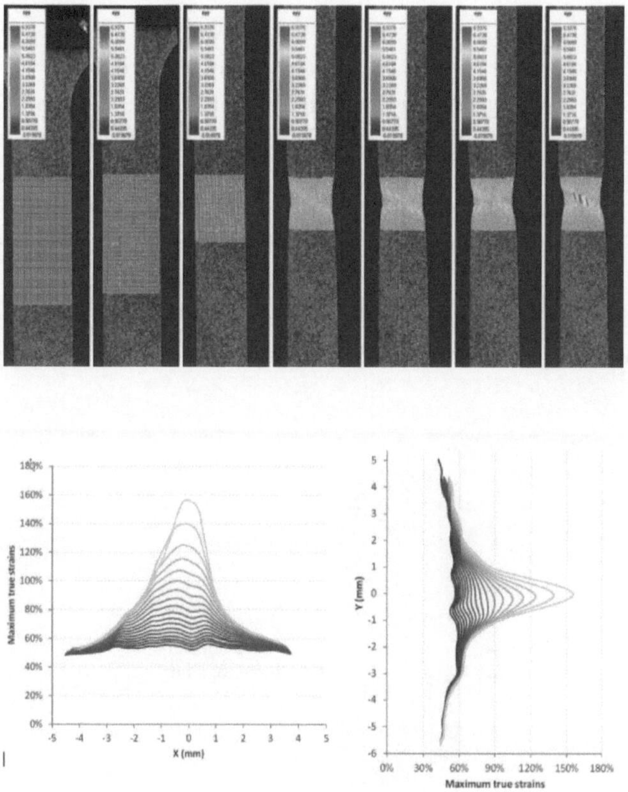

Figure 7.28 In the top, a dog–bone shape of a tensile specimen, aleatory black and white painted for the use of the DIC technique. It is clearly seen the localization of strain, also highlighted in the bottom graphs. From M. Gonzales and M. Alves, An inverse methodology for tuning material parameters in numerical modelling of mechanical structures, in MecSol2013, International Symposium on Solid Mechanics, Porto Alegre, Brazil.

mechanisms so that, when the torque and axial load are removed we are left with some permanent deformation. This is to say that there is a combination of stress states that lead the material to plastic flow, as already commented on Chapter 5. We need to establish the magnitude of these stress components that cause such a phenomenon.

To simplify matters, we postulate that the material is isotropic and has therefore the same behaviour in compression and in tension. Another postulate is that yielding does not depend on the hydrostatic press. This has been shown experimentally to be much closer to reality for metals.

To understand what the hydrostatic stress is, in the next figure we see a specimen that has a lateral notch and it is pulled. We can imagine that the bunch of hypothetical threads shown in the figure try to align themselves along the load direction. This is avoided though by internal forces acting in the material. These forces vary along the cross-section. It could be imagined that the material (threads) is held together by some external pressure, as if the specimen were surrounded by a liquid exerting pressure in all the body, hence the name hydrostatic stress, also referred to as mean stress. Of course that high pressures are generated when the notch radii becomes smaller. In passing, this loading mode causes material degradation, leading to failure. This is to say that the hydrostatic stress has a leading role in material failure criteria. However, for metals, various experimental evidence indicates that the hydrostatic stress does not affect yielding. For soil, on the other hand, it is easy to show that high surrounded pressure, *ie* negative hydrostatic stress, which can be obtained by squeezing outwardly a sand block, delays yielding.

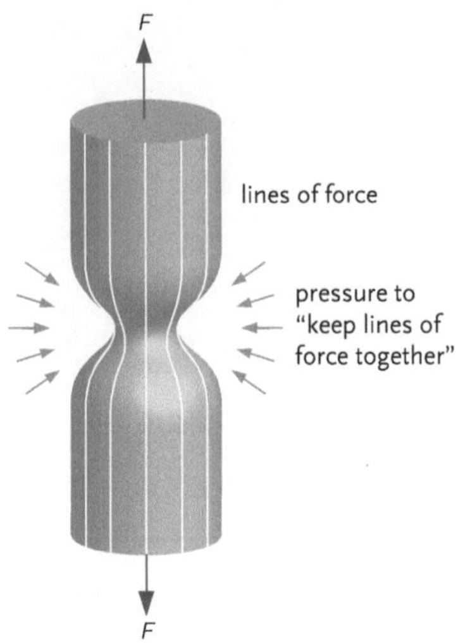

Figure 7.29 A simple representation of the hydrostatic stress.

We need to elaborate further the concept of hydrostatic stress and how to express it via the so called invariants of the stress tensor.

See I.H. Shames and F.A. Cozzarelli, Elastic and Inelastic Stress Analysis, Prentice Hall, 1992.

First of all, let us consider the generalized body in the next figure and pass an imaginary internal surface on it. We can imagine that the so formed two parts are held together by a traction vector $\mathbf{T}$ with components T_1, T_2 and T_3, each function of the Cartesian coordinates and time. $\mathbf{T}$ is a force intensity per unit area such that $\mathrm{d}\mathbf{F} = \mathbf{T}\mathrm{d}A$.

The normal to the area element we denote by the unit vector $\mathbf{n}$, which can have any orientation, even different from the direction of the traction vector. $\mathbf{n}$ has components $l = A_x/A$, $m = A_y/A$ and $n = A_z/A$. In the limit $A \to 0$, we can write

$$\mathbf{T} = \lim_{A\to 0} \frac{\mathbf{F}}{A}$$

and since $\mathbf{F}$ has the components F_x, F_y, F_z, it follows the stresses

$$\sigma_{xx} = T_x = \lim_{A_x \to 0} \frac{F_x}{A_x}; \sigma_{xy} = T_y = \lim_{A_x \to 0} \frac{F_y}{A_x}; \sigma_{xz} = T_z = \lim_{A_x \to 0} \frac{F_z}{A_x};$$

with A_x being the projection of A in the x direction. We can reason in the same way for the other area projections directions and generate another set of stress components, which can be arranged in the matrix representation

$$\boldsymbol{\sigma} = \begin{bmatrix} \sigma_{xx} & \sigma_{xy} & \sigma_{xz} \\ \sigma_{yx} & \sigma_{yy} & \sigma_{yz} \\ \sigma_{zx} & \sigma_{zy} & \sigma_{zz} \end{bmatrix}.$$

We can now apply Newton's law which, in the absence of body forces and having in mind the next figure, gives

$$T_1 A = \sigma_{xx} A_x + \sigma_{xy} A_y + \sigma_{xz} A_z,$$

or

$$T_1 = \sigma_{xx} l + \sigma_{xy} m + \sigma_{xz} n,$$

to be replicated for the other directions and leading to the so called Cauchy formula, the tensor $\boldsymbol{\sigma}$ being the Cauchy stress,

$$T_i = \sigma_{ij} n_j.$$

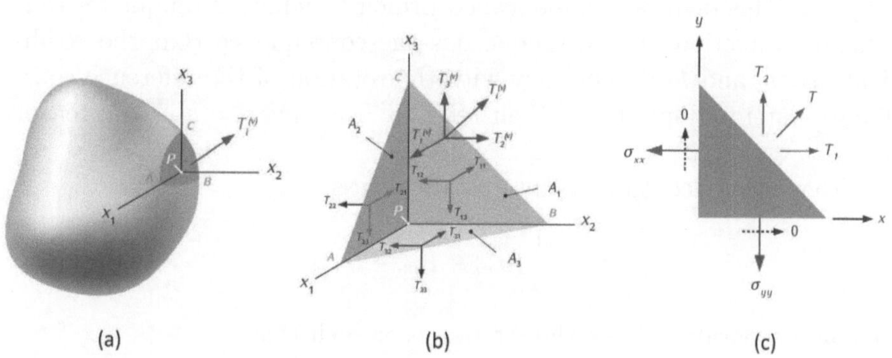

Figure 7.30 A boundary and the force–stress components.

It is easy to understand, having in mind the plane projection (c) in the above figure, that there should exist a direction of the plane such that the traction force is counter balanced only by normal stresses. We denote these stresses, σ, as principal and from the above equation it follows that

$$\sigma n_i = \sigma_{ij} n_j,$$

which, using the Kronecker notation, δ_{ij}, gives

$$(\sigma_{ij} - \boldsymbol{\sigma}\delta_{ij})n_j = 0.$$

A non trivial solution for this equation requires that the determinant equals to zero,

$$|\sigma_{ij} - \boldsymbol{\sigma}\delta_{ij}| = \boldsymbol{\sigma} = \begin{vmatrix} \sigma_{xx} - \boldsymbol{\sigma} & \sigma_{xy} & \sigma_{xz} \\ \sigma_{xy} & \sigma_{yy} - \boldsymbol{\sigma} & \sigma_{yz} \\ \sigma_{xz} & \sigma_{yz} & \sigma_{zz} - \boldsymbol{\sigma} \end{vmatrix} = 0,$$

leading to the cubic equation

$$\boldsymbol{\sigma}^3 - I_1\boldsymbol{\sigma}^2 + I_2\boldsymbol{\sigma} - I_3 = 0,$$

with

$$I_1 = \sigma_{xx} + \sigma_{yy} + \sigma_{zz},$$
$$I_2 = \sigma_{xx}\sigma_{yy} + \sigma_{xx}\sigma_{zz} + \sigma_{yy}\sigma_{zz} - \sigma_{xy}^2 - \sigma_{xz}^2 - \sigma_{yz}^2,$$
$$I_3 = \sigma_{xx}\sigma_{yy}\sigma_{zz} - \sigma_{xx}\sigma_{yz}^2 - \sigma_{yy}\sigma_{xz}^2 - \sigma_{zz}\sigma_{xy}^2 + 2\sigma_{xy}\sigma_{xz}\sigma_{yz}.$$

By solving this cubic equation we obtain the three values of the principal stresses. These stresses have a direction which can be obtained by solving the equation $(\sigma_{ij} - \boldsymbol{\sigma}\delta_{ij})n_j = 0$ together with the relation $n_i n_i = 1$. This defines a plane, called principal, which is unique for that principal direction. To be unique has the consequence that the coefficients I_1, I_2 and I_3 do not vary with the rotation of the reference coordinate system adopted. We then refer to I_i as invariants of the tensor $\boldsymbol{\sigma}$.

From the concept of the hydrostatic stress,

$$\sigma_h = \frac{1}{3}\left(\sigma_{xx} + \sigma_{yy} + \sigma_{zz}\right),$$

we could separate it from the stress tensor such that

$$\sigma_{ij} = s_{ij} + \frac{1}{3}\sigma_{kk}\delta_{ij}.$$

Note that the (negative) hydrostatic stress only compresses the material; it does not distort it. Hence, the distortion can only be caused by what is left from σ when removing σ_{kk}. Accordingly, we call $\mathbf{s}$ the deviatoric tensor. This tensor also has its invariants, here called J_1, J_2 and J_3. When calculating the invariants of $\mathbf{s}$ as we have done for I_i, we arrive at

$$J_1 = \text{tr}\mathbf{s} = s_{ii} = 0,$$

$$-J_2 = -\frac{1}{2}\text{tr}\mathbf{s}^2 = -\frac{1}{2}s_{ijj}s_{jij},$$

$$J_3 = \frac{1}{3}\text{tr}\mathbf{s}^3 = \frac{1}{3}s_{ij}s_{jk}s_{ki}.$$

Returning now to plasticity, the hydrostatic stress acts to squeeze the material, which hardly can be successful in changing the shape of a metal. It is rather easier to distort it. Hence, we can postulate that energy coming from the deviator tensor is the only responsible for yielding. So, our yield criterion is based on J_2 and it is referred to as the J_2 theory of plasticity.

Another implicit postulate in the development of yield criteria is the existence of a yield surface, a concept that we have already explored in Chapter 5. Here, the yield function

$$f(\sigma_{xx}, \sigma_{yy}, \sigma_{zz}, \sigma_{xy}, \sigma_{yz}, \sigma_{xz})$$

can indicate whether the material entered or not in the plastic regime, according to

$$\text{Elastic material behaviour} \rightarrow f < 0 \quad \text{or} \quad f = 0 \quad \text{and} \quad \frac{\mathrm{d}f}{\mathrm{d}t} < 0,$$

$$\text{Plastic material behaviour} \rightarrow f = 0 \quad \text{and} \quad \frac{\mathrm{d}f}{\mathrm{d}t} \geq 0.$$

The stress state in a point can have up to six components so it is difficult to graphically represent the yield function. This is not the case if we work on the principal stress space, which has at most three components of principal stresses, $\sigma_1, \sigma_2, \sigma_3$. It follows that f would be a function of $\sigma_1, \sigma_2, \sigma_3$. Recall now that f does not depend on the hydrostatic stress and as we move on the hydrostatic line in the next figure, the yield surface is the same.

One of the postulates in developing the yield surface is that the hydrostatic stress does not influence yielding. We have also seen that the hydrostatic stress can only expand or shrink the material, *ie* it does not cause any angular distortion. This can be further confirmed by the

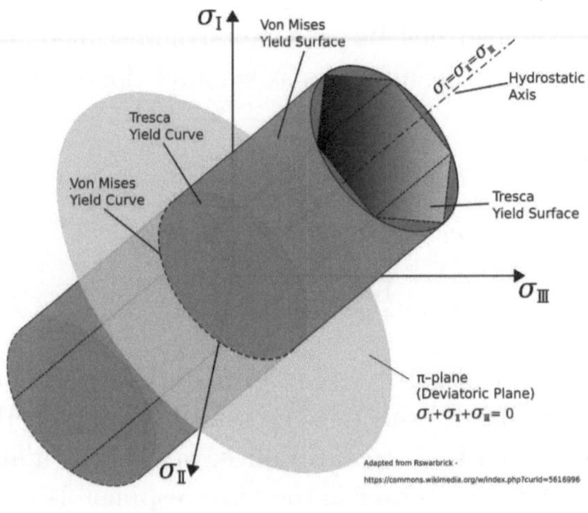

Figure 7.31 Yield surface and the deviatoric plane.

fact that the hydrostatic stress is not a function of any shear stress component. This fact is presented in the von Mises yield criterion, that indicates that plastic flow is caused only by the distortion strain energy, E_d, given by the difference between the total elastic strain energy density of deformation, E_t, and the energy associated with a pure hydrostatic stress state, E_h, *ie*

$$\begin{aligned} E_d &= E_t - E_h \\ &= \tfrac{1}{2E}\left[(\sigma_1^2 + \sigma_2^2 + \sigma_3^2) - 2\nu(\sigma_1\sigma_2 + \sigma_1\sigma_3 + \sigma_2\sigma_3\right] - \\ &\quad \tfrac{1-2\nu}{6E}(\sigma_1 + \sigma_2 + \sigma_3)^2 \\ &= \tfrac{1+\nu}{6E}\left[(\sigma_1 - \sigma_2)^2 + (\sigma_1 - \sigma_3)^2 + (\sigma_2 - \sigma_3)^2\right]. \end{aligned}$$

If we equate this energy with the one associated with an uniaxial stress state at yielding, *ie* $\sigma_1 = \sigma_y$, $\sigma_2 = \sigma_3 = 0$, it follows that

$$\frac{1+\nu}{6E}\left[(\sigma_1 - \sigma_2)^2 + (\sigma_1 - \sigma_3)^2 + (\sigma_2 - \sigma_3)^2\right] = \frac{1+\nu}{3E}\sigma_y^2,$$

or

$$\sigma_y = \sigma_{eq} = \sqrt{\frac{(\sigma_1 - \sigma_2)^2 + (\sigma_1 - \sigma_3)^2 + (\sigma_2 - \sigma_3)^2}{2}},$$

which is identical to the von Mises equivalent stress obtained before, but now written in the principal stress space.

An important observation from the above equations is that the distortion strain energy can also be expressed as

$$E_d = \frac{1+\nu}{E}J_2,$$

with J_2 being the second invariant of the stress tensor. Hence, at yielding, we can state that

$$\sigma_{eq} = \sqrt{3J_2}.$$

Note that the equivalent stress can also be written in terms of the deviatoric tensor as

$$\sigma_{eq} = \sqrt{\frac{3}{2}\dot{\mathbf{s}} : \dot{\mathbf{s}}},$$

with the contracted tensor product being defined as $\dot{\mathbf{s}} : \dot{\mathbf{s}} = s_x s_y + s_y s_y + \ldots = \sum_{i=1}^{n} \sum_{j=1}^{n} s_{ij} s_{ij}$.

7.8.1 Normality and consistency condition

To introduce the subject, let us take a bar and load it from its virgin state to a permanent deformation, point A in the next figure. Now increase both torsion and tension leading the bar to a strain state $(\varepsilon_{N_{A'}}, \varepsilon T_{A'})$. Apply a new loading pair (orange arrows) such that the new strain state is $\varepsilon(N_{B'}, \varepsilon T_{B'})$ and from there unload the bar (point O). Now, perform the loading path defined by the red arrows, opposite to the first one. After full unloading, we are left with a strain state different from the first loading cycle, even when at point O.

This only indicates that the way we load the specimen, *ie* the path load, leads to different final configuration, which of course needs to be taken into account. Had all this loading process occurred in the elastic limit of the material, the final result would be path loading independent. Our conclusion is that, in plasticity, we need to deal with increments of strain; the current state of stress and strain at a point in the material depends on the loading and on its history.

We postulate that the so called associated flow is such that the increment in the plastic strain tensor occurs in the direction normal to the yield surface, as represented in the principal stress space. This is indicated in the figure and allows us to write in a rate form,

$$\dot{\varepsilon}^p = \gamma \frac{\partial f}{\partial \sigma}.$$

Note here that the magnitude of the plastic strain rate is given by so called plastic multiplier γ, with the direction given by $\partial f / \partial \boldsymbol{\sigma}$.

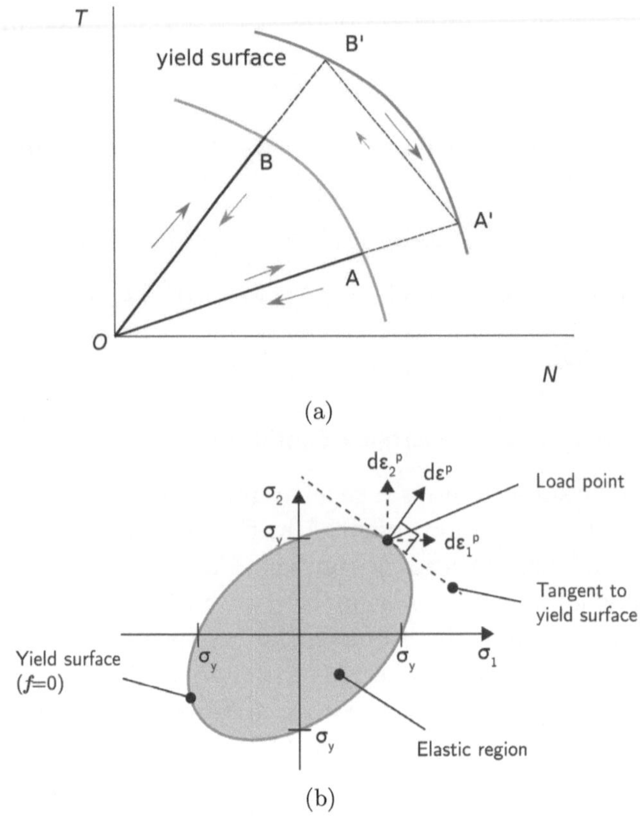

Figure 7.32 (a) A yield surface and the loading path of a bar. (b) a von Mises yield surface and the strain increment components.

Calculation of the plastic multiplier

This problem was adapted from F. Dunne and N. Petrinic, Introduction to Computational Plasticity, Oxford Press, 2005.

Prove that $\dot{\lambda} = \dot{\varepsilon}_{eq}$ for a von Mises material and a plane stress state. Our plane stress yield surface is given by

$$f = \sqrt{\frac{(\sigma_1 - \sigma_2)^2 + (\sigma_1)^2 + (\sigma_2)^2}{2}} - \sigma_y = 0,$$

with the derivatives

$$\frac{\partial f}{\partial \sigma_1} = \frac{2\sigma_1 - \sigma_2}{2\sqrt{\sigma_1^2 + \sigma_2^2 - \sigma_1\sigma_2}} = \frac{2\sigma_1 - \sigma_2}{2\sigma_{eq}} = \frac{3s_1}{2\sigma_{eq}}$$

and

$$\frac{\partial f}{\partial \sigma_2} = \frac{2\sigma_2 - \sigma_1}{2\sqrt{\sigma^2 + \sigma_2^2 - \sigma_1\sigma_2}} = \frac{2\sigma_2 - \sigma_1}{2\sigma_{eq}} = \frac{3s_2}{2\sigma_{eq}}.$$

Now,

$$\dot{\varepsilon}^p = \gamma \frac{\partial f}{\partial \boldsymbol{\sigma}} = \gamma \frac{3\boldsymbol{s}}{2\sigma_{eq}}$$

and with the previously obtained equivalent strain rate for small elastic strains

$$\dot{\varepsilon}_{eq} = \sqrt{\frac{2}{3}\dot{\boldsymbol{\varepsilon}} : \dot{\boldsymbol{\varepsilon}}},$$

we obtain

$$\dot{\varepsilon}_{eq} = \gamma \frac{\sqrt{\frac{3}{2}\boldsymbol{s} : \boldsymbol{s}}}{\sigma_{eq}} = \gamma$$

when using the equivalent stress definition.

We finally have that

$$\dot{\boldsymbol{\varepsilon}}^p = \frac{3\boldsymbol{s}}{2\sigma_{eq}}\dot{\varepsilon}_{eq}.$$

■

7.8.2 Isotropic strain hardening

We refer in general to hardening as the further strength enhancement that a material exhibits upon loading. This hardening can be due to straining of the material, strain hardening, or due to the speed one loads it, strain rate hardening. Both concepts have been seen before but we pay attention now at the strain hardening in connection with the yield surface.

As indicated in various equivalent stress–strain curves, strain hardening is represented by the ever increasing material strength, in contradistinction to the strain softening. For isotropic materials, the increase in strength in the principal direction 1 in the next figure due to a straining $\Delta\varepsilon_e$ means that the flow stress increases from σ_{y_1} to $\sigma_{y_1'}$. Conversely, in the principal direction 2, the flow stress increases from σ_{y_2} to $\sigma_{y_2'}$ for the same $\Delta\varepsilon_e$. This generalises to an in concert enlargement of the yield surface, here in the plane state case but also valid for the whole Mises cylinder.

For a general case, we have that the yield surface is written as

$$f(\sigma, \varepsilon_e) = \sigma_e - \sigma_y(\varepsilon_e),$$

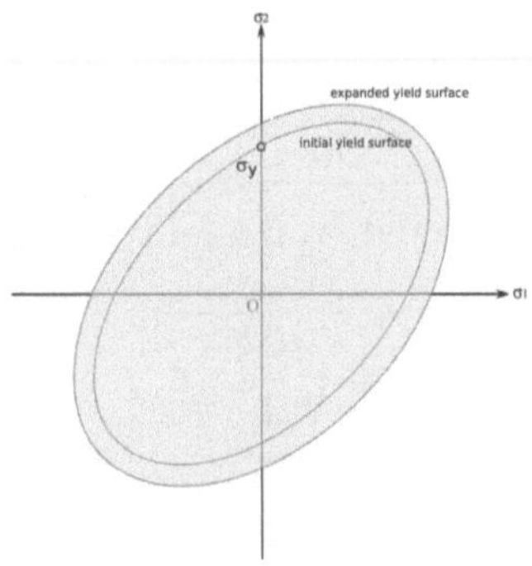

Figure 7.33 Strain hardening: expansion of the yield surface.

with the flow stress being given by the constitutive law. Clearly, for each hardening rule and loading mode, we have a specific plastic multiplier. We shall see in Chapter 9 the case of linear strain hardening in the context of the computational plasticity.

7.9 Material failure

Parts of this section are from Benchmark study of failure criteria for ship collision modeling using purpose-designed tensile specimen geometries, M.A.G. Calle, P. Verleysen and M. Alves, Marine Structures, 53, p 68–85, 2017 and from L. Driemeier, R.T. Moura, I.F. Machado and M. Alves, A bifailure specimen for accessing failure criteria performance, International Journal of Plasticity 71, p 62–86, 2015.

Material failure is a very complex phenomenon and to date there is no failure criteria valid for a wide range of parameters like triaxiality, temperature and strain rate. That is why the subject is constantly being researched and new criteria candidates are always being proposed.

We define failure as a local material separation visible at low magnification or even to the naked eye, as shown in the next figure. Note that plastic strains in a shaft would probably induce the machine to malfunction and one would claim a failure. This is indeed the case but here we are only concerned with failure in a crack sense.

We note first that it has been established in various experimental works that material failure is strongly influenced by triaxiality. As depicted in Figure 7.35, a wide range of triaxialities occur in various geometrical configurations and this is associated with the formation of shear bands, for low triaxiality values, and voids, for larger ones. But only knowing the triaxiality is not sufficient to define the failure locus.

Here we define triaxiality as

$$\eta = \frac{\sigma_h}{\sigma_{eq}}.$$

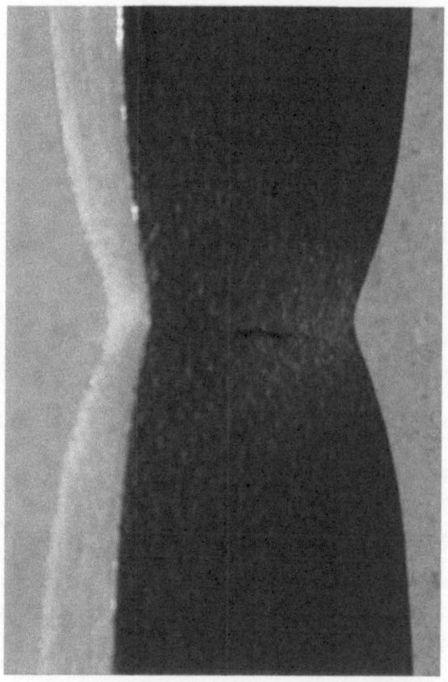

Figure 7.34 Failure in the necking zone of a rectangular cross-section tensile specimen.

Limiting ourselves here to some basic failure criteria for metals, let us start with the most common one, available in all finite elements codes, which is the accumulated plastic strain criterion. It is expressed by

$$\varepsilon_{fa} = \int_0^t \dot{\varepsilon}^p \mathrm{d}t.$$

The calculation of the accumulated plastic strain can easily be output from a FE code but the problem here is to establish the value of ε_{fa}. One simple idea is to measure the engineering strain that takes place on a tensile test at failure. We already know that it is not representative of a local value but for some rather complex loading cases this criterion ends up yielding reasonable results, despite its clear simplicity.

More sophisticated criteria are built upon the design and tests of different specimen geometries, as the one shown in Figure 7.36. They were pulled in traction until failure, resulting in a wide range of negative and positive triaxialities. Note, incidentally, the good experimental, left on each subfigure and as obtained by DIC technique, and numerical, right in each figure, match.

Now, when one evaluates, for instance, the plastic energy to failure, *ie* the area under the force–displacement (up to failure) curve, one notes in Figure 7.37 that, with some exceptions, the strain to failure criterion performs reasonably well.

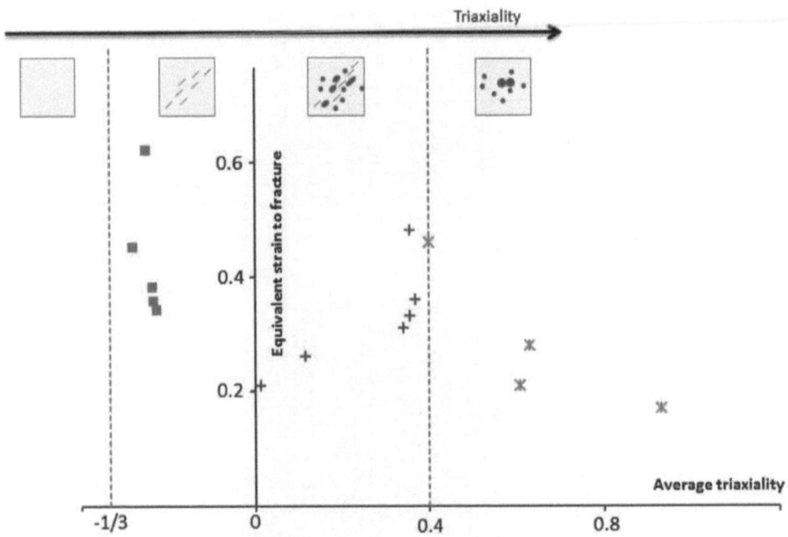

Figure 7.35 Influence of the triaxiality on the strain to failure. Experimental data from Y. Bao and T. Wierzbicki, On the fracture locus in the equivalent strain and stress triaxiality space. Int. J. Mech. Sci. 46, p 81–98, 2004.

Note that the failure strain criterion in Figure 7.37 was modified to

$$\varepsilon_{fa} = \varepsilon_u + \frac{t}{L_e}\varepsilon_n.$$

At failure, the uniform strain along the element length is ε_u. ε_n is the localized strain due to necking, t is the element thickness and L_e is the equivalent element length calculated as the square root of the element area. The figure depicts three other criteria, GL (Germanischer Lloyd), RTCL and BWH.

RTCL is based on Crockroft-Latham criterion, more adapted to low triaxialities, and on the Rice–Tracey criterion based on void growth, which is more appropriate for high triaxialities, here denoted as η. The RTCL criterion is then given by

$$f(\eta)_{RTCL} = \begin{cases} 0 & \eta \le -\frac{1}{3} \\ \frac{2+2\eta\sqrt{12-27\eta^2}}{3\eta+\sqrt{12-27\eta^2}} & -\frac{1}{3} < \eta < \frac{1}{3} \\ e^{\left(-\frac{1}{2}\right)}e^{\left(\frac{3}{2}\eta\right)} & \eta \ge \frac{1}{3} \end{cases}$$

From this function one obtains the damage parameter

$$D = \int_0^{\bar{\varepsilon}_f} f(\eta)_{RTCL} d\bar{\varepsilon},$$

which is then compared to a critical calibrated one.

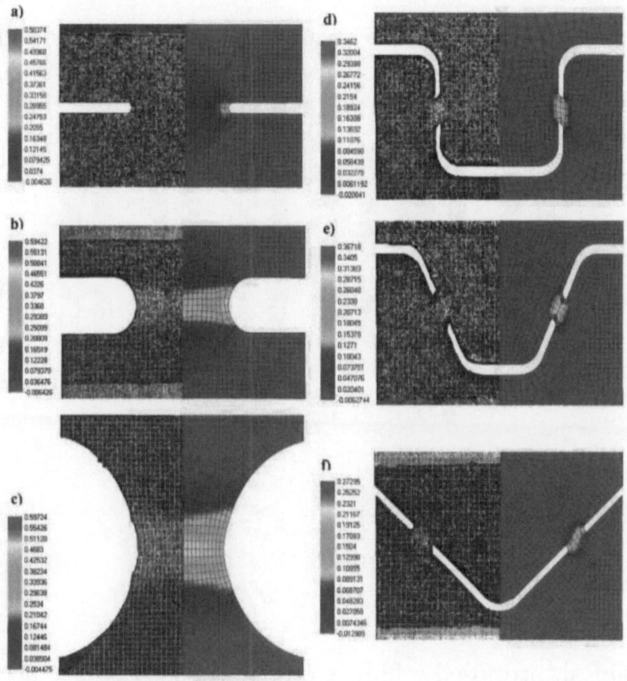

Figure 7.36 A wide range of triaxialities is obtained with these geometries. a) notch A, b) notch B, c) notch C, d) shear A, e) shear B and f) shear C samples at maximum load. Left in each figure is the strain field from DIC, to be compared to the one at right from FE analysis.

Interestingly that the BWH criterion, given by

$$\sigma_{1,f} = \begin{cases} \frac{2K}{\sqrt{3}} \left(\frac{1+\frac{\beta}{2}}{\sqrt{\beta^2+\beta+1}} \right) \left(\frac{\varepsilon_{1c}\left(\frac{t}{k}+1\right)}{\sqrt{3}(1+\beta)} \sqrt{\beta^2+\beta+1} \right)^n & \beta \leq 0 \\ \frac{2K}{\sqrt{3}} \frac{\left(\frac{1}{\sqrt{3}} \varepsilon_{1c} \left(\frac{t}{L_e}+1 \right) \right)^n}{\sqrt{1-\left(\frac{\beta}{2+\beta}\right)^2}} & \beta > 0 \end{cases}$$

is based on a stress, rather than strain, critical value. The β parameter is defined as the relation between the minor and major strain rates in the plane principal directions which can be approximated to the analogous relation between the principal strains as $\beta = \frac{\dot{\varepsilon}_2}{\dot{\varepsilon}_1}$. The BWH criterion is also dependent on the equivalent element length, L_e, with K and n being the strength coefficient and strain hardening exponent of the power hardening law $\sigma = K\varepsilon^n$, with ε_{1c} being a material constant.

As already anticipated, another well know criterion is that of Cockcroft and Latham, who suggest that fracture takes place when the plastic work associated with the maximum principal stress σ_1 reaches a

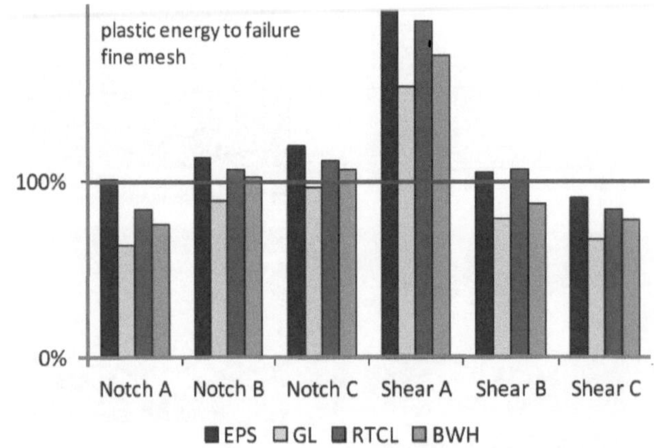

Figure 7.37 Plastic energy to failure of the tests in the specimens of the previous figure.

threshold value w_{fa}, such that

$$w_{fa} = \int_0^{\bar{\varepsilon}_f^p} \sigma_1 \mathrm{d}\bar{\varepsilon}^p \leq w_f^p \text{ for } \sigma_1 > 0.$$

Wilkins et al. proposed a failure criterion based on the accumulation of plastic deformation weighted by w_1 and w_2, which accounts for the hydrostatic and deviatoric load components, respectively. Failure occurs when the cumulative damage variable, D, reaches a critical value, D_c. It follows then that

$$D = \int_0^{\varepsilon_{fa}} \omega_1 \omega_2 \mathrm{d}\varepsilon_{eq} \leq D_c$$

with

$$\omega_1 = \frac{1}{(1 - a\eta)^\alpha} \quad \text{and} \quad \omega_2 = (2 - A)^\beta,$$

where

$$A = \max\left\{\frac{s_2}{s_1}, \frac{s_2}{s_3}\right\}.$$

$s_1 > s_2 > s_3$ are the principal deviatoric stresses and the material constants a, α and β are obtained from experimental tests that ideally should cover a wide range of triaxialities.

Note that, in the Wilkins model, damage is a function of the triaxiality. This is shown on the next figure that indicates that the model captures the fact that failure is, in a sense, a diverse phenomenon of plasticity alone. Clearly, for high (low) pressures, the limit of fracture tends to infinity (zero). The figure also shows the influence of β in the shape of the deviatoric part of the Wilkins mode. For $\beta = 0$ the circular profile in deviatoric plane is similar to the von Mises criterion.

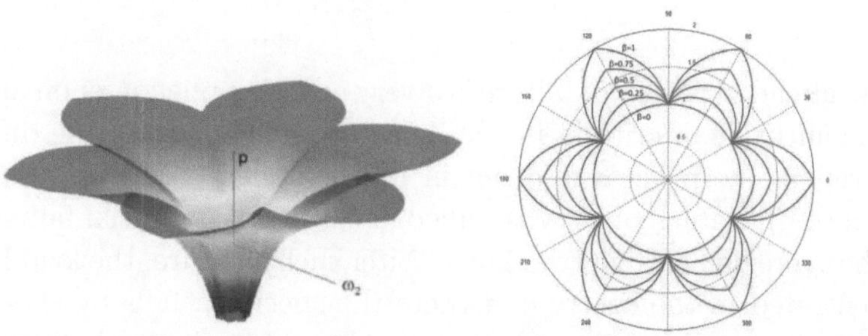

Figure 7.38 The failure envelope as predicted by the Wilkins model.

Johnson and Cook performed comprehensive tests using various materials, at different temperatures, strain rates and triaxialities and from them emerged that failure occurs when the accumulated plastic strain, $\bar{\varepsilon}^p$, reaches $\bar{\varepsilon}^p_{fa}$, with

$$
\begin{aligned}
\bar{\varepsilon}^p_{fa} &= \omega_p(\eta)\omega_{\dot{\varepsilon}}\left(\dot{\varepsilon}^p\right)\omega_T(T),\\
\omega_p &= C_1 + C_2 \exp\left(C_3\eta\right),\\
\omega_{\dot{\varepsilon}} &= 1 + C_4 \log\frac{\dot{\varepsilon}^p}{\dot{\varepsilon}_0},\\
\omega_T &= 1 + C_5\frac{T - T_0}{T_m - T_0},
\end{aligned}
$$

with the parameters already defined when presenting the Johnson–Cook constitutive equation.

Finally, we present the Xue–Wierzbicki failure model, stated as

$$
\bar{\varepsilon}^p_{fa} = D_1 e^{-D_2\eta} - \left(D_1 e^{-D_2\eta} - D_3 e^{-D_4\eta}\right)\left(1 - \xi^{1/n}\right)^n.
$$

With the failure strain, the damage parameter is calculated according to

$$
D = \int_0^{\bar{\varepsilon}^p_{fa}} \frac{d\bar{\varepsilon}^p}{\bar{\varepsilon}^p_{fa}}
$$

and by reaching a critical value, failure occurs.

In this failure criterion, D_i are fitting constants, n is the hardening material parameter and ξ is the Lode parameter given by

$$\xi = \frac{3\sqrt{3}}{2}\frac{J_3}{J_2^{3/2}}.$$

As already indicated, failure is a very complex phenomenon and there are a multitude of criteria in the literature. A glimpse of the difficulties involved in the process is given in the next figure. There, a specimen with a complicate geometry is pulled. This causes an initial failure in the notches, a region of high triaxiality. After such a failure, the load becomes concentrated in the centre and there the specimen fails by shearing, at low triaxiality. This failure process is reflected in the load–displacement curve, which exhibits a strong jump.

Now, when simulating this failure process via Finite Elements, with some of the criteria given above, we obtain the curves in the figure. Clearly, it was not possible to truely represent the experimental curve, despite the complexity of the John–Cook (J-C) and Xue–Wierzbicki (X-W) criteria.

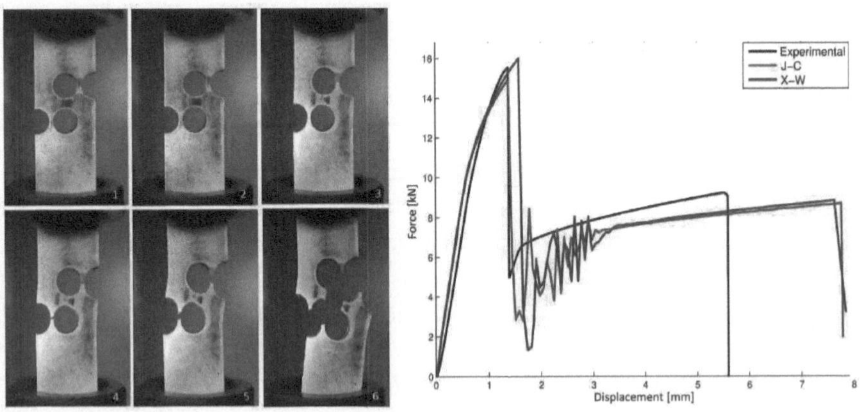

Figure 7.39 (a) A double failure specimen and a (b) comparison between experimental load–displacement curve and the J-C and X-W failure criteria.

7.10 Closure

We have seen in this chapter how crucial it is in any analysis, theoretical or numerical, to know the material behaviour. This knowledge is gained via mechanical tests, some of which are described here. From these test results, it is possible to develop and fit constitutive laws. These mainly take into account strain, strain rate and temperature.

The mechanical tests also give an indication of plastic flow and this is important in the sense that, once plasticity is in play, the system under analysis will permanently change its shape. Such a plasticity limit is established by the yield criteria and a glimpse on this subject was offered to the reader in this chapter.

While paving the way to finite elements, to be seen in the next two chapters, we presented some concepts on the so called normality and consistency conditions. From there we explored material failure, highlighting the importance and the difficulty of this topic.

7.11 Problems

1. In an EWM what is the length of the pressure pulse generated by a striker of length L? Discuss the influence of determining the rising of the pulse, ie when the pulse is first detected.
2. For an EWM, discuss the benefits of: a) having a long bar, b) to place the gauges in the middle of the bar, c) to use a long striker, d) to measure the strains in the bar–material interface.
3. Prove that $\sigma = s(1+e)$ and $\varepsilon = \ln(1+e)$ is valid in the presence of volume conservation.
4. Perform a literature review on the correction coefficient f_c for rectangular cross–section specimens.
5. Show that $\dot{\varepsilon}_s = \frac{\dot{e}_s}{1+e_s}$.
6. For the cylinder under tension show that $\varepsilon_{eq} = 2\ln\left(\frac{d_0}{d}\right)$. Is this an engineering strain measure?
7. A poor country government has decided to allow importation of only good crash rated cars. Your car company works against this determination since it is a major exporter of bad rated cars to this country. You are invited to an open panel where this issue will be discussed. What would you answer to the question: Why do you sell bad cars in our country while yours allow only good performance cars? How could you act to push forward better car design and manufacturing? Would you see this issue as a contributing or a detrimental factor to the quality of your work?

CHAPTER 8

Linear Finite Elements Analysis

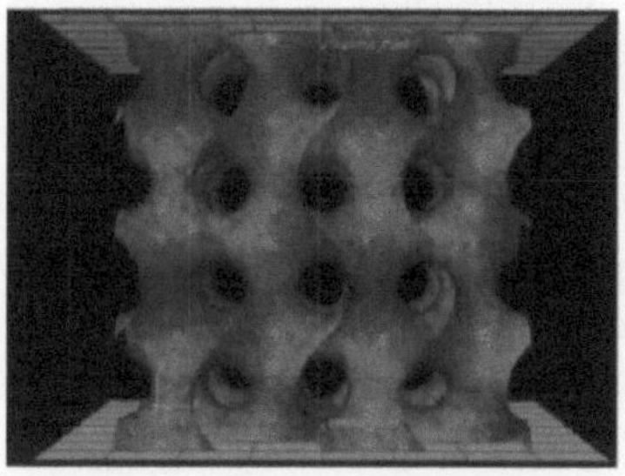

Figure 8.1 A gyroid represented by shell finite elements.

The majority of the results presented so far are based on models that are prone to be solved via closed form mathematical expressions. This has many advantages. The study of these models may disclose important features of the phenomenon. Also, a sensitivity analysis, *ie* how a given variable affects the overall response of the model, can be performed in a somewhat simple fashion. However, these models are in many ways limited in representing structures of complex shapes, and/or made of unusual material behaviour and/or subjected to loads of complex time profiles. Science and technology have overcome these limitations by using a dual approach, where theoretical fundamentals of solid mechanics were, so to say, translated to computer language. Chiefly among these

DOI: 10.1201/9781003485292-8

techniques is the Finite Element Method, which will now be presented in its various aspects.

We start by exploring the simple rod finite element method, focusing on its use in the propagation of waves in bars. We take the opportunity to present explicit and implicit methods of time integration. We advance further by studying finite elements in the context of beams, introducing the important concepts of boundary conditions application, coordinate transformation and finite element assembly. An axisymmetric finite element is also given and the topic is enhanced by presenting isoparametric formulation and numerical integration.

8.1 The finite element method in dynamics: explicit and implicit

In the Finite Element Method, FEM, one uses the so called shape functions to interpolate the variable of interest, here displacement, in the domain, *ie* in the structure. To obtain the shape function for a given variable, say the axial displacement in a bar, u, assume that it can be approximate to the linear function

$$u = ax + b.$$

Consider now that, at $x = 0$ and at $x = L_e$, the displacements are u_{e_1} and u_{e_2}, respectively. This gives $b = u_{e_1}$ and $a = (u_{e_2} - u_{e_1})/L_e$ so that, in matrix form,

$$\boldsymbol{u} = \bar{\boldsymbol{N}}\boldsymbol{u}_e = [\bar{N}_1 \quad \bar{N}_2]\left\{\boldsymbol{u}_{e_1}\boldsymbol{u}_{e_2}\right\}^T,$$

with $\bar{\boldsymbol{N}}$, the shape function matrix, being formed by

$$\bar{N}_1 = 1 - \frac{x}{L_e}, \qquad \bar{N}_2 = \frac{x}{L_e}$$

and depicted in the next figure.

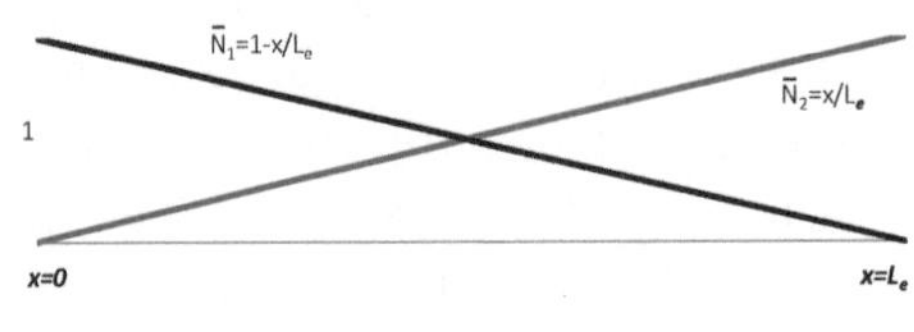

Figure 8.2 Shape function for a linear bar finite element.

■ DISPLACEMENT CALCULATION WITH INTERPOLATION FUNCTIONS

Obtain the displacement in the bar at $x = 3$ assuming a linear interpolation for the variable.

As seen in the next figure, the displacement can be easily obtained from

$$u = \left(1 - \frac{3}{5}\right) 7 + \frac{3}{5} 13 = 10.6.$$

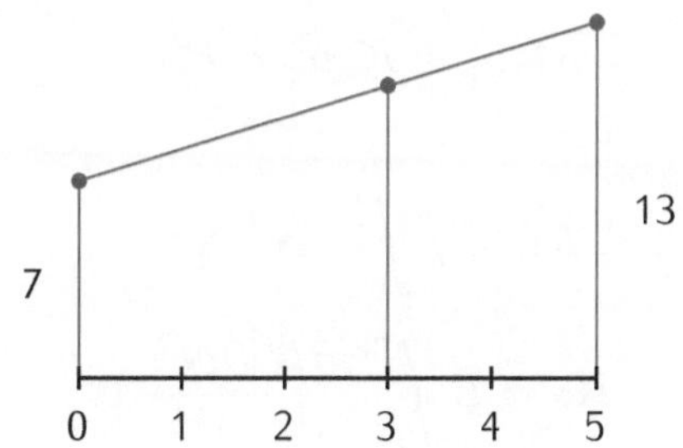

Figure 8.3 The displacement in a bar using the shape functions.

■

With the finite element method we seek the magnitude of a given variable, usually the displacement, at the nodes. They are judiciously positioned to geometrically well represent the body under analysis.

For the uniaxial case, the forced wave propagation equation reads

$$\frac{E}{\rho} \frac{\partial^2 u(x,t)}{\partial x^2} + \frac{1}{\rho A} f(x,t) = \frac{\partial^2 u(x,t)}{\partial t^2},$$

ρ being the density and f the distributed load.

Let us now multiply it by a virtual displacement, δu, and integrate in $x \in [0, L_e]$, to obtain

$$\int_0^{L_e} E \frac{\partial^2 u}{\partial x^2} . \delta u dx + \int_0^{L_e} \frac{f}{A} . \delta u dx = \int_0^{L_e} \rho \frac{\partial^2 u}{\partial t^2} . \delta u dx.$$

Using the interpolation functions via $u = \bar{\boldsymbol{N}} \boldsymbol{u}_e$ and reducing the order of the x derivative via integration by parts, *ie* $\int \boldsymbol{h} d\boldsymbol{g} = \boldsymbol{hg} - \int \boldsymbol{g} d\boldsymbol{h}$,

We use here the dot product. Note that $\boldsymbol{A}.\boldsymbol{B} = \boldsymbol{B}^T\boldsymbol{A}$ and $(\boldsymbol{A}\boldsymbol{B})^T = \boldsymbol{B}^T\boldsymbol{A}^T$.

with $\boldsymbol{h} = \bar{\boldsymbol{N}}$ and $\mathrm{d}\boldsymbol{g} = \partial^2\boldsymbol{u}/\partial x^2\mathrm{d}x$, we obtain

$$\rho\int_0^{L_e}\bar{\boldsymbol{N}}^T\bar{\boldsymbol{N}}\mathrm{d}x\frac{\partial^2\boldsymbol{u}_e}{\partial t^2} + E\int_0^{L_e}\frac{\partial\bar{\boldsymbol{N}}}{\partial x}\frac{\partial\bar{\boldsymbol{N}}^T}{\partial x}\mathrm{d}x\boldsymbol{u}_e = \int_0^{L_e}\frac{f}{A}\bar{\boldsymbol{N}}^T\mathrm{d}x$$

$$+E\bar{\boldsymbol{N}}^T\frac{\partial\bar{\boldsymbol{N}}}{\partial x}\boldsymbol{u}_e.$$

The last term above is zero because it is evaluated at the boundaries $x = 0$ and $x = L_e$, so it follows that

$$\boldsymbol{M}_e\ddot{\boldsymbol{u}}_e + \boldsymbol{K}_e\boldsymbol{u}_e = \boldsymbol{R}_e^{ext},$$

where

$$\boldsymbol{M}_e = \rho\int_0^{L_e}\bar{\boldsymbol{N}}^T\bar{\boldsymbol{N}}\mathrm{d}x,$$

$$\boldsymbol{K}_e = E\int_0^{L_e}\frac{\partial\bar{\boldsymbol{N}}}{\partial x}\frac{\partial\bar{\boldsymbol{N}}^T}{\partial x}\mathrm{d}x$$

and

$$\boldsymbol{R}_e^{ext} = \int_0^{L_e}\frac{f}{A}\bar{\boldsymbol{N}}^T\mathrm{d}x.$$

Using our shape functions for the bar finite element we have,

$$\boldsymbol{M}_e = \frac{\rho AL}{6}\begin{bmatrix} 2 & 1 \\ 1 & 2 \end{bmatrix} \quad \text{and} \quad \boldsymbol{K}_e = \frac{EA}{L}\begin{bmatrix} 1 & -1 \\ -1 & 1 \end{bmatrix}$$

and

$$\boldsymbol{R}_e^{ext} = \frac{fL}{2}[1 \quad 1]^T.$$

$\boldsymbol{M}_e$ is the mass matrix. It is called consistent because it was obtained using the shape functions. An alternative would be to concentrate the finite element mass in the diagonals of the matrix,

$$\boldsymbol{M}_e = \rho AL\begin{bmatrix} 1/2 & 0 \\ 0 & 1/2 \end{bmatrix}.$$

Several remarks are opportune here. First of all, the above procedure allowed us to reduce the partial wave differential governing equation in a set of algebraic equations, whose unknowns are the coefficients of a polynomial representing, in this case, the displacement of nodes in the bar. Clearly, it is of computational advantage to solve an algebraic system of equations rather than a differential equation. For more complicated problems, ruled by a set of differential partial equations, it is too difficult, if at all possible, to solve them. In contrast, it is always possible to solve the associate algebraic equations. This is one of the major advantages of the finite element method.

The equation of motion obtained above needs to be solved at various time intervals so one can obtain the time evolution of each node displacement. If we know the time lapse, we can obtain velocity and acceleration of all the nodes, then having a clear picture of the structure configuration in time and space.

Note that the above formulation was presented for one finite element only. In practice, it is necessary to represent the rod by various finite elements. We need to assemble these elements to obtain the equation of motion

$$\boldsymbol{M}\ddot{\boldsymbol{u}}(t) + \boldsymbol{K}\boldsymbol{u}(t) = \boldsymbol{R}^{ext}(t),$$

with each degree of freedom in the global stiffness and mass matrices receiving contribution from two adjacent elements, as explained in the next figure. A more general assembling of elements will be presented later.

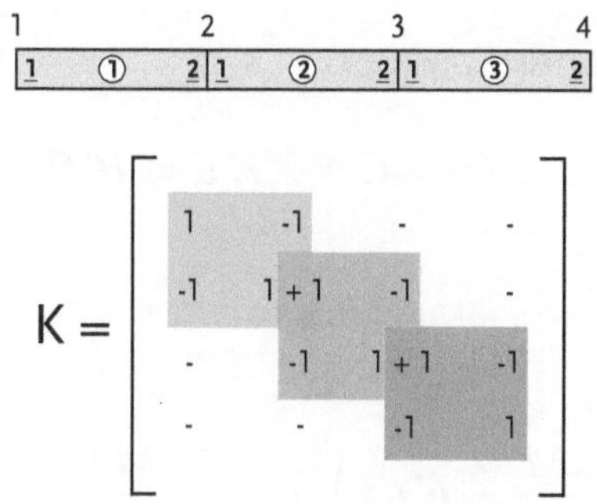

Figure 8.4 Assembling of a rod global stiffness matrix.

After the global matrix is assembled, it is necessary to impose the boundary condition. One way of doing so is by filling with zeros the row and column of the stiffness and mass matrix in the respective position where we do not want the node to move. We also make the diagonal term of the stiffness and mass matrices equal to 1 and the corresponding line in the vector force to have the value 0. This makes the corresponding degree of freedom (displacement) to be fixed in space, representing the sought boundary condition.

In the static analysis, the unknown vector $\boldsymbol{u}$ is obtained by, generally speaking, multiplying the inverse of the stiffness matrix by the load vector, $\boldsymbol{R}^{ext}$, which requires the computer demanding task of inverting $\boldsymbol{K}$. This strategy in the dynamic case is not necessary. For impact problems, instead, we time integrate the equation of motion seen above.

The problem can be solved using other procedures rather than assembling the whole stiffness matrix and inverting it.

There are two general classes of time integration, giving rise to implicit and explicit integration procedures. Explicit integration has become the standard in solving structural impact problems so attention is now given to this formulation.

If we expand in the Taylor series the displacement $\boldsymbol{u}$ we obtain

The Taylor series is given by $f(a+t) = f(a) + (\mathrm{d}f(a)/\mathrm{d}x)t/1! + (\mathrm{d}^2 f(a)/\mathrm{d}x^2)t^2/2! + \ldots$ and the present method is called central difference, in counter–distinction to the half–step central difference. See also R.D. Cook, D.S. Malkus, M.E. Plesha and R.J. Witt, Concepts and Applications of Finite Element Analysis, John Wiley, 2002.

$$\boldsymbol{u}_{n+1} = \boldsymbol{u}_n + \Delta t \dot{\boldsymbol{u}}_n + \frac{\Delta t^2}{2} \ddot{\boldsymbol{u}}_n + \ldots$$

and

$$\boldsymbol{u}_{n-1} = \boldsymbol{u}_n - \Delta t \dot{\boldsymbol{u}}_n + \frac{\Delta t^2}{2} \ddot{\boldsymbol{u}}_n + \ldots,$$

from which

$$\dot{\boldsymbol{u}}_n = \frac{\boldsymbol{u}_{n+1} - \boldsymbol{u}_{n-1}}{2\Delta t} \quad \text{and} \quad \ddot{\boldsymbol{u}}_n = \frac{\boldsymbol{u}_{n+1} + \boldsymbol{u}_{n-1} - 2\boldsymbol{u}_n}{\Delta t^2},$$

allowing the equation of motion

$$\boldsymbol{M}\ddot{\boldsymbol{u}} + \boldsymbol{C}\dot{\boldsymbol{u}} + \boldsymbol{K}\boldsymbol{u} = \boldsymbol{R}^{ext}$$

to be written as

$$\boldsymbol{u}_{n+1} = \left(\frac{\boldsymbol{M}}{\Delta t^2} + \frac{\boldsymbol{C}}{2\Delta t}\right)^{-1}$$
$$\left\{\boldsymbol{R}_n^{ext} + \left(\frac{2\boldsymbol{M}}{\Delta t^2} - \boldsymbol{K}\boldsymbol{u}_n\right)\boldsymbol{u}_n - \left(\frac{\boldsymbol{M}}{\Delta t^2} - \frac{\boldsymbol{C}}{2\Delta t}\right)\boldsymbol{u}_{n-1}\right\}.$$

Note the introduction of the matrix $\boldsymbol{C}$. It represents the structural damping and it takes into account dissipative forces in a structure. It allows the representation of motion which dies away, as it happens in the free vibration of any structure.

One important aspect to observe in the above equation is that it calculates the displacement of the model in increment $n+1$ from information of what happened in the previous increments n and $n-1$ (although velocity and acceleration calculation require the current displacement configuration). This is to say that the sought configuration is explicitly obtained. But this is not the only reason for the current time integration method to be called explicit.

In the above equation we need to invert $\boldsymbol{M}$, which is trivial if the mass matrix is diagonal, as in the case of the rod finite element developed in this chapter. The mass matrix of more complex finite elements is usually sparse and not easily invertible. If the mass matrix is diagonal, the system of equations to be solved is uncoupled, assuming the use of internal forces $\boldsymbol{R}_{int}$ in place of the stiffness matrix. The fact that the system is uncoupled allow us the calculation of one node displacement independent of the others, which is of computational advantage. Hence, it only makes sense to talk about explicit finite element method when the mass matrix is diagonal.

Some efforts have been made to change the basis of the interpolation function to obtain a diagonal stiffness matrix. See M.L. Bittencourt, M.G. Vazquez and T.G. Vazquez. Construction of shape functions for the h– and p– versions of the FEM using tensorial product. International Journal for Numerical Methods in Engineering, 71, p. 529–563, 2007.

Again, to keep this idea of diagonal matrix, the calculation of the damping matrix $\boldsymbol{C}$ can be made by

$$\boldsymbol{C} = \alpha \boldsymbol{M} + \beta \boldsymbol{K},$$

with α and β being constants that depend on the structure, frequency and material.

Note that, in the evaluation of $\boldsymbol{u}_1$, it is necessary to calculate

$$\boldsymbol{u}_{-1} = \boldsymbol{u}_0 - \Delta t\{\dot{\boldsymbol{u}}\}_0 + \frac{\Delta t^2}{2}\{\ddot{\boldsymbol{u}}\}_0$$

and

$$\ddot{\boldsymbol{u}}_0 = \boldsymbol{M}^{-1}\left(\boldsymbol{R}_0^{ext} - \boldsymbol{K}\boldsymbol{u}_0 - \boldsymbol{C}\dot{\boldsymbol{u}}_0\right),$$

which come from the Taylor series and from the governing equation for the rod, where damping was included.

In the Taylor expansion, terms of order higher than 2 were disregarded so that the primary error is proportional to Δt^2, suggesting that the smaller the time increment the smaller the error (halving the time increment quarters the error). The time increment is also bounded by the required stability of the explicit method. Ideally, we want to make Δt large so as to finish the analysis quickly. However, if the time increment is too large, the solution may unbound orders of magnitude. One simple way to calculate the maximum time increment which assures stability, Δt_{cr}, is using

$$\Delta t_{cr} = \frac{l}{c},$$

where l is the smallest finite element size (or edge) in the mesh and $c = \sqrt{E/\rho}$ is the elastic wave speed in the structure. This equation reveals two simple ways to increase the time increment and hence to quicken the analysis. First, we can avoid too small finite elements. Second, we can artificially use very high values for the material density. This mass scaling is only possible, however, if the inertia forces are not important, as in a quasi-static analysis or low velocity impact events.

The fact that there is a time limit increment is a characteristic of the explicit method and we refer to it as a conditioned stable method, *ie* only under certain conditions is the method stable. In simpler programs, like the one to be shown here, the time increment is kept constant throughout the whole analysis. However, depending on the structure response, it is of interest to increase the time increment as the structure response is found to be stable, which can be made, so to say, automatically.

There are many situations where a structure is not loaded by a force. Consider for instance the case where one displaces the end of a rule and let it go, with the other end being held against a table. This is the case of an imposed displacement. It is also possible to impose an initial velocity, *eg* via an explosion, as we discussed in Chapter 5.

■ A BAR UNDER A STEP FORCE

A rod is suddenly compressed by a step force, as illustrated in the next figure. Obtain the bar response via the finite element method at its middle span and compare the solution with the theoretical one.

This example was run in a dedicated MatLab based on the theory above. The solution shows some oscillations but it should be noted that no damping is taken into account here.

The problem was suggested as a means to explore different time increments in the solution. Defining the Courant number as

$$C_n = \frac{\Delta t_{actual}}{\Delta t_{cr}},$$

it is possible to explore how values slightly different, and smaller, from 1 can improve the response when compared with the theoretical one. For the present program, somewhat large variations of C_n do not affect

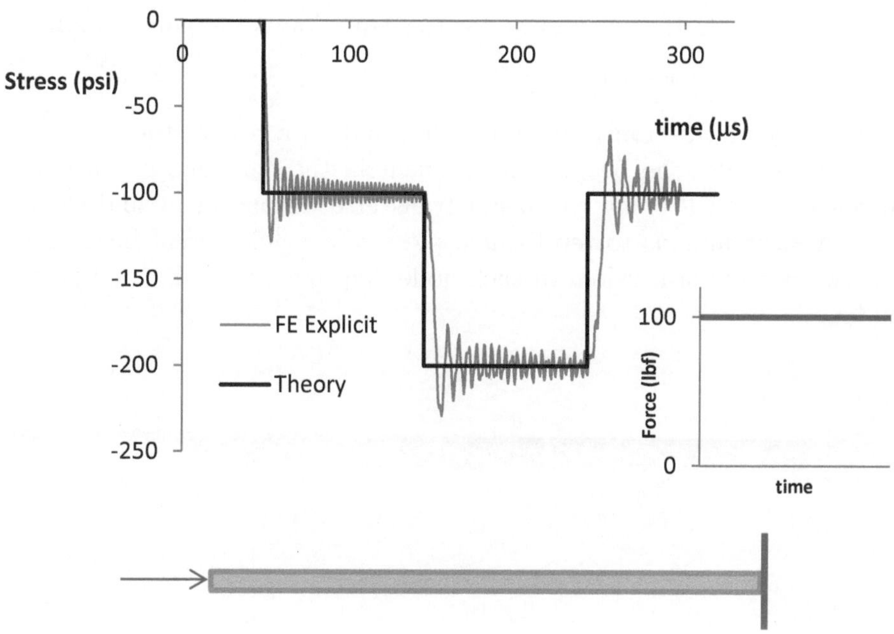

Figure 8.5 A bar under a step load of 100 lbf and the stress profile with time at the middle span. $L = 20$ in, $A = 1$ in^2, $E = 30$ Mpsi, $\rho = 7.4\text{E-}4$ lb.s^2/in^4. Example extracted from R.D. Cook et al., Concepts and Applications of Finite Element Analysis.

much the stress values. A more refined discretization does improve the response.

■

8.2 Finite elements for beams

We further develop the FEM technique to the more complex beam finite element. This element allows us to solve many practical problems, such as frame design. For these structures, it is necessary to use coordinate transformation and to assemble the finite elements, leading to the concept of node connectivity. Let us first see the case of static loads.

8.2.1 Quasi–static loads

Let us start with the beam equilibrium equation for a linear elastic material with no inertia effects

$$EI\frac{\mathrm{d}^4 w(x)}{\mathrm{d}x^4} = p(x),$$

with I being the moment of inertia of the beam cross-section.

We can verify that a solution for this equation is the polynomium

$$w(x) = a_1 + a_2 x + a_3 x^2 + a_4 x^3 = \lfloor 1\ x\ x^2\ x^3 \rfloor \lfloor a_1\ a_2\ a_3\ a_4 \rfloor^T = \lfloor X \rfloor [a]^T,$$

with $[\,]^T$ being the transpose of a matrix and $\lfloor\,\rfloor$ a row vector.

Now consider the beam finite element in the next figure, with the unkowns w and θ being the beam transverse displacement and slope. This element intends to represent a piece of a whole actual beam and has two degrees of freedom in each node, $[w_1\ \theta_1\ w_2\ \theta_2]$, which ought to be found.

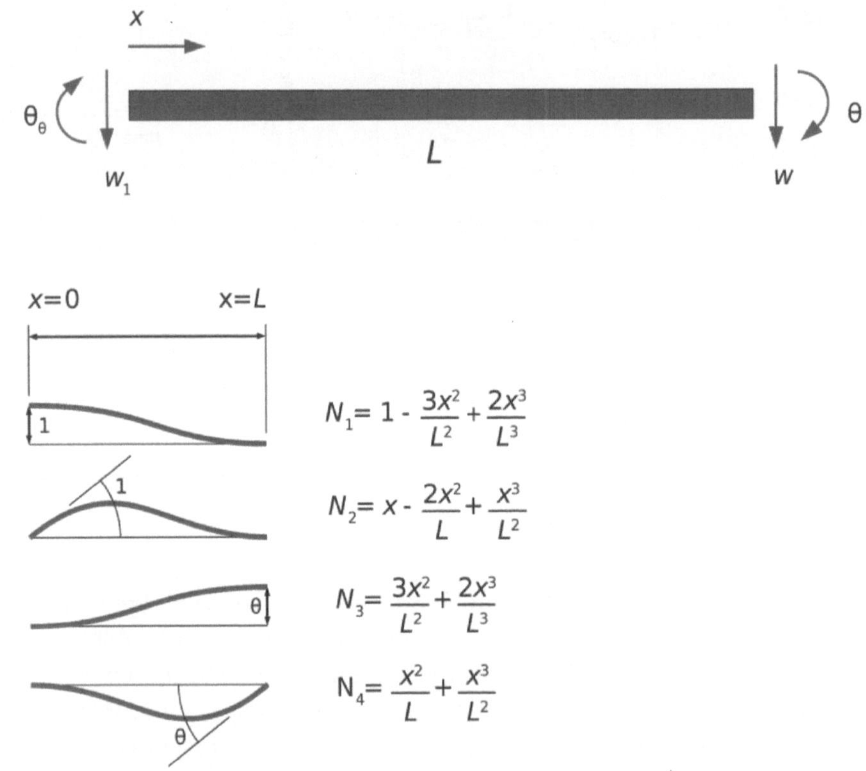

Figure 8.6 A beam finite element and its shape functions (Hermite polynomium).

From the above expression, we can obtain the values of w and θ at $x = 0$ and at $x = L_e$, represented by $\{w_e\} = [A]\{a\}$, or

$$\begin{Bmatrix} w_1 \\ \theta_1 \\ w_2 \\ \theta_2 \end{Bmatrix} = \begin{bmatrix} 1 & 0 & 0 & 0 \\ 0 & 1 & 0 & 0 \\ 1 & L_e & L_e^2 & L_e^3 \\ 0 & 1 & 2L_e & 3L_e^2 \end{bmatrix} \begin{Bmatrix} a_1 \\ a_2 \\ a_3 \\ a_4 \end{Bmatrix}.$$

Now, since $\{w\} = [X]\{a\}$ it follows that

$$\{w\} = [X][A]^{-1}\{w_e\} = [\bar{N}]\{w_e\}.$$

The matrix $[\bar{N}] = [X][A]^{-1}$ is formed by a row of functions in x. They are the shape functions, summarized in the previous figure. The shape functions are of fundamental importance to the performance of the finite element method and to the correct representation of the displacement field. Note that the shape functions for this beam element attain a value of unity in one node and zero in the other. This function behaviour is a characteristic of the so called Lagrange polynomials. Our polynomials here have the additional characteristic that their derivative attain the unit value in the respective node where the amplitude is one, we call them Hermite polynomial.

Let us now return to the beam quasi–static equilibrium equation (with $M = EI\kappa$) and multiply it by a small displacement, δw, which does not cause any stress on the beam (hence, called virtual displacement). Let us also integrate the resulting equation along the beam length, so that

$$\int_{L_e} \frac{\partial^2 M}{\partial x^2}\delta w \mathrm{d}x = \int_{L_e} p\delta w \mathrm{d}x.$$

Integration by parts yields

Use $u = w$, followed by $u = \mathrm{d}w/\mathrm{d}x$ in $\int u\mathrm{d}v = [uv] - \int v\mathrm{d}u$.

$$\underbrace{\int_{L_e} M.\delta\kappa \mathrm{d}x}_{\substack{\text{Internal energy}\\ \text{due to a change}\\ \text{in the curvature}}} = \underbrace{\int_{L_e} p.\delta w \mathrm{d}x}_{\substack{\text{External work}\\ \text{due to distributed}\\ \text{loads}}} + \underbrace{\left[Q\delta w - M\frac{\mathrm{d}\delta w}{\mathrm{d}x}\right]}_{\substack{\text{External work}\\ \text{due to concentrated}\\ \text{loads}}}.$$

A small variation in the beam transverse virtual displacement causes a small variation in the virtual curvature, $\delta\kappa$, so that

$$\delta\kappa = -\frac{\partial^2 \delta w}{\partial x^2} = \frac{\partial^2([\bar{N}]\{\delta w_e\})}{\partial x^2} = \frac{\partial^2[\bar{N}]}{\partial x^2}\{\delta w_e\},$$

which can be substituted in the above functional, together with $\{\delta w\} = [\bar{N}]\{\delta w_e\}$ and $\{\mathrm{d}\delta w\}/\mathrm{d}x = \mathrm{d}[\bar{N}]/\mathrm{d}x\ \{\delta w_e\}$, to give

Note that we use $M = EI\kappa$ and the internal product of two matrices since the functional is an energy statement (work).

$$\int_{L_e} EI\frac{\partial^2[\bar{N}]}{\partial x^2}.\frac{\partial^2[\bar{N}]}{\partial x^2}\{\delta w_e\}\mathrm{d}x =$$

$$\int_{L_e} p.[\bar{N}]\{\delta w_e\}\mathrm{d}x + \left[Q.[\bar{N}]\{\delta w_e\} - M.\frac{\mathrm{d}[\bar{N}]}{\mathrm{d}x}\{\delta w_e\}\right].$$

Disregarding the concentrated loads, this expression can be written as

$$[K_e]\{w_e\} = \{F_e\},$$

or

$$\underbrace{\frac{EI}{L_e^3}\begin{bmatrix} 12 & 6L_e & -12 & 6L_e \\ 6L_e & 4L_e^2 & -6L_e & 2L_e^2 \\ -12 & -6L_e & 12 & -6L_e \\ 6L_e & 2L_e^2 & -6L_e & 4L_e^2 \end{bmatrix}}_{\text{beam stiffness matrix}} \underbrace{\begin{Bmatrix} w_1 \\ \theta_1 \\ w_2 \\ \theta_2 \end{Bmatrix}}_{\text{nodes d.o.f.}} = \underbrace{\begin{Bmatrix} F_1 \\ M_1 \\ F_2 \\ M_2 \end{Bmatrix}}_{\text{load vector}},$$

with $[K_e]$ being the stiffness matrix of the beam finite element and $\{F_i\}$, $\{M_i\}$ the force and bending moments in the nodes, respectively.

If we compare the above equation with the beam equilibrium equation, we will again notice that the former is an algebraic equation (easy to solve) and the latter a differential equation (difficult to solve). It is also interesting to note that the distributed load, p, was transformed in concentrated loads acting in the nodes via $\int_{L_e} [\bar{N}]^T p dx$.

Consideration of normal forces

For a structure as the one shown in the next figure, it is necessary to consider the cooperation of normal forces to the functional. This can be done by allowing a third degree of freedom for the beam finite element in its axial direction.

An increase in the degrees of freedom, obviously increases the order of the local and global stiffness matrix. Also, more degrees of freedom are necessary to represent a structure in space rather than the plane case here.

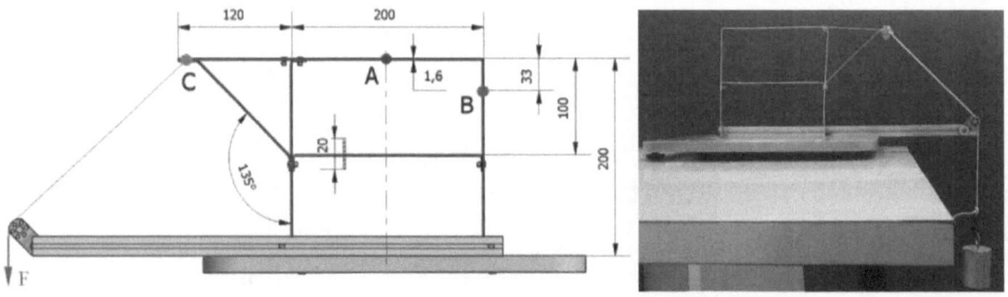

Figure 8.7 A more complex beam–like structure.

The displacement field in this direction, $\{u\}$, can be assumed linear according to

$$\{u\} = a_5 + a_6 x = [X]\{a\}.$$

At the extremes of the element, the displacements are u_1 and u_2, such that

$$\begin{Bmatrix} u_1 \\ u_2 \end{Bmatrix} = \begin{bmatrix} 1 & 0 \\ 1 & L_e \end{bmatrix} \begin{Bmatrix} a_5 \\ a_6 \end{Bmatrix} = [A]\{a\},$$

giving

$$\{u\} = [\bar{N}] \left\{ \begin{array}{c} u_1 \\ u_2 \end{array} \right\}$$

with

$$[\bar{N}] = \{x\}[A]^{-1} = [1 - x/L_e \quad x/L_e]$$

being the shape function matrix.

The functional $\frac{1}{2}\int_0^{L_e} \sigma\varepsilon A \mathrm{d}x$ becomes

$$\frac{EA}{2}\int_0^{L_e} \{u_1\ u_2\}\frac{\mathrm{d}[\bar{N}]^T}{\mathrm{d}x}\frac{\mathrm{d}[\bar{N}]}{\mathrm{d}x}\{u_1\ u_2\}\mathrm{d}x$$

since $\varepsilon = \partial u/\partial x = \mathrm{d}[\bar{N}]/\mathrm{d}x\,\{u_1\ u_2\}$. This axial degree of freedom should be incorporated in the right position of the stiffness matrix and load force, yielding the more complete beam finite element translated to the following algebraic system of equations

Note that torsion, which naturally arises in 3D frames, is not considered here.

$$\frac{E}{L_e}\begin{bmatrix} A & 0 & 0 & -A & 0 & 0 \\ 0 & \frac{12I}{L_e^2} & \frac{6I}{L_e} & 0 & -\frac{12I}{L_e^2} & \frac{6I}{L_e} \\ 0 & \frac{6I}{L_e} & 4I & 0 & -\frac{6I}{L_e^2} & 2I \\ -A & 0 & 0 & A & 0 & 0 \\ 0 & -\frac{12I}{L_e^2} & -\frac{6I}{L_e} & 0 & \frac{12I}{L_e^2} & -\frac{6I}{L_e} \\ 0 & \frac{6I}{L_e} & 2I & 0 & -\frac{6I}{L_e} & 4I \end{bmatrix}\left\{\begin{array}{c} u_1 \\ w_1 \\ \theta_1 \\ u_2 \\ w_2 \\ \theta_2 \end{array}\right\} = \left\{\begin{array}{c} N_1 \\ F_1 \\ M_1 \\ N_2 \\ F_2 \\ M_2 \end{array}\right\}.$$

Applying boundary conditions and loads

Let us now consider the hyperstatic problem of the beam in the next figure. The beam is represented by a single finite element. With this problem we will show how to impose boundary conditions and to apply concentrated loads to a finite element model.

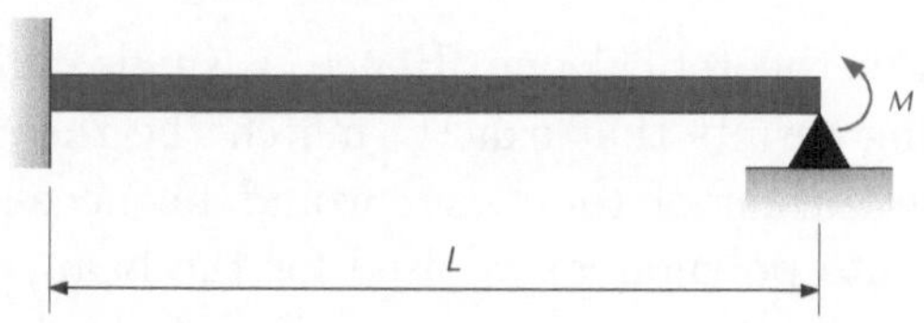

Figure 8.8 A hyperstatic beam problem solved using a single FE.

Since the distributed loads, the transverse shear force and the normal force are all zero, the load vector reduces to

$$\{F_e\} = \int_{L_e} p[\bar{N}]\mathrm{d}x + \left[Q[\bar{N}] - M\frac{\mathrm{d}[\bar{N}]}{\mathrm{d}x}\right]$$
$$= -M\frac{\mathrm{d}[\bar{N}]}{\mathrm{d}x} = \lfloor 0 \quad 0 \quad 0 \quad -M \rfloor^T.$$

If a given boundary condition is given by $w_m = 0$, we set in the finite element stiffness matrix $K_{e_{mm}} = 1$, $K_{e_{mj}} = K_{e_{im}} = 0$, $F_{e_j} = 0$, $F_{e_m} = 1$, for $n = 1..6, j = 1..6, m \neq i$ or j.

In our problem here we have, disregarding axial forces, $w_1 = \theta_1 = w_2 = 0$ so that, the local stiffness matrix reads

$$\begin{bmatrix} 1 & 0 & 0 & 0 \\ 0 & 1 & 0 & 0 \\ 0 & 0 & 1 & 0 \\ 0 & 0 & 0 & 4EI/L \end{bmatrix} \begin{Bmatrix} w_1 \\ \theta_1 \\ w_2 \\ \theta_2 \end{Bmatrix} = \begin{Bmatrix} 0 \\ 0 \\ 0 \\ -M \end{Bmatrix},$$

giving

$$\theta_2 = -\frac{ML}{4EI}.$$

It is interesting to note that professional finite element codes depict the deformed configuration of a structure using the information of the nodes displacement. For this problem and mesh, such a procedure would be of no avail since the nodes do not displace. However, the beam profile comes from

$$\{w\} = [\bar{N}]\{w_e\} = \lfloor \bar{N}_1 \quad \bar{N}_2 \quad \bar{N}_3 \quad \bar{N}_4 \rfloor \lfloor 0 \quad 0 \quad 0 \quad \theta_2 \rfloor^T = \frac{M}{4EI}\left(x^2 - \frac{x^3}{L}\right),$$

which matches the theoretical one. Hence, a single finite element was capable of yielding results that exactly match the theoretical ones and no further discretization of the beam would be necessary. This is so because the Hermite polynomial we used for the beam finite element is in itself a solution to the governing equation of the beam, so forming what is called a complete basis.

Coordinate transformation

We have seen that the normal forces in one finite element do not act in the same direction as the ones in the other finite elements. The same occurs for other forces and bending momenta. Therefore, they cannot be added directly. Rather, they should be referenced to the same global coordinate system. We then need to perform what is called a coordinate transformation so as to bring all the vectors in the same direction and only after this transformation can we solve the algebraic equations.

Now consider the next figure. Here, the vector v has the components

$$v_X = v\cos\beta \qquad \text{and} \qquad v_Y = v\sin\beta.$$

It also has the components

$$v_x = v\cos(\beta-\alpha) \qquad \text{and} \qquad v_y = v\sin(\beta-\alpha),$$

or

$$v_x = v(\cos\beta\cos\alpha + \sin\beta\sin\alpha)$$

and

$$v_y = v(\sin\beta\cos\alpha - \cos\beta\sin\alpha),$$

or, in matrix form,

$$\{v_{xy}\} = [T]\{v_{XY}\} \quad \text{and} \quad \{v_{XY}\} = [T]^{-1}\{v_{xy}\},$$

with

$$T = \begin{bmatrix} \cos\alpha & \sin\alpha \\ -\sin\alpha & \cos\alpha \end{bmatrix}.$$

By inspection, $[T]^{-1} = [T]^{\mathrm{T}}$, *ie* the coordinate transformation matrix $[T]$ is orthogonal.

The comments above regarding force are also valid for any other vector. In fact, we need to operate on each local stiffness (and mass) matrix so as to orientate all the degrees of freedom according to the global coordinate system. To this end, consider that $[T]$ is a matrix that transform a vector (or matrix) $\{F_e\}$ into $\{F_e\}'$, *ie* it rotates the vector from one cartesian coordinate system to the other, according to $\{F_e\}' = [T]\{F_e\}$. The finite element equation $[K_e]\{w_e\} = \{F_e\}$, written in a rotated reference system, is $[K_e]'\{w_e\}' = \{F_e\}'$ such that

Remember that there are as many local stiffness matrix as finite elements.

$$[K_e]'\{w_e\}' = [T]\{F_e\} = [T][K_e]\{w_e\} = [T][K_e][T]^{-1}\{w_e\}',$$

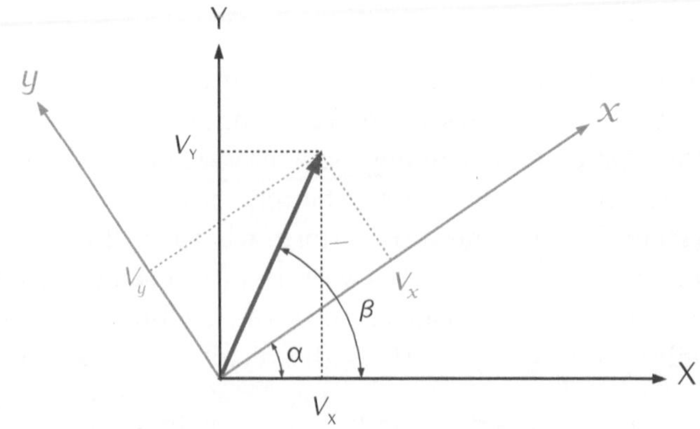

Figure 8.9 A vector represented by two different coordinate systems.

or

$$[K_e]' = [T]^{\mathrm{T}}[K_e][T],$$

where advantage was taken from the fact that the matrix $[T]$ is an orthogonal matrix. Hence, it is necessary to perform the above operation for all matrix and vectors before assembling the global equilibrium equation.

Assembling of elements

We have seen for rods a simple way to connect the finite elements along a bar, leading us to work with a global stiffness matrix, $[K]$, and the respective global vector of displacements, $\{w\}$, and global vector of forces, $\{F\}$. In more complicate structures, see next figure, we need to indicate what are the nodes of the finite elements and how they are connected, which is done using the connective matrix, $[cm]$.

The information in the matrix $[cm]$ is used to locate the contribution of each node to the global matrix. One basic way to perform such an operation for a beam finite element with three degree of freedoms in each node is given by the algorithm below.

Listing 8.1 A basic way to assemble elements and to generate a global stifness matrix for a 2D beam element with no axial force.

```
%x,y global coord
gc(1,1)=0;%x
gc(1,2)=0;%y
gc(2,1)=10;gc(2,2)=0;gc(3,1)=30;gc(3,2)=0;gc(4,1)=10;gc(4,2)=-15;
%row n stores nodes of element n
%conective matrix
```

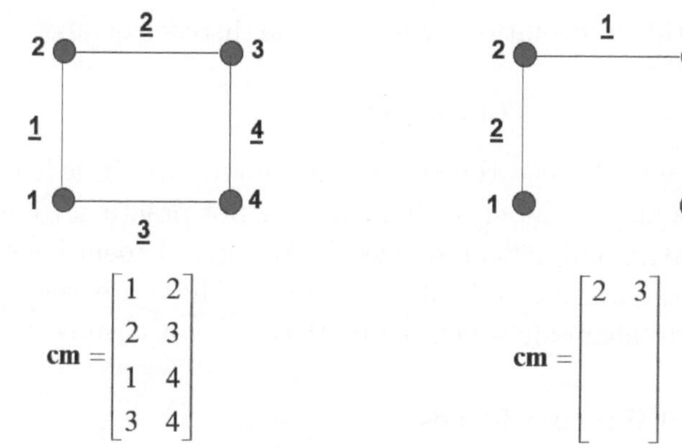

Figure 8.10 A connective matrix for two structures.

```
cm(1,1)=1;cm(1,2)=2;cm(2,1)=2;cm(2,2)=4;cm(3,1)=2;cm(3,2)=3;
....
k_global=zeros(dofg,dofg);
for el_n=1:nel
        nd(1)=cm(el_i,1);%node 1 of element el_1
        nd(2)=cm(el_i,2);%node 2 of element el_1
        x1=gc(nd(1),1);%x global node 1 of element el_i
        y1=gc(nd(1),2);%y global node 1 of element el_i
        x2=gc(nd(2),1);%x global node 2 of element el_i
        y2=gc(nd(2),2);%y global node 1 of element el_i
        dx(el_i)=x2-x1; dy(el_i)=y2-y1;
        le(el_i)=sqrt(dx(el_i)^2+dy(el_i)^2);
        c(el_i)=dx(el_i)/le(el_i);s(el_i)=dy(el_i)/le(el_i);
%transformation matrix
     T(1,1)=c(el_i);T(1,2)=s(el_i);T(2,1)=-s(el_i);T(2,2)=c(el_i);
     T(3,3)=1;T(4,4)=c(el_i);T(4,5)=s(el_i);T(5,4)=-s(el_i);T(5,5)=c(el_i);T(6,6)=1;
     c1=b*h*Em/le(el_i);c2=Em*b*h^3/12/le(el_i)^3;
     k_e(1,1)=c1;                    k_e(1,4)=-c1;
     k_e(2,2)=12*c2;                 k_e(2,3)=6*c2*le(el_i);
     k_e(2,5)=-12*c2;                k_e(2,6)=6*c2*le(el_i);
     k_e(3,3)=4*c2*le(el_i)^2;       k_e(3,5)=-6*c2*le(el_i);
     k_e(3,6)=2*c2*le(el_i)^2;       k_e(4,4)=c1;
     k_e(5,5)=12*c2;                 k_e(5,6)=-6*c2*le(el_i);
     k_e(6,6)=4*c2*le(el_i)^2;
     k_e(:,1)=k_e(1,:);              k_e(:,2)=k_e(2,:);
     k_e(:,3)=k_e(3,:);              k_e(:,4)=k_e(4,:);
     k_e(:,5)=k_e(5,:);              k_e(:,6)=k_e(6,:);
     g(1)=3*cm(el_i,1)-2;
     g(2)=g(1)+1;
     g(3)=g(1)+2;
     g(4)=3*cm(el_i,2)-2;
     g(5)=g(4)+1;
     g(6)=g(4)+2;
     k_aux=zeros(dofg,dofg);
     for n=1:6
        for j=1:6
            k_aux(g(i),g(j))=k_e(i,j);
        end
     end
     k_global=k_global+k_aux;
end
```

Having all this information, it follows the displacement of the nodes via

$$\{w\} = [K]^{-1}\{F\}.$$

Note that we could read the above equation as: obtain w from K and F. In fact, for a large number of elements it is not practical to invert the full stiffness matrix and instead one works at finite element level, *ie* with the local stiffness matrix and internal forces. There are very powerful methods to solve algebraic systems but they are not covered here.

Reactions and internal forces

The knowledge of the reactions at the supports of a structure is obviously very useful. It is also useful to know the forces acting in the nodes so that one can be sure of the equilibrium state of the structure as well as to build diagrams for the internal bending moment, transverse shear and normal forces.

The calculation of the reaction forces can be obtained by multiplying the global stiffness matrix by the calculated displacements. Here, the global stiffness matrix is not reduced by the boundary conditions: it needs to be the full matrix. A simple, but probably inefficient procedure, is to evaluate $[\mathbf{K}]_{\text{global}}$, reduce it by applying the boundary conditions, obtain the displacements, recalculate $[\mathbf{K}]_{\text{global}}$ and without applying the boundary conditions, multiply it by the displacements. The resulting force vector will contain the (negative) reaction forces.

If we take the example of the hyperstatic beam, the reaction force is given by

$$\frac{EI}{L^3}\begin{bmatrix} 12 & 6L & -12 & 6L \\ & 4L^2 & -6L & 2L^2 \\ & & 12 & -6L \\ & & & 4L^2 \end{bmatrix}\begin{Bmatrix} 0 \\ 0 \\ 0 \\ -\frac{ML}{4EI} \end{Bmatrix} = \begin{Bmatrix} R_A \\ M_A \\ R_B \\ -M \end{Bmatrix} \rightarrow \begin{matrix} R_A = -\frac{3M}{2L} \\ M_A = -\frac{M}{2} \\ R_B = \frac{3M}{2L} \end{matrix}$$

By knowing the internal forces, it is possible to verify if the structure is in equilibrium. This is a way to check if the calculations are correct; when adding all the applied loads, *ie* forces, reactions and internal forces acting in a node, they must be zero. We shall see in the sequence that, for non–linear problems, the fact that this summation is not always zero establishes a loop in the program until equilibrium is satisfied.

8.2.2 Dynamic loading: forced response

Beam dynamics can be efficiently analysed via the FEM and to this end we need to add the inertia term

$$m\frac{\mathrm{d}^2w}{\mathrm{d}t^2},$$

to our discretized equation. Noting that $\{w\} = [\bar{N}]\{w_e\}$ and that $\mathrm{d}^2\{w\}/\mathrm{d}t^2 = [\bar{N}]\mathrm{d}^2\{w_e\}/\mathrm{d}t^2$, since the shape function matrix is time independent, we can write the inertia term in the functional as

$$\int_{L_e} \delta w^{\mathrm{T}} \rho \mathrm{d}^2 w/\mathrm{d}t^2 \mathrm{d}V = \delta w_e \int_{L_e} \rho A[\bar{N}]^{\mathrm{T}}[\bar{N}]\mathrm{d}x\mathrm{d}\{w_e\}/\mathrm{d}t^2,$$

where m is the mass per unit length of the beam. Performing the above integration we obtain the mass matrix for a beam finite element. We can add in this procedure the contribution of the axial inertia and this results in the beam mass matrix

$$[M_e] = \frac{mL}{420}\begin{bmatrix} 140 & 0 & 0 & 70 & 0 & 0 \\ & 156 & 22L_e & 0 & 54 & -13L_e \\ & & 4L_e^2 & 0 & 13L_e & -3L_e^2 \\ & & & 140 & 0 & 0 \\ & & & & 156 & -22L_e \\ & & & & & 4L_e^2 \end{bmatrix}.$$

Note that this mass matrix was obtained directly from the shape functions and this lends to it the name consistent. Another and more important mass matrix is the diagonal one, which can be formed in many ways. One idea is to equally distribute the total mass of the element along the diagonal. Another way, by lumping the element mass, was investigated by Hinton, Rock and Zienkiewicz who suggested the HRZ matrix for beam elements without axial forces as

$$M = \frac{mL_e}{78}\lceil\ 39 \quad L_e^2 \quad 39 \quad L_e^2\ \rfloor.$$

while a diagonal mass matrix preserving the inertia axial contribution reads

$$M = mL_e\lceil\ 1/2 \quad 39/78 \quad L_e^2/78 \quad 1/2 \quad 39/78 \quad L_e^2/78\ \rfloor.$$

The consistent mass matrix is positive definite but the diagonal mass matrix may be semi–definite in the presence of nil diagonal terms. Generally speaking, the use of a consistent mass matrix leads to more accurate results for bending problems. The diagonal matrix gives lower bound values for natural frequencies. Note that the mass matrix needs to be rotated using the finite element local coordinate transformation matrix.

We have seen analytically how to solve some basic problems of forced vibration of beams. In the context of the finite element method, the forced response of a structure can be obtained by discretizing in time both the input force and the structure response. Hence, for each time increment, starting from $t = 0$, we have to input the load at the various time increments. Using the formulation regarding implicit and explicit time integration, we obtain the displacement of each node of the structure at those times. The solution marches on time and eventually the dynamic load comes to an end. To the structure though kinetic energy was imparted, it carries on moving, as captured by the continuous time integration of the dynamic problem, until a steady state regime is achieved or the motion is damped out.

Damping is included in the analyses by the damping matrix

$$[C] = \alpha[M] + \beta[K],$$

with α and β being constants that best represent the natural damping of a structure. Although called constants, they are somewhat frequency dependent so care should be exercised here.

Now, we can run in Finite Elements the problem of the falling beam on the support, seen before analytically in Chapter 4. As depicted in the next figure, the numerical results match quite well the theoretical ones. The figure also offers a solution in a commercial code. Here, a consistent mass matrix was used for the in-house program.

8.2.3 Modal analysis

Modal analysis means the determination of the natural frequencies of a structure and the associated vibration modes. There are different experimental techniques used to perform a modal analysis and a simple one is to continuously excite a structure with an harmonic motion whose frequency is changed until the structure maximizes its motion. By doing such a tuning we are in fact measuring the so called resonant frequency,

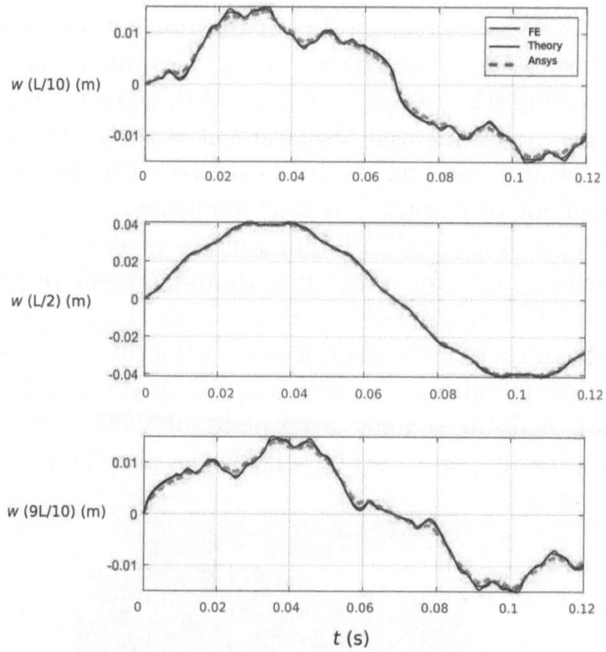

Figure 8.11 Finite element versus theoretical solution for the problem of a beam falling on a support.

which is very close to the natural frequency for metallic structures with low damping.

In the context of the finite element method, a modal analysis is based on the equation of motion

$$[\boldsymbol{M}]\{\ddot{\boldsymbol{D}}\} + [\boldsymbol{C}]\dot{\boldsymbol{D}} + [\boldsymbol{K}]\{\boldsymbol{D}\} = \{\boldsymbol{R}^{\text{ext}}\},$$

Now, since we are dealing with modal analysis, the structure will respond with an harmonic motion of the type

$$\{\boldsymbol{D}\} = \{\bar{\boldsymbol{D}}\}\sin\omega t \rightarrow \{\ddot{\boldsymbol{D}}\} = -\omega^2\{\bar{\boldsymbol{D}}\}\sin\omega t$$

which when substituted in the equation of motion gives, damping being disregarded,

$$\underbrace{([\boldsymbol{K}] - \omega^2[\boldsymbol{M}])}_{\text{dynamic stiffness matrix}} \{\bar{\boldsymbol{D}}\} = \{\boldsymbol{0}\}$$

or

$$[\boldsymbol{M}]^{-1}[\boldsymbol{K}]\{\bar{\boldsymbol{D}}\} = \omega^2\{\bar{\boldsymbol{D}}\}.$$

This means that we need to find a scalar ω and a vector $\{\bar{\boldsymbol{D}}\}$ that satisfies the above equation. This is performed by different algorithms not studied here. What is relevant to our purpose is that the eigenvalues of the problem above is the square of the natural frequencies of our structure possessing stiffness $[\boldsymbol{K}]$ and mass $[\boldsymbol{M}]$, while the corresponding eigenvectors are the normalized displacement of the nodes at that given frequency. If these displacements are imagined to be frozen we have a picture of the mode of vibration at that frequency.

As an example of the realization of a modal analysis, the next figure shows some experiments for a free–free aluminium square plate. There, the plate is set to vibrate by a shaker, whose frequency is tuned to maximize the motion of the sand that covers the plate. The sand moves to regions of low amplitude, so revealing the mode of deformation for that frequency. A finite element using plate, or shell, or plane, or solid elements can be easily performed by software, as also shown in the figure.

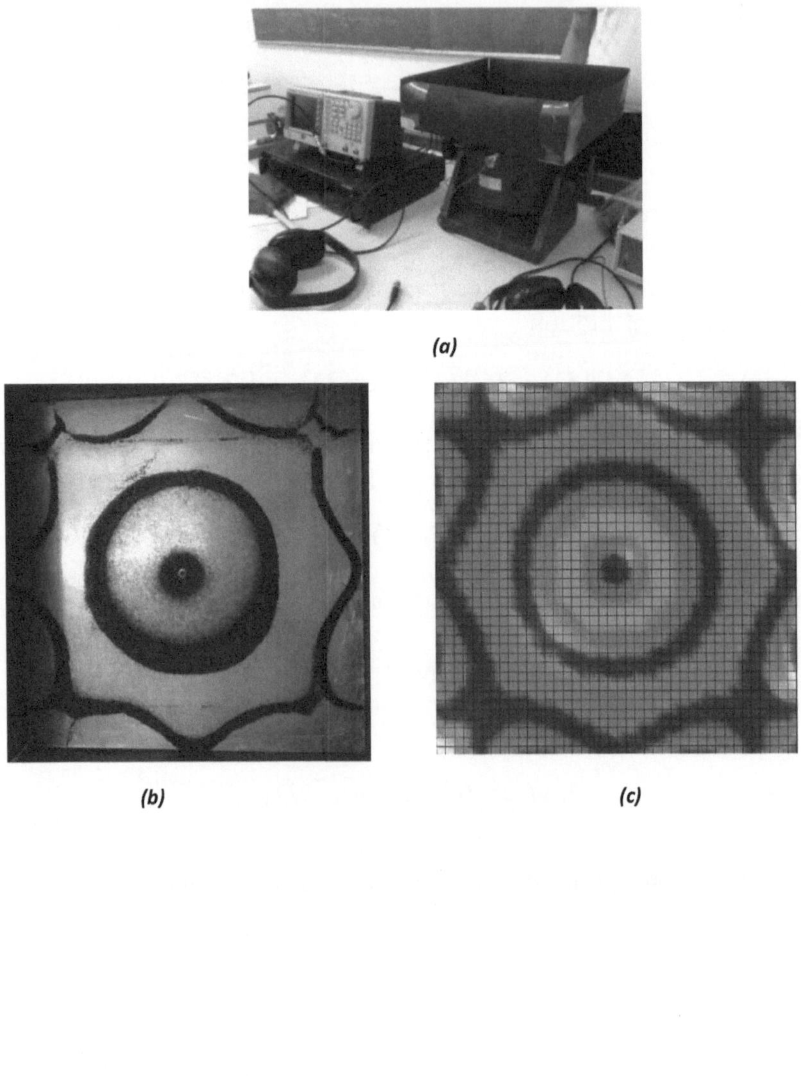

Figure 8.12 A modal analysis of a free rectangular plate. (a) Experimental apparatus with a wave generator, power amplifier, shaker and the plate surrounded by a black cardboard to avoid sand spill. (b) Sand pattern for the fourth mode of vibration at 741 Hz. (c) Finite element model at 716 Hz.

8.3 An axisymmetric finite element

There are many structures that behave axisymmetrically, like the cylindrical tubes under axial compression studied in Chapter 6. Such behaviour allows for the use of analytical and numerical models that are bi–dimensional, hence simpler than the more complete 3D models. It is opportune then to explore a linear axisymmetric finite element, which can be used to analyse dynamic elastic buckling of cylindrical shells. Also, by studying this element we will introduce the important concepts of isoparametric finite element and numerical integration. This will serve as a basis for the more complete three dimensional finite element.

Non–axisymmetric loads are possible to be dealt with these elements but we will not consider them here.

8.3.1 Basic formulas

When developing the beam finite element we obtained its stiffness matrix by applying the principle of virtual work. Here, we follow the same approach but instead of working with bending moment and curvature, we use stresses and strains.

Let us pause a bit here the study of axisymmetric shells and develop a finite element on a plane. We shall see in the sequence that to transform this element to an axisymmetric one is just a matter of using a different constitutive matrix.

For the differential (quasi–static) plane element of the next figure, equilibrium in the x and y directions will give

$$\frac{\partial \sigma_{xx}}{\partial x} + \frac{\partial \sigma_{xy}}{\partial y} + F_x = 0$$

$$\frac{\partial \sigma_{xy}}{\partial x} + \frac{\partial \sigma_{yy}}{\partial y} + F_y = 0$$

called a strong form of equilibrium since it is valid at every point of the domain.

Note that $\mathbf{F}$ is a body force vector per unity volume and can include, via the d'Alembert principle, inertia forces, *ie* $-\rho\mathbf{a}$, where $\mathbf{a}$ is the acceleration vector.

For the sake of completeness, these equations can be expanded to the z direction, giving the full equilibrium equations for a solid as

$$\frac{\partial \sigma_{xx}}{\partial x} + \frac{\partial \sigma_{xy}}{\partial y} + \frac{\partial \sigma_{xz}}{\partial z} + F_x = 0$$

$$\frac{\partial \sigma_{xy}}{\partial x} + \frac{\partial \sigma_{yy}}{\partial y} + \frac{\partial \sigma_{yz}}{\partial z} + F_y = 0$$

$$\frac{\partial \sigma_{xz}}{\partial x} + \frac{\partial \sigma_{yz}}{\partial y} + \frac{\partial \sigma_{zz}}{\partial z} + F_z = 0$$

or

$$\{\partial\}^T\{\sigma\} + \{\mathbf{F}\} = \{\mathbf{0}\},$$

with

$$\{\partial\} = \begin{bmatrix} \partial/\partial x & 0 \\ 0 & \partial/\partial y \\ \partial/\partial y & \partial/\partial x \end{bmatrix}$$

and

$$\{\partial\} = \begin{bmatrix} \partial/\partial x & 0 & 0 \\ 0 & \partial/\partial y & 0 \\ 0 & 0 & \partial/\partial z \\ \partial/\partial y & \partial/\partial x & 0 \\ 0 & \partial/\partial z & \partial/\partial y \\ \partial/\partial z & 0 & \partial/\partial x \end{bmatrix},$$

for the 2-D and 3-D cases, respectively.

These equilibrium equations are complemented by the boundary conditions, *ie* prescribed displacement and stresses on the sides (or surfaces) of the body. Surface tractions, $\mathbf{\Phi}$, are forces per surface area and can always be decomposed parallel to the normal coordinate axes. Hence, for the Cartesian coordinates x, y and z, the surface tractions are

$$\{\mathbf{\Phi}\} = \{\Phi_x \quad \Phi_y \quad \Phi_z\}$$

with

$$\phi_x = l\sigma_{xx} + m\sigma_{xy} + n\sigma_{xz}\phi_y = l\sigma_{xy} + m\sigma_{yy} + n\sigma_{yz}\phi_z = l\sigma_{xz} + m\sigma_{yz} + n\sigma_{zz},$$

being l, m, n the direction cosines of a vector normal to the surface under consideration.

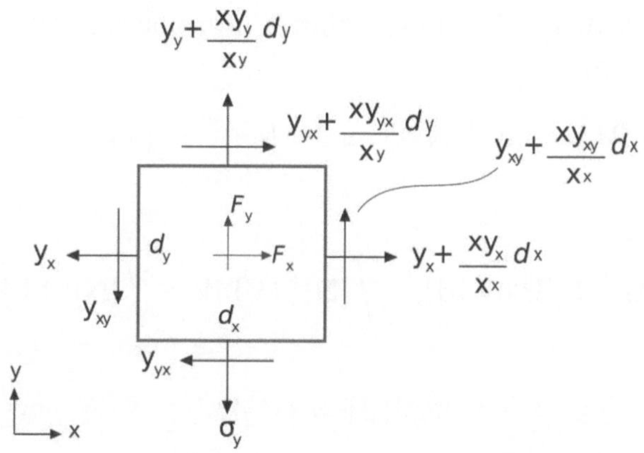

Figure 8.13 A differential plane element.

In contradistinction to the strong form of equilibrium, we have the weak form, so called because it is of an integral nature, valid in an average sense. The weak form is most appropriate for the finite element approach and it is obtained by multiplying the strong form by a virtual displacement, $\{\delta\mathbf{u}\}$ and integrating the result by parts, leading to

$$\int \{\delta\varepsilon\}^T\{\sigma\}\mathrm{d}V = \int \{\delta\mathbf{u}\}^T\{\mathbf{F}\}\mathrm{d}V + \int \{\delta\mathbf{u}\}^T\{\mathbf{\Phi}\}\mathrm{d}S,$$

Note that here $\{\mathbf{u}\}$ is a generalized vector with components u, v, w, later to be specialized to the axisymmetric variables.

We now interpolate the continuous displacement field in our domain by using the shape functions and the displacements at the nodes via

$$\{\mathbf{u}\} = [\mathbf{N}]\{\mathbf{d}\}$$

and since in the linear theory adopted in this section the strains are the first derivative of the displacements,

$$\left\{ \begin{array}{c} \varepsilon_x \\ \varepsilon_y \\ \varepsilon_{xy} \end{array} \right\} = \left[\begin{array}{cc} \frac{\partial}{\partial x} & 0 \\ 0 & \frac{\partial}{\partial y} \\ \frac{\partial}{\partial y} & \frac{\partial}{\partial x} \end{array} \right] \left\{ \begin{array}{c} u \\ v \end{array} \right\},$$

it follows that

$$\{\varepsilon\} = [\mathbf{B}]\{\mathbf{d}\} \quad \text{with} \quad [\mathbf{B}] = [\partial][\mathbf{N}],$$

with temperature effects, residual stresses and residual strains all being disregarded.

Applying it to the functional above, interpolating the virtual displacement field by

$$\{\delta \mathbf{u}\}^T = \{\delta \mathbf{d}\}^T[\mathbf{N}]^T \quad \text{and} \quad \{\delta \varepsilon\}^T = \{\delta \mathbf{d}\}^T[\mathbf{B}]^T,$$

we obtain

$$\{\delta \mathbf{d}\}^T \left(\int [\mathbf{B}]^T[\mathbf{E}][\mathbf{B}]\mathrm{d}V\{\mathbf{d}\} - \int [\mathbf{N}]^T\{\mathbf{F}\}\mathrm{d}V - \int [\mathbf{N}]^T\{\boldsymbol{\Phi}\}\mathrm{d}S \right) = 0,$$

or

$$[\mathbf{K}_e]\{\mathbf{d}\} = \{\mathbf{R}_e\},$$

with

$$[\mathbf{K}_e] = \int [\mathbf{B}]^T[\mathbf{E}][\mathbf{B}]\mathrm{d}V$$

and

$$\{\mathbf{R}_e\} = \int [\mathbf{N}]^T\{\mathbf{F}\}\mathrm{d}V + \int [\mathbf{N}]^T\{\boldsymbol{\Phi}\}\mathrm{d}S.$$

Note that the type of plane state problem we can solve with this procedure is in our hands: we can handle linear plane stress, plane strain and their corresponding axisymmetric cases, only being necessary to adopt the corresponding elastic isotropic constitutive relation that comes from Hooke's law and represented by the tensor $[\mathbf{E}]$.

In the case of tube under axial load in the z direction, axisymmetric case, the plane strain state is described by

$$\sigma_z = \nu(\sigma_r + \sigma_\theta), \qquad \sigma_{\theta z} = \sigma_{rz} = 0,$$

and Hooke's law reads

$$\begin{Bmatrix} \sigma_r \\ \sigma_\theta \\ \sigma_{r\theta} \end{Bmatrix} = \underbrace{\frac{E}{(1+\nu)(1-2\nu)} \begin{bmatrix} 1-\nu & \nu & 0 \\ \nu & 1-\nu & 0 \\ 0 & 0 & 1 \end{bmatrix}}_{[\mathbf{E}]} \begin{Bmatrix} \varepsilon_r \\ \varepsilon_\theta \\ \varepsilon_{r\theta} \end{Bmatrix}.$$

8.3.2 Rectangular finite element

As an application of the previous equations, let us develop the formulation of the linear rectangular finite element shown in the next figure.

One way to obtain its shape functions is by inspection, or by applying the Lagrange's interpolation formula

$$N_k = \frac{(x_1 - x)(x_2 - x)\ldots[x_k - x]\ldots(x_n - x)}{(x_1 - x_k)(x_2 - x_k)\ldots[x_k - x_k]\ldots(x_n - x_k)},$$

with the terms within the brackets being omitted in order to obtain the kth shape function. In either case, the shape functions are

$$N_1 = \frac{(a-x)(b-y)}{4ab}, \quad N_2 = \frac{(a+x)(b-y)}{4ab},$$

$$N_3 = \frac{(a+x)(b+y)}{4ab}, \quad N_4 = \frac{(a-x)(b+y)}{4ab},$$

which allow us to relate the displacement field in the domain of the finite element via $\{\mathbf{u}\} = [\boldsymbol{N}]\{\mathbf{d}\}$, with

$$[\mathbf{N}] = \begin{bmatrix} N_1 & 0 & N_2 & 0 & N_3 & 0 & N_4 & 0 \\ 0 & N_1 & 0 & N_2 & 0 & N_3 & 0 & N_4 \end{bmatrix}.$$

Now, since $[\mathbf{B}] = [\boldsymbol{\partial}][\mathbf{N}]$ is follows that

$$\mathbf{B} = \frac{1}{4ab}\begin{bmatrix} -b+y & 0 & b-y & 0 & b+y & 0 & -b-y & 0 \\ 0 & -a+x & 0 & -a-x & 0 & a+x & 0 & a-x \\ -a+x & -b+y & -a-x & b-y & a+x & b+y & a-x & -b-y \end{bmatrix},$$

from which it results the stiffness matrix

$$[\mathbf{K}_e] = \int_{-b}^{b}\int_{-a}^{a} [\mathbf{B}]^T[\mathbf{E}][\mathbf{B}]t\mathrm{d}x\mathrm{d}y,$$

with

$$[\varepsilon] = [\mathbf{B}]\{\mathbf{d}\}, \qquad \{\boldsymbol{\sigma}\} = [\mathbf{E}][\mathbf{B}]\{\mathbf{d}\},$$

t being the element thickness.

One limitation of this finite element is that it exhibits shear locking, *ie* shear strains appear in a pure bending loading case, absorbing energy and requiring values of bending moment larger than the correct one.

This behaviour is quantified in R.D. Cook, D.S. Malkus, M.E. Plesha and R.J. Witt, Concepts and Applications of Finite Element Analysis, 4th. ed., Wiley, 2002, from which this section is based on.

Depending on the problem at hand, it is possible to improve an analysis by using a larger number of finite elements. This enrichment is referred to a h. Another way is to increase the degree of the shape functions and when this p enrichment is applied to the linear rectangular element we will have as a next step a quadratic one, which does not exhibit shear

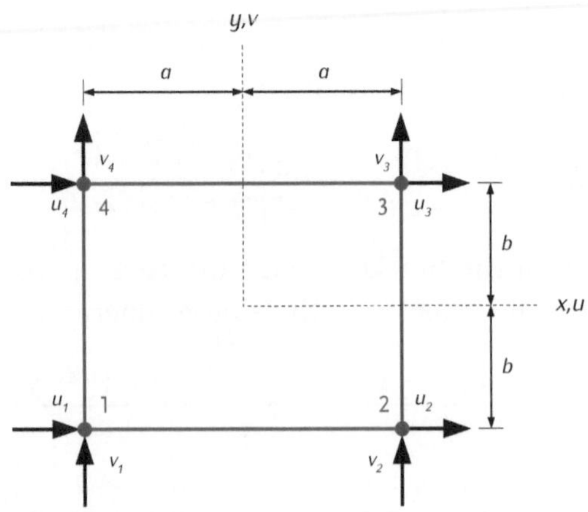

Figure 8.14 A bilinear rectangular finite element.

locking. We shall argue though that, for impact analysis, linear shape functions have been used complemented by h enrichment, *ie* fine meshes.

Now consider the conical shell in the next figure, with its longitudinal cross-section being represented on one side by the rectangular finite elements seen above and the other side represented by distorted rectangular finite elements. It is clear that nonregular domains require another type of finite element, explored next under the so called isoparametric formulation.

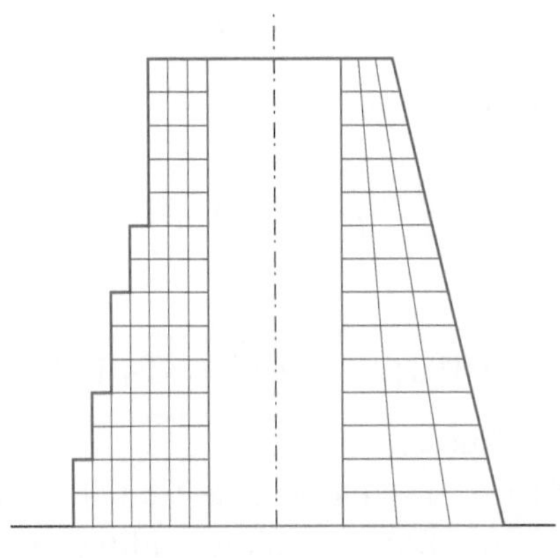

Figure 8.15 A conical shell cross–section modeled by rectangular finite elements and by isoparametric ones.

8.3.3 Isoparametric formulation

The key idea behind the isoparametric formulation is the transformation of coordinates and the fact that both the displacement field (or any other scalar field, such as temperature) within the element and its geometry are interpolated using the same shape functions. We adopt here a coordinated system r, s whose origin in the physical space does not need to be in the geometrical centre of the element. In the r, s space though, the origin is the center of a square element such that $-1 \leq r \leq 1$ and $-1 \leq s \leq 1$. Note that the displacement field given by u and v is referred to as the Cartesian system $x - y$. We can enrich the element by allowing mid–nodes. Such a quadratic formulation, not detailed here, allows the element to have curved sides. Hence, as shown in the next figure, both physical geometric representation, of straight and curved characteristics on the left side, can be represented by isoparametric finite elements.

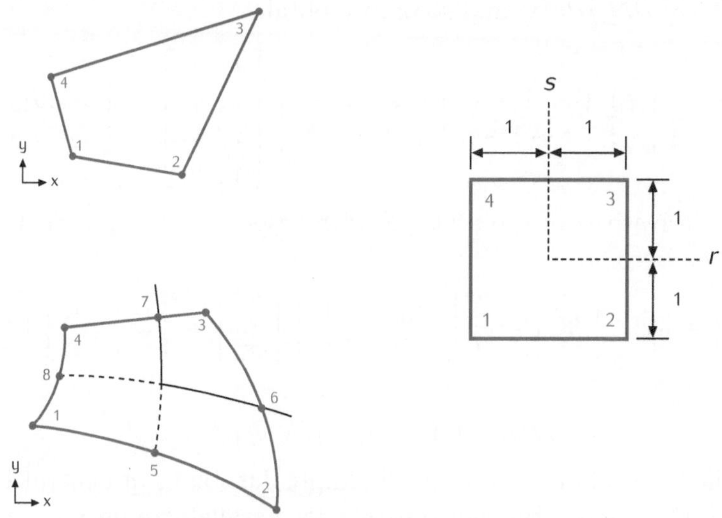

Figure 8.16 Straight and curved geometrical physical domains represented by a isoparametric finite element.

The shape functions for the isoparametric bilinear quadrilateral finite element are the same as seen for the rectangular element, but setting $a = b = 1$ and $x = r$ and $y = s$. The displacement field is, as before, $\{\mathbf{u}\} = [\mathbf{N}]\{\mathbf{d}\}$ and the physical coordinates are interpolated via $\{\mathbf{x}\} = [\mathbf{N}]\{\mathbf{c}\}$, with

$$\{\mathbf{d}\} = [\, u_1 \; v_1 \; u_2 \; v_2 \; u_3 \; v_3 \; u_4 \; v_4 \,]^T,$$

$$\{\mathbf{x}\} = [\, x \; y \,]^T,$$

and

$$\{\mathbf{c}\} = [\, x_1 \; y_1 \; x_2 \; y_2 \; x_3 \; y_3 \; x_4 \; y_4 \,]^T.$$

Now, we need to evaluate $[\mathbf{B}]$, which requires the calculation of the derivative operator $[\partial]$. But it is not possible to perform directly, for instance, the derivative $\partial/\partial x$, since our shape functions have r and s as arguments. To move from one basis, $r-s$, to the other, $x-y$, we define the operator

$$[\mathbf{J}] = \begin{bmatrix} \frac{\partial x}{\partial r} & \frac{\partial y}{\partial r} \\ \frac{\partial x}{\partial s} & \frac{\partial y}{\partial s} \end{bmatrix}$$

and write, from the chain rule,

$$\begin{bmatrix} \frac{\partial S}{\partial r} \\ \frac{\partial S}{\partial s} \end{bmatrix} = [\mathbf{J}] \left\{ \begin{matrix} \frac{\partial S}{\partial x} \\ \frac{\partial S}{\partial y} \end{matrix} \right\},$$

with S being any scalar. For our quadrilateral finite element and using $\partial x/\partial r = (\sum \partial \boldsymbol{N}_i/\partial r)x_i$ and so on, we obtain

$$[\mathbf{J}] = \frac{1}{4}\begin{bmatrix} -1+s & 1-s & 1+s & -1-s \\ -1+r & -1-r & 1+r & 1-r \end{bmatrix} \begin{bmatrix} x_1 & y_1 \\ x_2 & y_2 \\ x_3 & y_3 \\ x_4 & y_4 \end{bmatrix} = \begin{bmatrix} J_{11} & J_{12} \\ J_{21} & J_{22} \end{bmatrix}.$$

The derivatives of the scalar S with respect to the physical coordinates then become

$$\begin{bmatrix} \frac{\partial S}{\partial x} \\ \frac{\partial S}{\partial y} \end{bmatrix} = [\mathbf{J}]^{-1} \begin{bmatrix} \frac{\partial S}{\partial r} \\ \frac{\partial S}{\partial s} \end{bmatrix} = \frac{1}{J}\begin{bmatrix} J_{22} & -J_{12} \\ -J_{21} & J_{11} \end{bmatrix} \begin{bmatrix} \frac{\partial S}{\partial r} \\ \frac{\partial S}{\partial s} \end{bmatrix} = \begin{bmatrix} \Gamma_{11} & \Gamma_{12} \\ \Gamma_{21} & \Gamma_{22} \end{bmatrix} \begin{bmatrix} \frac{\partial S}{\partial r} \\ \frac{\partial S}{\partial s} \end{bmatrix},$$

with

$$J = \det[\mathbf{J}] = J_{11}J_{22} - J_{21}J_{12}.$$

$\boldsymbol{J}$ is called Jacobian matrix and J simply Jacobian, in general being a function of r and s, although constant for parallelograms.

We still need to build up the matrix $[\mathbf{B}]$ for our isoparametric quadrilateral finite element and to this end we notice that

$$\left\{ \begin{matrix} \varepsilon_x \\ \varepsilon_y \\ \varepsilon_{xy} \end{matrix} \right\} = \begin{bmatrix} 1 & 0 & 0 & 0 \\ 0 & 0 & 0 & 1 \\ 0 & 1 & 1 & 0 \end{bmatrix} \left\{ \begin{matrix} \frac{\partial u}{\partial x} \\ \frac{\partial u}{\partial y} \\ \frac{\partial v}{\partial x} \\ \frac{\partial v}{\partial y} \end{matrix} \right\} = \begin{bmatrix} 1 & 0 & 0 & 0 \\ 0 & 0 & 0 & 1 \\ 0 & 1 & 1 & 0 \end{bmatrix} \times$$

$$\begin{bmatrix} \Gamma_{11} & \Gamma_{12} & 0 & 0 \\ \Gamma_{21} & \Gamma_{22} & 0 & 0 \\ 0 & 0 & \Gamma_{11} & \Gamma_{12} \\ 0 & 0 & \Gamma_{21} & \Gamma_{22} \end{bmatrix} \begin{bmatrix} \frac{\partial N_1}{\partial r} & 0 & \frac{\partial N_2}{\partial r} & 0 & \frac{\partial N_3}{\partial r} & 0 & \frac{\partial N_4}{\partial r} & 0 \\ \frac{\partial N_1}{\partial s} & 0 & \frac{\partial N_2}{\partial s} & 0 & \frac{\partial N_3}{\partial s} & 0 & \frac{\partial N_4}{\partial s} & 0 \\ 0 & \frac{\partial N_1}{\partial r} & 0 & \frac{\partial N_2}{\partial r} & 0 & \frac{\partial N_3}{\partial r} & 0 & \frac{\partial N_4}{\partial r} \\ 0 & \frac{\partial N_1}{\partial s} & 0 & \frac{\partial N_2}{\partial s} & 0 & \frac{\partial N_3}{\partial s} & 0 & \frac{\partial N_4}{\partial s} \end{bmatrix} \left\{ \begin{matrix} u_1 \\ v_1 \\ u_2 \\ v_2 \\ u_3 \\ v_3 \\ u_4 \\ v_4 \end{matrix} \right\},$$

so [**B**] is what precedes the nodal vector {**d**}. Now, the stiffness matrix comes from the integration

$$[\mathbf{K}_e] = \int\int [\mathbf{B}]^T[\mathbf{E}][\mathbf{B}]t dx dy = \int_{-1}^{1}\int_{-1}^{1} [\mathbf{B}]^T[\mathbf{E}][\mathbf{B}]tJ dr ds.$$

A quick look at the above equation reveals that there is a cross product of the derivatives of the shape functions that need to be integrated. Also, in most cases, the Jacobian is not a constant and all this renders the above equation quite difficult to be analytically integrated. Hence the preference to perform the above integration numerically, to be seen next.

■ JACOBIAN CALCULATION

Evaluate the Jacobian for the finite elements in the figure.

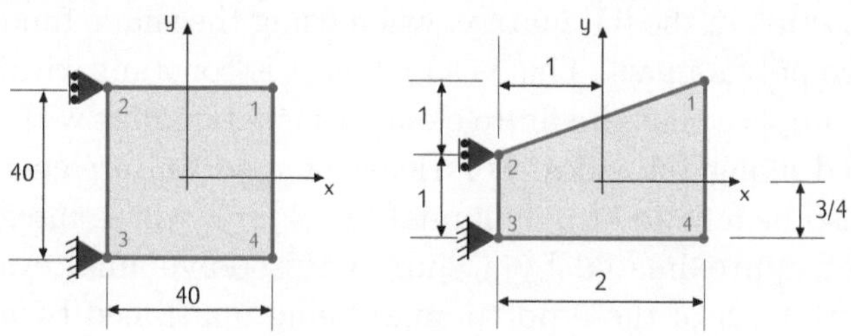

Figure 8.17 Quadrilateral finite elements.

For element 1,

$$x = 20r \quad \text{and} \quad y = 20s$$

so that

$$J = \begin{bmatrix} 20 & 0 \\ 0 & 20 \end{bmatrix}$$

is constant.

For element 2, we use the shape functions to express the coordinates x and y by

$$x = N_1u_1 + N_2u_2 + N_3u_3 + N_4u_4$$

and

$$y = N_1v_1 + N_2v_2 + N_3v_3 + N_4v_4$$

with

$$N_1 = \frac{1}{4}(1+r)(1+s) \quad N_2 = \frac{1}{4}(1-r)(1+s)$$

$$N_3 = \frac{1}{4}(1-r)(1-s) \quad N_4 = \frac{1}{4}(1+r)(1-s),$$

where u_i and v_i are the vertices coordinates in the figure. From the Jacobian definition it follows that

$$J = \frac{1}{4}\begin{bmatrix} 4 & 1+s \\ 0 & 3+r \end{bmatrix},$$

which is not constant. ■

8.3.4 Numerical integration

The various terms in the [**B**] matrix, when using the shape functions seen here, [**N**], are polynomials. This is so when J is constant within the element, which implies that the finite element is rectangular with the nodes evenly spaced along the sides. For richer shaped finite elements, these terms can also be a ratio of polynomials, say $f(r)$, whose integration we assume to be approximated by a sum of this polynomial evaluated at points r_i, with each of these polynomial being multiplied by a factor (a weight) w_i. Both r_i and w_i have to be determined and we then have

$$\int_{-1}^{1} f(r)\mathrm{d}r = \sum_{i=0}^{p} w_i f(r_i) = w_0 f(r_0) + w_1 f(r_1) + \ldots + w_p f(r_p).$$

In the unidimensional case, we can write our polynomial as

$$f(r) = a_0 + a_1 r + a_2 r^2 + \ldots + a_n r^n,$$

whose integration is

$$\int_{-1}^{1} f(r)\mathrm{d}r = 2a_0 + \frac{2}{3}a_2 + \ldots + \frac{a_n}{n+1}\left[1 - (-1)^{n+1}\right].$$

Now, $f(r_0) = a_0 + a_1 r_0 + \ldots + a_n r_0^n$, $f(r_1) = a_0 + a_1 r_1 + \ldots + a_n r_1^n$, so the sum above has the terms $w_0 a_0$, $w_1 a_0$, $w_2 a_0$, $\ldots w_p$, which equal what multiplies a_0, *ie* 2. Repeating this procedure we arrive at the system of equations

$$\begin{aligned} & w_0 + w_1 + w_2 + ... + w_p = 2 \\ & w_0 r_0 + w_1 r_1 + ... + w_p r_p = 0 \\ & \vdots \\ & w_0 r_0^n + w_1 r_1^n + ... + w_p r_p^n = \tfrac{1}{n+1}\left[1 - (-1)^{n+1}\right], \end{aligned}$$

which has $n+1$ equations and $2(p+1)$ unknowns, giving $n = 2(p+1)-1$. This is to say that, if there are $p+1$ terms in the sum, then we can integrate exactly a polynomial of degree n.

This numerical integration procedure, called Gauss quadrature, can be readily expanded to double and triple integrals and for our quadrilateral element we have

$$\int_{-1}^{1}\int_{-1}^{1} f(r,s)\,\mathrm{d}r\mathrm{d}s = \sum_{i,j=0}^{p} w_{ij} f(r_i, s_j).$$

■ INTEGRATION OF A POLYNOMIAL

Obtain the weights and integration points necessary to integrate exactly a polynomial of degree 3.

For $n = 3$, we have $p+1 = 2$, from which

$$\begin{aligned} & w_0 + w_1 = 2 \\ & w_0 r_0 + w_1 r_1 = 0 \\ & w_0 r_0^2 + w_1 r_1^2 = \tfrac{2}{3} \\ & w_0 r_0^3 + w_1 r_1^3 = 0 \end{aligned}$$

and

$$\begin{aligned} & w_0 = w_1 = 1 \\ & r_1 = -r_0 = \tfrac{1}{\sqrt{3}} \end{aligned}$$

■

For our bilinear quadrilateral finite element we adopt 4 integration points at the positions shown in the next figure, whose value are $\pm 1/\sqrt{3}$, with the weights being 1.0.

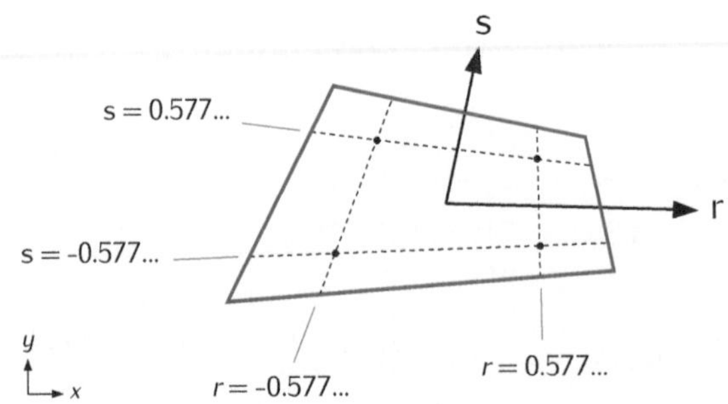

Figure 8.18 Integration points for the bilinear quadratic finite element.

It is sometimes useful to use reduced integration. This means that we approximate our integrals for the bilinear quadrilateral element by

$$\int_{-1}^{1}\int_{-1}^{1} f(r,s)\mathrm{d}r\mathrm{d}s \approx 4f(0,0)$$

and a genuine question would be if to reduce the order of quadrature implies loosing accuracy. It is indeed possible to use a low order quadrature and even so to have more accurate FE results and reduced computation time. However, care should be exercised here since such a procedure introduces spurious modes.

We refer to full integration when the quadrature rule used yields the exact integration of the stiffness matrix coefficients. This occurs if the finite element is rectangular with side nodes evenly spaced. For distorted finite elements, we refer to full integration when exact integration of the stiffness matrix coefficients would be possible if the element were undistorted.

8.3.5 Stress calculation

Stresses are one of the most sought variables in a structural analysis but it is a secondary quantity in the FEM discussed here. Stress comes from the derivative of the displacements, *ie* the strains. We note that the displacement field is continuous but this may not be the case for the strains, leading also to discontinuities on the stress field at the nodes. Hence, to a common node of two finite elements, it is usually attributed two or more stress values, which are averaged in part to give to the analyst a more pleasant visual output. We can use this difference of stresses in a shared node as an indicative of the mesh refinement: finer mesh will yield less discontinuities of the strain–stress field, so we can think of these discontinuities as a measure of error or as a warning to the analyst.

The common procedure to calculate stresses within an element is to use the stress magnitude in the Gauss points and to extrapolate to the nodes via the same shape functions used to interpolate the displacement field. Let the integration points have the coordinates r and s for a quadrilateral element, which are related to the dimensionless coordinates $\xi = r/\sqrt{3}$ and $\zeta = s/\sqrt{3}$. We now evaluate the shape functions at r and s, *eg* $N_1 = (1-r)(1-s)/4$, and extrapolate the stresses calculated there to the nodes via

$$\sigma = N_i \sigma_i,$$

which gives

$$\sigma_C = 0.1340\sigma_1 - 0.5\sigma_2 + 1.866\sigma_3 - 0.5\sigma_4.$$

Note that the stress at C, when evaluated directly would give σ_3, hence not transferring any influence of the stresses levels of the other Gauss points. It seems then that the extrapolation method is more interesting as much as it considers the stresses at all integration points. This technique also applies to solid elements, with a third coordinate calculated via $\zeta = t/\sqrt{3}$.

Other ways to obtain the stresses are not discussed here but we point out that it is sometimes misleading to judge the correctness of an (in house) implemented finite element by comparing its stress output with a reference (professional) code.

8.4 Closure

We have given an overview of the linear finite element analysis. The subject is a vast one and we endeavoured to present dynamic analysis, with the necessary distinction between implicit and explicit analyses. We also introduced the assembly procedure for the common case of a multitude of finite elements, as well as coordinate transformation. Little can be advanced in this field without the use of isoparametric finite element, which was here introduced side by side with the Jacobian and the numerical integration.

Now having these concepts in mind we can advance further to nonlinear finite element analysis, as explored next.

8.5 Problems

1. For the problem in the next figure, show that

$$\frac{EI}{L^3}\begin{bmatrix} 1 & 0 & 0 & 0 & 0 & 0 \\ 0 & 1 & 0 & 0 & 0 & 0 \\ 0 & 0 & 12+12 & 6L+6L & 0 & 6L \\ 0 & 0 & 6L+6L & 4L^2+4L^2 & 0 & 2L^2 \\ 0 & 0 & 0 & 0 & 1 & 0 \\ 0 & 0 & 6L & 2L^2 & 0 & 4L^2 \end{bmatrix}\begin{bmatrix} w_1 \\ \theta_1 \\ w_2 \\ \theta_2 \\ w_3 \\ \theta_3 \end{bmatrix} = \frac{pL}{2}\begin{bmatrix} 0 \\ 0 \\ 1 \\ L/6 \\ 1 \\ -L/6 \end{bmatrix}$$

when discretising the beam with two finite elements. Assume $L_1 = L_2$ and no axial load.

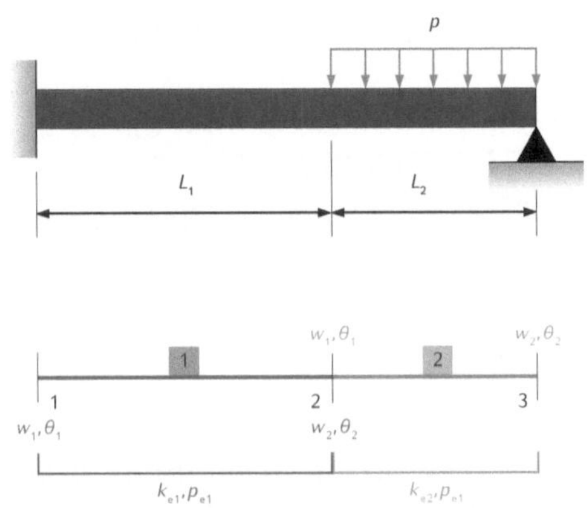

Figure 8.19 A beam discretized by two finite elements.

2. Why is it necessary to rotate the mass matrix when gravity is not being considered in a given model?

3. Write a FE program for rods so it can handle bars of different properties and damping. Study the case of a cantilever rod under the action of a harmonic wave. Plot the strain profile along time and along the length of the bars.

4. Obtain the weak form of equilibrium for a solid from the strong one and show that

$$\int \{\delta\varepsilon\}^T\{\boldsymbol{\sigma}\}\mathrm{d}V = \int \{\delta\mathbf{u}\}^T\{\mathbf{F}\}\mathrm{d}V + \int \{\delta\mathbf{u}\}^T\{\boldsymbol{\Phi}\}\mathrm{d}S.$$

5. Deduce the expression for the rectangular finite element **B** matrix.

6. What are the main differences between the Gauss and Lobatto numerical integration procedures?

7. Given that finite element codes can perform a much more detailed analyses of the buckling of a shell, what would be the reason, if any, to study this phenomenon analytically?

8. Your company advocates the use of composite car structures on the basis that, being light, it implies in less fuel consumption. Studies in the literature indicate that the manufacturing and recycling energy for this car outbalance the apparent environment friendly advantages of the composite structure. In a interview panel you are exposed to these two scenarios and a journalist asks you if you are aware that your company misleads the public. How would you react?

CHAPTER 9

Non-linear Finite Elements Analysis

with Larissa Driemeier

Figure 9.1 Simulation of a carbon–fiber tube under axial impact.

The various phenomena seen here lend to the Impact Engineering a clear nonlinear character. Material behaviour, large deformation, coupling of stresses, inertia effects, etc... require, for more complex structures, an approach based on finite elements, with these nonlinear features being incorporated in the formulation.

To present in a more clear way these nonlinear aspects of the finite element method we set aside the inertia effects, treated anyway in the time integration methods presented in the previous chapter, and focus in large displacement and deformation, as well as in computational plasticity.

DOI: 10.1201/9781003485292-9

Nonlinearitics of geometrical nature are presented via a nonlinear truss and beam finite element. The solution procedure require to implement the Newton–Raphson procedure, also detailed here. We advance further into the plasticity realm in the context of finite element, both uni and tri dimensional, when the radial return mapping algorithm is presented. Basic concepts of kinematics are given next, such as gradient of deformation and stress rates. These concepts are believed to equip the reader with some background to further study this rich subject of computational mechanics.

9.1 Non-linear truss element

We have seen in Chapter 7 the unidimensional strain family and the conjugate stress counter parts. We have also explored the advantages of the natural strain definition, associated with the true stress definition. It is this strain–stress pair that is going to be used in this section to develop a nonlinear truss element. Because we want to explore geometrical nonlinear aspects of the Finite Element formulation, we will restrict ourselves to the static analysis here in this section.

This section is based on a Report by Dr. G.M. Barbosa, from GMSIE-USP, 2003.

A truss element is like a rod, in the sense that it sustains only axial loads. The 3D truss finite element is depicted in the next figure, where it can be appreciated the local, r, and the global, x, y, z coordinate systems as well as the degree of freedoms of the nodes 1, u_1, v_1, w_1, and 2, u_2, v_2, w_2. Forces act only in these nodes, in the directions of the displacements.

We start by stating that the sum of all internal and external virtual work in the truss element should be zero,

$$\delta W_{\text{I}} + \delta W_e = 0,$$

which gives

$$\int_V \delta\varepsilon\sigma dV = \int_V \Phi f^B dV + \int_S \Phi f^S dS + \sum_i \delta p_i R_i,$$

where we set the body and area forces, f^B and f^S as zero, leaving only the loads at the nodes, R_i, whose vector form is

$$\boldsymbol{F} = [R_1 \quad R_2 \quad R_3 \quad R_4 \quad R_5 \quad R_6]^{\text{T}},$$

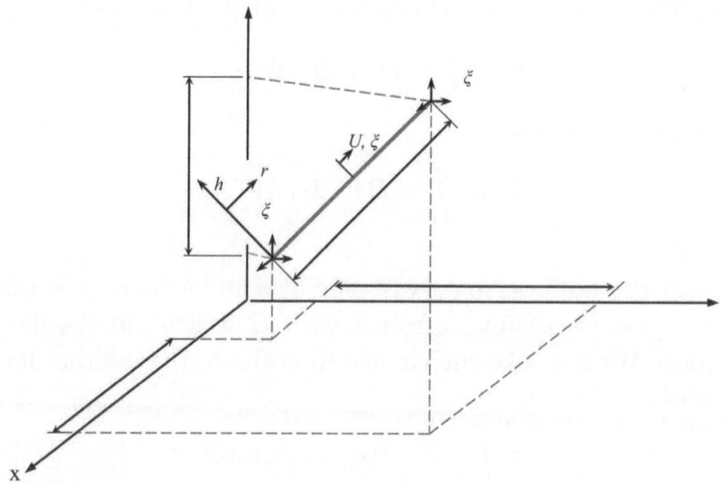

Figure 9.2 A truss finite element.

associated to the node displacements in the vector form of

$$\boldsymbol{p} = \begin{bmatrix} u_1 & v_1 & w_1 & u_2 & v_2 & w_2 \end{bmatrix}^{\mathrm{T}}.$$

The virtual external work can then be written as

$$\delta W_E = \sum_i \delta p_i R_i = \boldsymbol{F}^{\mathrm{T}} \delta \boldsymbol{p}$$

and the displacement along the truss element as

$$U = \boldsymbol{H}\boldsymbol{p},$$

from which we obtain the strain and the nominal stress along the element

$$\varepsilon = \frac{\partial U}{\partial r} = \frac{\partial \boldsymbol{H}}{\partial r}\boldsymbol{p} \quad \Rightarrow \quad \varepsilon = \boldsymbol{B}\boldsymbol{p}$$

and

$$\sigma = E\boldsymbol{B}\boldsymbol{p},$$

respectively, with $\boldsymbol{H}$ to be defined later.

The internal virtual work reads

$$\delta W_I = \delta p \int_{V^e} \left(\boldsymbol{B}^{\mathrm{T}} E \boldsymbol{B}\boldsymbol{p}\right) \mathrm{d}V^e$$

and by equating to the external one results in the force vector

$$\boldsymbol{F} = \int_{V^e} \left(\boldsymbol{B}^{\mathrm{T}} \boldsymbol{E} \boldsymbol{B}\right) \mathrm{d}V^e \boldsymbol{p}$$

and in the stiffness matrix

$$\boldsymbol{K} = \int_{V^e} \left(\boldsymbol{B}^{\mathrm{T}} \boldsymbol{E} \boldsymbol{B}\right) \mathrm{d}V^e$$

since $\boldsymbol{F} = \boldsymbol{K}\boldsymbol{p}$.

Note that this stiffness matrix is valid also for the nonlinear case and for that we need to occupy ourselves with $\boldsymbol{B}$ written in the deformed configuration. We can take the chance to evaluate the internal forces in the elements as

$$\boldsymbol{F}^{int} = \int_{V^e} \boldsymbol{B}^{\mathrm{T}} \sigma \mathrm{d}V^e = A_n L_n \boldsymbol{B}^{\mathrm{T}} \sigma,$$

or

$$\boldsymbol{F}^{int} = N L_n \boldsymbol{B}^{\mathrm{T}},$$

with N, A_n and L_n being the nominal normal force and the current area and length of the element, respectively. Of course that, for small deformation, $L_n \approx L$ and $A_n \approx A_0$.

For the finite element, we adopt the linear set of shape functions, $h_1 = \frac{1}{2}(-\xi + 1)$ and $h_2 = \frac{1}{2}(\xi + 1)$ and we define

$$\boldsymbol{h} = \frac{1}{2}\left[1 - \xi \;\; 1 + \xi\right],$$

so that the geometry of the element and its displacement at any position can be defined, according to the isoparametric formulation, as

$$x = \tfrac{1}{2}\left[1 - \xi \;\; 1 + \xi\right] \begin{pmatrix} x_1 \\ x_2 \end{pmatrix} = \textstyle\sum_{i=1}^{2} h_i x_i$$

$$y = \tfrac{1}{2}[1 - \xi \quad 1 + \xi] \begin{pmatrix} y_1 \\ y_2 \end{pmatrix} = \textstyle\sum_{i=1}^{2} h_i y_i$$

$$z = \tfrac{1}{2}[1 - \xi \quad 1 + \xi] \begin{pmatrix} z_1 \\ z_2 \end{pmatrix} = \textstyle\sum_{i=1}^{2} h_i z_i$$

and

$$u = \tfrac{1}{2}\left[1 - \xi \;\; 1 + \xi\right] \begin{pmatrix} u_1 \\ u_2 \end{pmatrix} = \textstyle\sum_{i=1}^{2} h_i u_i$$

$$v = \tfrac{1}{2}\left[1 - \xi \;\; 1 + \xi\right] \begin{pmatrix} v_1 \\ v_2 \end{pmatrix} = \textstyle\sum_{i=1}^{2} h_i v_i$$

$$w = \tfrac{1}{2}\left[1 - \xi \;\; 1 + \xi\right] \begin{pmatrix} w_1 \\ w_2 \end{pmatrix} = \textstyle\sum_{i=1}^{2} h_i w_i$$

Now, the axial displacement, U, is

$$U = \frac{ux_{21} + vy_{21} + wz_{21}}{L} =$$

$$\frac{1}{2l}\begin{bmatrix} x_{21} & y_{21} & z_{21} \end{bmatrix} \begin{bmatrix} 1-\xi & 0 & 0 & 1+\xi & 0 & 0 \\ 0 & 1-\xi & 0 & 0 & 1+\xi & 0 \\ 0 & 0 & 1-\xi & 0 & 0 & 1+\xi \end{bmatrix} \begin{bmatrix} u_1 \\ v_1 \\ w_1 \\ u_2 \\ v_2 \\ w_2 \end{bmatrix}$$

or

$$U = \boldsymbol{H}\boldsymbol{p},$$

from which

$$\boldsymbol{H} = \frac{1}{2l}\begin{bmatrix} x_{21} & y_{21} & z_{21} \end{bmatrix} \begin{bmatrix} 1-\xi & 0 & 0 & 1+\xi & 0 & 0 \\ 0 & 1-\xi & 0 & 0 & 1+\xi & 0 \\ 0 & 0 & 1-\xi & 0 & 0 & 1+\xi \end{bmatrix},$$

when defining

$$x_{21} = x_2 - x_1, \quad y_{21} = y_2 - y_1, \quad z_{21} = z_2 - z_1.$$

It is possible to obtain the $\boldsymbol{B}$ matrix,

$$\boldsymbol{B} = \frac{\partial \boldsymbol{H}}{\partial r} = \frac{\partial \boldsymbol{H}}{\partial \xi}\frac{\partial \xi}{\partial r} =$$

$$\frac{1}{L^2}\begin{bmatrix} -x_{21} & -y_{21} & -z_{21} & x_{21} & y_{21} & z_{21} \end{bmatrix} = \frac{1}{L^2}\boldsymbol{b}^{\mathrm{T}}(x),$$

with

$$\boldsymbol{b}^{\mathrm{T}}(x) = [-x_{21} \quad -y_{21} \quad -z_{21} \quad x_{21} \quad y_{21} \quad z_{21}],$$

from which the stiffness matrix becomes

$$\boldsymbol{K} = \int_{V^e} \boldsymbol{B}^{\mathrm{T}} E \boldsymbol{B} \mathrm{d}V = EA\frac{1}{L^3}\boldsymbol{b}(x)\boldsymbol{b}^{\mathrm{T}}(x).$$

The matrix $\boldsymbol{B}$ was obtained before as $\boldsymbol{b}^{\mathrm{T}}(x)$ but for a nonlinear analysis the element length L is no longer constant and becomes L_n. Indeed,

$$\boldsymbol{B}^{\mathrm{T}} = \frac{\partial \varepsilon}{\partial \boldsymbol{p}} = \frac{L}{L_n}\frac{1}{L}\frac{\partial L_n}{\partial \boldsymbol{p}} = \frac{1}{L_n}\frac{\partial L_n}{\partial \boldsymbol{p}}$$

and

$$L_n^2 = (\boldsymbol{x}_{21} + \boldsymbol{p}_{21})^{\mathrm{T}} (\boldsymbol{x}_{21} + \boldsymbol{p}_{21})$$

with

$$\boldsymbol{x}_{21} = \begin{bmatrix} x_{21} \\ y_{21} \\ z_{21} \end{bmatrix} \quad \boldsymbol{p}_{21} = \begin{bmatrix} u_{21} \\ v_{21} \\ w_{21} \end{bmatrix},$$

from which

$$L_n^2 = (\boldsymbol{x}_{21} + \boldsymbol{p}_{21})^{\mathrm{T}} (\boldsymbol{x}_{21} + \boldsymbol{p}_{21}),$$

or

$$L_n^2 = (\boldsymbol{x} + \boldsymbol{p})^{\mathrm{T}} \boldsymbol{A} (\boldsymbol{x} + \boldsymbol{p})$$

when adopting $\boldsymbol{x} = [x_1 \quad y_1 \quad z_1 \quad x_2 \quad y_2 \quad z_2]$ and

$$A = \begin{bmatrix} 1 & 0 & 0 & -1 & 0 & 0 \\ 0 & 1 & 0 & 0 & -1 & 0 \\ 0 & 0 & 1 & 0 & 0 & -1 \\ -1 & 0 & 0 & 1 & 0 & 0 \\ 0 & -1 & 0 & 0 & 1 & 0 \\ 0 & 0 & -1 & 0 & 0 & 1 \end{bmatrix}.$$

Since $L_n^2 = \boldsymbol{x}^{\mathrm{T}} \boldsymbol{A} \boldsymbol{x} + \boldsymbol{p}^{\mathrm{T}} \boldsymbol{A} \boldsymbol{p} + \boldsymbol{x}^{\mathrm{T}} \boldsymbol{A} \boldsymbol{p} + \boldsymbol{p}^{\mathrm{T}} \boldsymbol{A} \boldsymbol{x}$, we can differentiate it to obtain

$$\boldsymbol{B}^{\mathrm{T}} = \frac{\boldsymbol{A}(\boldsymbol{x} + \boldsymbol{p})}{(\boldsymbol{x} + \boldsymbol{p})^{\mathrm{T}} \boldsymbol{A} (\boldsymbol{x} + \boldsymbol{p})}.$$

We will see next the Newton–Raphson procedure of solving nonlinear equations, where we will show the need of working with the tangent matrix, $\boldsymbol{K}_T$. For the truss nonlinear finite element, its tangent matrix can be written as

$$\boldsymbol{K}_T = -\frac{\partial \left(\boldsymbol{F}^{ext} - \boldsymbol{F}^{int}\right)}{\partial \boldsymbol{p}} = \frac{\partial \boldsymbol{F}^{int}}{\partial \boldsymbol{p}} = \frac{\partial \left(\sigma A L \boldsymbol{B}^T\right)}{\partial \boldsymbol{p}}$$

with the internal force vector being

$$\boldsymbol{F}^{int} = \sigma \frac{A_0}{L} \lambda^{-(1+2\nu)} \boldsymbol{A} \boldsymbol{x}'$$

when adopting $\boldsymbol{x}' = \boldsymbol{x} + \boldsymbol{p}$, since

$$\frac{A_n}{A_0} = \left(\frac{L_n}{L}\right)^{-2\nu} = \lambda^{-2\nu},$$

ν being the Poisson ratio.

We can now write the tangent matrix as

$$\boldsymbol{K}_T = \frac{\partial \boldsymbol{F}^{int}}{\partial \boldsymbol{p}} =$$
$$= \left(\frac{\partial \sigma}{\partial \boldsymbol{p}}\right)\frac{A_n}{L}\lambda^{-1}\boldsymbol{A}\boldsymbol{x}' + \boldsymbol{A}\left(\frac{\partial \boldsymbol{x}'}{\partial \boldsymbol{p}}\right)\frac{A_n}{L}\sigma\lambda^{-1} + \left(\frac{\partial \lambda^{-(1+2\nu)}}{\partial \boldsymbol{p}}\right)\frac{A_0}{L}\sigma \boldsymbol{A}\boldsymbol{x}'$$

or

$$\boldsymbol{K}_T = \frac{A_0}{L^3}\lambda^{-(3+2\nu)}(E-(1+2\nu)\sigma)\boldsymbol{C}(x')\,\boldsymbol{C}(x')' + \frac{A_0}{L}\sigma\lambda^{-(l+2\nu)}\boldsymbol{A},$$

with $\boldsymbol{C}(\boldsymbol{x}') = \boldsymbol{A}\boldsymbol{x}'$.

9.2 Newton–Raphson procedure

In order to become acquainted with the Newton–Raphson procedure, let us consider the truss assemble of the next figure. The load–displacement behaviour, with the variables depicted in the figure, is given by

$$F = 2E(w-H)\left[1 - \frac{1}{L}\sqrt{\frac{L^2+H^2}{1+\left(\frac{w-H}{L}\right)^2}}\right],$$

which clearly is a non–linear equation, as plotted in the figure for $E = 1$, $H = 0.08$m and $L = 0.1$m.

The problem posed is to obtain the displacement of the joining node for different loads. Of course this cannot be obtained explicitly for the equation above so a numerical procedure is required.

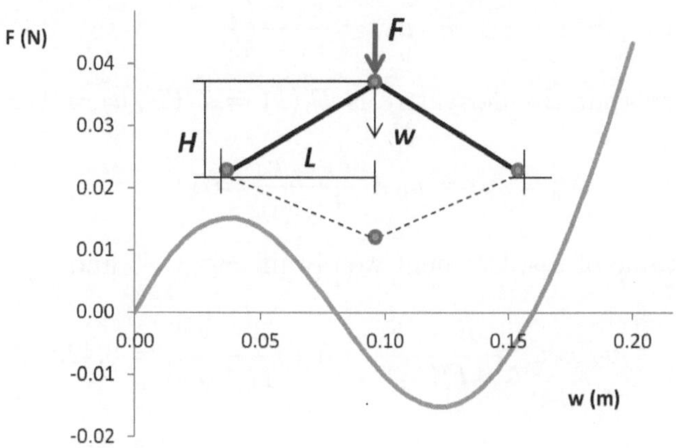

Figure 9.3 A truss configuration and its force–displacement behaviour.

For the sake of explanation, we abandon the equation above and instead we deal with the similar equation

$$F = A\sqrt{u},$$

with $A = 1$ and appropriate units. This equation is plotted in the next figure. We want to obtain the displacement u_s associated with the load $F_s = 3$. This load is desired to be reached in one step. Once the right value of u, here quite obviously $u_s = 9$, is found, we can progress to the next load step, if required, so covering different parts of the load–displacement curve.

Note that step here is not a single increment in time but rather an increment in the load.

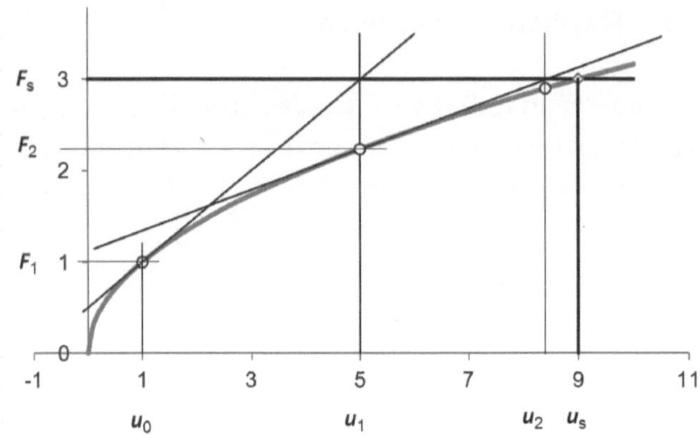

Figure 9.4 A graphical representation of the Newton–Raphson method for $F = \sqrt{u}$.

To develop the calculations, we need to guess an initial value for the displacement, here $u_0 = 1$, from which $F_1 = 1$. Now, from the figure it is clear that

$$F'(u_0) = \frac{F_s - F_1}{u_1 - u_0}$$

but we know that the derivative is $F_1'(1) = 1/(2\sqrt{u}) = 1/2$. It follows then that

$$u_1 = u_0 + \frac{F_s - F_1}{F'(u_0)} = 5.$$

With this value of displacement we obtain $F_2 = \sqrt{5}$ and

$$u_2 = u_1 + \frac{F_s - F_2}{F'(u_1)} = 5 + \frac{3 - \sqrt{5}}{1/(2\sqrt{5})} = 8.42.$$

This procedure is repeated until the difference between F_s and F_n is within an established tolerance, which can be as small as, say, 10E-6.

■ USE OF THE NEWTON–RAPHSON METHOD

Study the equation

$$F = u^2 \sin(u)$$

and obtain the displacements associated to $F = 200$, $F = 400$ and $F = -400$.

Of course we cannot obtain explicitly the values for u for a given F, so we solve this equation numerically using the Newton–Raphson scheme.

The equation is plotted in the next figure and the algorithm used requires that we give an initial value for u. Around that value, the solution is searched so that the error is minimised. Note that we work here with the function $F - u^2 \sin(u) = 0$. The following list indicates the solution evolution.

Listing 9.1 The evolution of the NR solution for $F - u^2 \sin(u) = 0$.

```
F=200   u0=13                                F=200   u0=22
it      u              error                 it      u              error
0       13             1                     0       22             1
1       13.785181      0.785180995           1       21.5782484     0.421751603
2       14.02173067    0.236549675           2       21.54608226    0.032166137
3       14.1147874     0.09305673            3       21.54573472    0.00034754
4       14.13996019    0.02517279            4       21.54573468    0.00000004
5       14.14220777    0.00224758
6       14.14222612    0.00001835
7       14.14222612    0

F=400   u0=13                                F=-400  u0=19
it      u              error                 it      u              error
0       13             1                     0       19       1
1       15.00259353    2.002593531           1       17.74770131    1.252298692
2       13.33008963    1.672503901           2       16.67242308    1.075278228
3       15.21800107    1.88791144            3       17.5958695     0.92344642
4       13.68648355    1.53151752            4       15.9190076     1.6768619
5       15.86457136    2.17808781            5       17.28233421    1.36332661
6       14.13217428    1.73239708            6       20.30718648    3.02485227
7       20.97689354    6.84471926            7       10.99149023    9.31569625
8       20.84288549    0.13400805            8       23.41299031    12.42150008
9       20.81590589    0.0269796             9       22.29987478    1.11311553
10      20.81469093    0.00121496            10      22.81062736    0.51075258
11      20.81468845    0.00000248            11      22.86146672    0.05083936
12      20.81468845    0                     12      22.86259308    0.00112636
13      22.86259366    0.00000058
```

■

Interaction is the loop that takes place in the code as many times as necessary to assure that the applied loads of a given problem result in internal forces in the structure in equilibrium with the external ones. Of course that the difference between the external and internal force vector

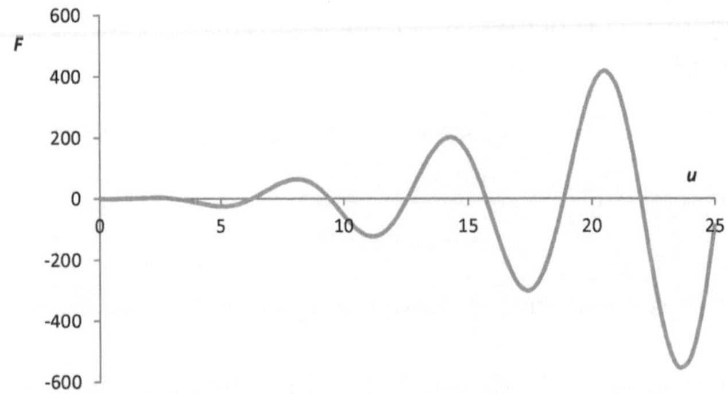

Figure 9.5 Plot of $F = u^2 \sin u$.

is not exactly zero but a rest remains. We can use a tolerance based on some kind of norm, *eg* the Euclidean norm

$$\sqrt{\frac{\sum_{m=1}^{N} |u_m^n - u_m^{n-1}|^2}{\sum_{m=1}^{N} |u_m^n|^2}} \leq tol,$$

which, when achieved, drives the solution procedure to the next step.

Now, we can generalise a bit more on the above procedure by noting that

$$f(x_2) = f(x_1) + T(x_1)(x_2 - x_1),$$

with

$$T(x_1) = \frac{\mathrm{d}f(x_1)}{\mathrm{d}x}.$$

This gives

$$x_2 = x_1 - \frac{f(x_1)}{T(x_1)}$$

since we target $f(x_2) = 0$.

This procedure leads to

$$x_{r+1} = x_r - \frac{f(x_r)}{T(x_r)}.$$

In the context of the finite element method, we have the equation

$$\mathbf{K}(\mathbf{u})\mathbf{u} = \mathbf{F},$$

which can be expressed in the form $f(x) = 0$ when writing

$$\mathbf{r}(\mathbf{u}) = \mathbf{K}(\mathbf{u})\mathbf{u} - \mathbf{F},$$

from which it follows that

$$\mathbf{u_{r+1}} = \mathbf{u_r} - \mathbf{T}^{-1}(\mathbf{u_r})\mathbf{r}(\mathbf{u_r}),$$

with the tangent matrix being given by

$$\mathbf{T}(\mathbf{u_r}) = \frac{\partial \mathbf{r}(\mathbf{u_r})}{\partial \mathbf{u_r}} = \mathbf{K}(\mathbf{u_r}) + \frac{\partial \mathbf{K}(\mathbf{u_r})}{\partial \mathbf{u}}\mathbf{u_r}.$$

One difficulty is to derive the tangent stiffness matrix, $\mathbf{T}(\mathbf{u_r})$. For a two dimensional problem, as an easy example, we can write the residual $\mathbf{r}$ as

$$\begin{Bmatrix} r_1 \\ r_2 \end{Bmatrix} = \begin{bmatrix} K_{11} & K_{12} \\ K_{21} & K_{22} \end{bmatrix} \begin{Bmatrix} u_1 \\ u_2 \end{Bmatrix} - \begin{Bmatrix} F_1 \\ F_2 \end{Bmatrix},$$

which allows us to write

$$\frac{\partial r_1}{\partial u_1} = \left(\frac{\partial K_{11}}{\partial u_1}u_1 + K_{11}\right) + \frac{\partial K_{12}}{\partial u_1}u_2.$$

In general,

$$T_{ij} = \frac{\partial r_i}{\partial u_j} = K_{ij} + \sum_{m=1}^{n} \frac{\partial K_{im}}{\partial u_j}u_m.$$

9.3 A non–linear beam finite element

We develop now a finite element suitable to deal with geometrical nonlinearities that arise in beams when their displacements are moderate, say equal to the beam thickness. By doing so, we obtain the stiffness matrix to this problem, which happens to be a function of the unknowns of the problem, *ie* the displacements themselves. This characterizes a nonlinear problem: the stiffness depends on the displacement, which depends on the stiffness. A robust method for solving the resulting system of equations is the Newton–Raphson procedure, which requires the tangent stiffness matrix, obtained by the derivative of the stiffness matrix.

This section, including the notation, is based on J.N. Reddy, An Introduction to Nonlinear Finite Element Analysis, Oxford Press, 2004.

Although we deal in this book with dynamic problems, the finite element of this section is meant to be applied only to quasi–static problems. We insist on this element because it presents important concepts that are also used in dynamic problems, like linearization and tangent matrix.

In contrast with the previous developing of a linear beam finite element, whose starting point was the equilibrium equation, we here use the principle of the virtual displacements. It states that a consequence of a body to be in equilibrium is that the external work, W_{ext}, as performed by the loads in moving through imaginary and infinitesimal displacement, hence called a virtual displacement, δw, must be equal to the internal work, W_{int}, *ie* the work made by the internal forces that arise as an internal reaction of the structure to the loading. Mathematically, we write

$$\delta W_{int} = \delta W_{ext} \rightarrow \int_V \delta\varepsilon_{ij}\sigma_{ij}\mathrm{d}V = \int q\delta w\mathrm{d}x + \int f\delta u\mathrm{d}x + \sum_{i=1}^{6} Q_i\delta\Delta,$$

with the integrals valid in the domain $[x_a, x_b]$.

Note that the present formulation allows for in plane load (along the beam length), f, for distributed transverse load, q, and for concentrated loads, Q_i, as given by bending moments, axial and transverse forces acting in the extremes of our beam element (see next figure).

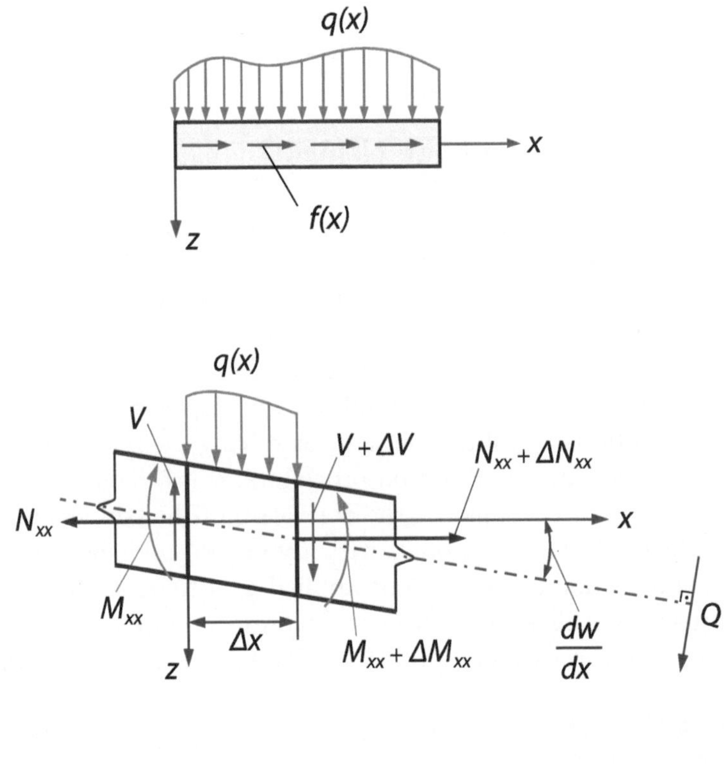

Figure 9.6 A beam element and its finite element representation.

We now need to work on the variation of each term in the above equation and, for that, the finite strain variation is given by

$$\delta\varepsilon_x = \delta\left[\frac{\mathrm{d}u}{\mathrm{d}x} + \frac{1}{2}\left(\frac{\mathrm{d}w}{\mathrm{d}x}\right)^2 + z\kappa\right] = \frac{\mathrm{d}\delta u}{\mathrm{d}x} + \frac{\mathrm{d}w}{\mathrm{d}x}\frac{\mathrm{d}\delta w}{\mathrm{d}x} - z\frac{\mathrm{d}^2\delta w}{\mathrm{d}x^2},$$

which, times the stress σ_x, results in

$$\delta W_{int} = \int\left[\left(\frac{\mathrm{d}\delta u}{\mathrm{d}x} + \frac{\mathrm{d}w}{\mathrm{d}x}\frac{\mathrm{d}\delta w}{\mathrm{d}x}\right)N - z\frac{\mathrm{d}^2\delta w}{\mathrm{d}x^2}M\right],$$

If we now collect the terms involving δu and δw and use the expressions for the external work, we obtain the so called weak forms

$$\int\frac{\mathrm{d}\delta u}{\mathrm{d}x}N = \int\delta u f\mathrm{d}x + Q_1\delta\Delta_1 + Q_4\delta\Delta_4$$

and

$$\int\left[\frac{\mathrm{d}\delta w}{\mathrm{d}x}\frac{\mathrm{d}w}{\mathrm{d}x}N - \frac{\mathrm{d}^2\delta w}{\mathrm{d}x^2}M\right]\mathrm{d}x =$$

$$= \int\delta w q\mathrm{d}x + Q_2\delta\Delta_2 + Q_3\delta\Delta_3 + Q_5\delta\Delta_5 + Q_6\delta\Delta_6.$$

An interesting remark here is that, from the above functional, *ie* from this weak form, we can obtain the equilibrium equation for the beam, as we have done using the so called vector approach. To this end, we need to eliminate the differentiation in the virtual displacements, which can be done by integration by parts of the above equations.

When the equilibrium equations obtained by the weak form and by the vector approach are the same, we refer to a consistent set of equilibrium equations, meaning that the assumed kinematic simplifications are consistent with the assumed simplification in the vector components of transverse shear force, normal force and bending moment.

Now that we have a consistent set of equilibrium equations, we can obtain the governing equation for the beam by joining the kinematic and stress fields via the material constitutive law, assumed here to be the linear one, *ie* $\sigma = E\varepsilon$. This will render the normal force and the bending moment as

$$N = \int E\varepsilon\mathrm{d}A = A_1\left[\frac{\mathrm{d}\delta u}{\mathrm{d}x} + \frac{1}{2}\left(\frac{\mathrm{d}\delta w}{\mathrm{d}x}\right)^2\right]$$

and

$$M=\int E\varepsilon z\mathrm{d}A=-A_2\frac{\mathrm{d}^2\delta w}{\mathrm{d}x^2},$$

with

$$A_1=\int E\mathrm{d}A,\qquad A_2=\int z^2\mathrm{d}A,$$

since $\int Ez\mathrm{d}A=0$, because the x axis is made to coincide with the geometrical centroidal axis. Note that, while the bending moment depends only on the curvature, the normal force depends not only on the axial stretching but also on the squared slope (close to zero for small displacements).

Note that the shear stress is not written here using the shear modulus and strain since, consistent with the adopted kinematic field, the transverse shear force is not an independent variable but, rather, it is a consequence of the bending moment distribution.

Our weak form expressions then become

$$\int A_1\frac{\mathrm{d}\delta u}{\mathrm{d}x}\left[\frac{\mathrm{d}u}{\mathrm{d}x}+\frac{1}{2}\left(\frac{\mathrm{d}w}{\mathrm{d}x}\right)^2\right]\mathrm{d}x=\int f\delta u\mathrm{d}x+Q_1\delta u(x_a)+Q_4\delta u(x_b)$$

and

$$\int\left\{A_1\frac{\mathrm{d}\delta w}{\mathrm{d}x}\frac{\mathrm{d}w}{\mathrm{d}x}\left[\frac{\mathrm{d}u}{\mathrm{d}x}+\frac{1}{2}\left(\frac{\mathrm{d}w}{\mathrm{d}x}\right)^2\right]+A_2\frac{\mathrm{d}\delta^2 w}{\mathrm{d}x^2}\frac{\mathrm{d}^2 w}{\mathrm{d}x^2}\right\}\mathrm{d}x=$$
$$\int q\delta w\mathrm{d}x+Q_2\delta w(x_a)+Q_3\delta\theta(x_a)+Q_5\delta w(x_b)+Q_6\delta\theta(x_b).$$

Only at this stage the fundamental idea of the finite element method is invoked, *ie* to use shape functions to transform the above integral equations in algebraic equations. To this end, we express the unknowns of this statement using a set of shape functions N_{a_i} for the axial displacement, the Lagrange family, and a set N_i for the transverse displacement, the Hermitian family, so that

$$u=u_1N_{a_1}+u_2N_{a_2}\quad\text{and}\quad w=w_1N_1+\theta_1N_2+w_2N_3+\theta_2N_4.$$

Lack of symmetry of the stiffness matrix also occurs in contact problems in the presence of friction.

Following the Galerkin method and Reddy's book, we substitute next the variations δu and δw by the shape functions N_{a_1} and N_1 in the weak forms. By so proceeding for the other shape functions N_{a_2}, N_2, N_3 and N_4 we obtain the non–symmetric stiffness matrix

$$
\begin{bmatrix}
\int A_1 N'_{a_1} N'_{a_1} \mathrm{d}x & \int A_1 N'_{a_1} N'_{a_2} \mathrm{d}x & \int \frac{A_1}{2} w' N'_{a_1} N'_1 \mathrm{d}x & \int \frac{A_1}{2} w' N'_{a_1} N'_2 \mathrm{d}x & \int \frac{A_1}{2} w' N'_{a_1} N'_3 \mathrm{d}x & \int \frac{A_1}{2} w' N'_{a_1} N'_4 \mathrm{d}x \\
\int A_1 N'_{a_2} N'_{a_1} \mathrm{d}x & \int A_1 N'_{a_2} N'_{a_2} \mathrm{d}x & \int \frac{A_1}{2} w' N'_{a_2} N'_1 \mathrm{d}x & \int \frac{A_1}{2} w' N'_{a_2} N'_2 \mathrm{d}x & \int \frac{A_1}{2} w' N'_{a_3} N'_3 \mathrm{d}x & \int \frac{A_1}{2} w' N'_{a_4} N'_4 \mathrm{d}x \\
\int A_1 w' N'_{a_1} N'_1 \mathrm{d}x & \int A_1 w' N'_{a_2} N'_1 \mathrm{d}x & \int A_2 N''_1 N''_1 \mathrm{d}x + \int A_3 N'_1 N'_1 \mathrm{d}x & \int A_2 N''_1 N''_2 \mathrm{d}x + \int A_3 N'_1 N'_2 \mathrm{d}x & \int A_2 N''_1 N''_3 \mathrm{d}x + \int A_3 N'_1 N'_3 \mathrm{d}x & \int A_2 N''_1 N''_4 \mathrm{d}x + \int A_3 N'_1 N'_4 \mathrm{d}x \\
\int A_1 w' N'_{a_1} N'_2 \mathrm{d}x & \int A_1 w' N'_{a_2} N'_2 \mathrm{d}x & \int A_2 N''_2 N''_1 \mathrm{d}x + \int A_3 N'_2 N'_1 \mathrm{d}x & \int A_2 N''_2 N''_2 \mathrm{d}x + \int A_3 N'_2 N'_2 \mathrm{d}x & \int A_2 N''_2 N''_3 \mathrm{d}x + \int A_3 N'_2 N'_3 \mathrm{d}x & \int A_2 N''_2 N''_4 \mathrm{d}x + \int A_3 N'_2 N'_4 \mathrm{d}x \\
\int A_1 w' N'_{a_1} N'_3 \mathrm{d}x & \int A_1 w' N'_{a_2} N'_3 \mathrm{d}x & \int A_2 N''_3 N''_1 \mathrm{d}x + \int A_3 N'_3 N'_1 \mathrm{d}x & \int A_2 N''_3 N''_2 \mathrm{d}x + \int A_3 N'_3 N'_2 \mathrm{d}x & \int A_2 N''_3 N''_3 \mathrm{d}x + \int A_3 N'_3 N'_3 \mathrm{d}x & \int A_2 N''_3 N''_4 \mathrm{d}x + \int A_3 N'_3 N'_4 \mathrm{d}x \\
\int A_1 w' N'_{a_1} N'_4 \mathrm{d}x & \int A_1 w' N'_{a_2} N'_4 \mathrm{d}x & \int A_2 N''_4 N''_1 \mathrm{d}x + \int A_3 N'_4 N'_1 \mathrm{d}x & \int A_2 N''_4 N''_2 \mathrm{d}x + \int A_3 N'_4 N'_2 \mathrm{d}x & \int A_2 N''_4 N''_3 \mathrm{d}x + \int A_3 N'_4 N'_3 \mathrm{d}x & \int A_2 N''_4 N''_4 \mathrm{d}x + \int A_3 N'_4 N'_4 \mathrm{d}x
\end{bmatrix}
\begin{Bmatrix} u_1 \\ u_2 \\ \bar{\Delta}_1 \\ \bar{\Delta}_2 \\ \bar{\Delta}_3 \\ \bar{\Delta}_4 \end{Bmatrix}
=
\begin{Bmatrix} \int f N_{a_1} \mathrm{d}x + \hat{Q}_1 \\ \int f N_{a_2} \mathrm{d}x + \hat{Q}_2 \\ \int q N_1 \mathrm{d}x + \bar{Q}_1 \\ \int q N_2 \mathrm{d}x + \bar{Q}_2 \\ \int q N_3 \mathrm{d}x + \bar{Q}_3 \\ \int q N_4 \mathrm{d}x + \bar{Q}_4 \end{Bmatrix},
$$

which can be concisely expressed as

$$
\sum_{j=1}^{2} K_{ij}^{11} u_j + \sum_{J=1}^{4} K_{iJ}^{12} \bar{\Delta}_J = F_i^1
$$

and

$$
\sum_{j=1}^{2} K_{Ij}^{21} u_j + \sum_{J=1}^{4} K_{IJ}^{22} \bar{\Delta}_J = F_I^2,
$$

with

$$
\begin{aligned}
K_{ij}^{11} &= \int A_1 N'_{a_i} N'_{a_j} \mathrm{d}x, \\
K_{iJ}^{12} &= \int \frac{A_1}{2} w' N'_{a_i} N'_J \mathrm{d}x, \\
K_{Ij}^{21} &= \int A_1 w' N'_{a_j} N'_I \mathrm{d}x, \\
K_{IJ}^{22} &= \int A_2 N''_I N''_J \mathrm{d}x + \int \frac{A_1}{2} (w')^2 N'_I N'_J \mathrm{d}x, \\
F_i^1 &= \int f N_{a_i} \mathrm{d}x + \hat{Q}_i, \\
F_I^2 &= \int q N_I \mathrm{d}x + \bar{Q}_I,
\end{aligned}
$$

where we changed the variables for easy indexing to $\bar{\Delta}_1 = w(x_a)$, $\bar{\Delta}_2 = \theta(x_a)$, $\bar{\Delta}_3 = w(x_b)$, $\bar{\Delta}_4 = \theta(x_b)$, $\hat{Q}_1 = Q_1$, $\hat{Q}_2 = Q_4$, $\bar{Q}_1 = Q_2$, $\bar{Q}_2 = Q_3$,

$\bar{Q}_3 = Q_5$, $\bar{Q}_4 = Q_6$. Also, $(\mathrm{d}/\mathrm{d}x) = (')$, $(\mathrm{d}^2/\mathrm{d}x^2) = ('')$, $A_3 = (A_1/2)(w'')^2$ and the lower cases indexes i, j vary from 1 to 2 and upper case ones, I, J vary from 1 to 4.

One important question here is how to deal with the non–linear term $\frac{1}{2}\left(\frac{\partial w}{\partial x}\right)^2$ of our weak form. If we square this term using the shape functions we obtain a squared degree of freedom, which will not allow us to write the global equilibrium of our finite element model as $K(\bar{\Delta})\bar{\Delta} = F$. To avoid this problem, we linearise the functional and calculate w' from the previous iteration,

$$w' = N_I'\bar{\Delta}.$$

Of course this linearisation process will imply a lack of equilibrium in the beam, as represented by the unbalanced external minus internal load. Such unbalanced load can be reapplied in a next iteration, leading to a closer approximation of the correct beam displacement. A generalization of this process and the technique underlying it is, essentially, the Newton–Raphson method, which requires the tangent matrix.

9.3.1 Tangent matrix

The tangent stiffness matrix for the non–linear beam finite element can be obtained, using the previous notation for the stiffness matrix, as follows.

The sub–matrix T^{11} comes from

$$T_{ij}^{11} = K_{ij}^{11} + \frac{\partial K_{ip}^{11}}{\partial u_j} u_p + \frac{\partial K_{iP}^{12}}{\partial u_j} \bar{\Delta}_P = K_{ij}^{11}, \quad p = 1, 2 : P = 1, 2, 3, 4,$$

since $\partial K_{ij}^{11}/\partial u_k = \partial K_{ij}^{12}/\partial u_k = 0$.

Following similar procedures and noting that

$$\frac{\partial w'}{\partial \bar{\Delta}_J} = \frac{\partial}{\partial \bar{\Delta}_J}\left(\sum_{K=1}^{4} \bar{\Delta}_K N_K'\right) = N_J',$$

we obtain

$$T_{iJ}^{12} = K_{Ji}^{21}.$$

Finally, algebraic manipulations leads to

$$T_{IJ}^{22} = K_{IJ}^{22} + \int A_1[u' + (w')^2] N_I' N_J' \mathrm{d}x.$$

Note that we need here to deal with the integration of the product of shape functions, whose analytical integration, if possible, can be troublesome.

Hence, we rely on the numerical integration techniques seen in Chapter 8.

Also, it is important to note that a boundary condition hinged–hinged implies that there is no axial strain when in–plane motion is allowed so that $u' + (w')^2/2$ must be zero. But this is difficult to achieve when considering the degree of the polynomials used to interpolate the axial and transverse displacement, leading to high values of the stiffness matrix, so characterizing what is called membrane locking. As suggested by Reddy, a way to deal with this situation is to use reduce integration in the calculation of the coefficients K_{ij}^{12}, K_{ij}^{21}, T_{ij}^{12}, T_{ij}^{21} and of the non–linear terms in K^{22} and T_{ij}^{22}.

9.4 One-dimensional computational plasticity

We have seen here that there are many situations where the a structure or component is permanently deformed. This analysis can be performed via the finite element method, now under the broad scope of computational plasticity. In this section we will study the uni–dimensional plasticity case, presenting its basic algorithm in the context of static loading.

This section and some of their figures are based on J.C. Simo and T.J.R. Hughes, Computational Inelasticity, Springer, 1998.

When a structure is loaded beyond its elastic limit, part of it or the whole will experience plastic deformation. The total strain field generated can be captured via, for instance, image sensors and it is formed by elastic and plastic strains. Their magnitude depend on the stress level, as illustrated in the next figure.

The figure presents also the actual material behaviour by a spring model. In this unity length and unity area material model, elastic strains are restricted to act in the spring and have a magnitude ε^e. Plastic strains of magnitude ε^p act on the Coulombic friction device with the constant flow stress $\sigma_y > 0$, such that the total strain is

$$\varepsilon = \varepsilon^e + \varepsilon^p.$$

The applied stress, σ, acting on the spring is given by

$$\sigma = E\varepsilon^e = E(\varepsilon - \varepsilon^p).$$

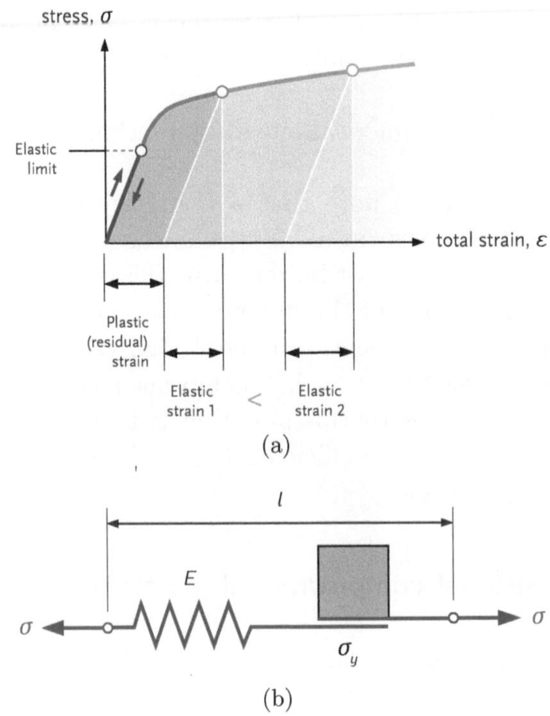

Figure 9.7 A unidimensional material model and its idealization.

Obviously, a change in the frictional device configuration will occur only if the plastic strain changes with time, *ie* $\mathrm{d}\varepsilon^p/\mathrm{d}t = \dot{\varepsilon}^p \neq 0$, so we can write

$$f(\sigma) = |\sigma| - \sigma_y \leq 0 \rightarrow \dot{\varepsilon}^p = 0,$$

which we refer to as a yield condition. Observe that this is a unidimensional stress condition, related to the bending moment M_0 in the case of beams. Also, the absolute value of the stress σ in the frictional device cannot be greater than σ_y. If the absolute value of the applied stress is less than the flow stress, it implies that $\dot{\varepsilon}^p = 0$, and the stress regime is elastic.

Calling the absolute value of the slip rate as $\gamma \geq 0$, we can write

$$\dot{\varepsilon}^p = \gamma \geq 0 \quad \text{if} \quad \sigma = \sigma_y > 0$$

and

$$\dot{\varepsilon}^p = -\gamma \leq 0 \quad \text{if} \quad \sigma = -\sigma_y < 0,$$

which is the same as

$$\dot{\varepsilon}^p = \gamma \text{sign}(\sigma) \quad \text{iff} \quad f(\sigma) = |\sigma| - \sigma_y = 0,$$

with the function sign being 1 if $\sigma > 0$, or -1 if $\sigma < 0$. The set of points $f(\sigma)$ defines a yield surface, which reduces to the two extreme points $(\sigma_y, -\sigma_y)$ in the present unidimensional case.

We now need to determine whether the absolute value of the slip rate is zero or positive, since $\gamma \geq 0$. To this end, we make use of the loading–unloading conditions. First, we know that $\gamma \geq 0$ and that $f(\sigma) \leq 0$. Also, $\gamma = 0$ when $f(\sigma) < 0$. For $\dot{\varepsilon}^p \neq 0$, $\gamma > 0$ if $f(\sigma) = 0$, implying that

$$\gamma f(\sigma) = 0,$$

meaning that plastic flow (here, slip) only occurs on the yield surface under an admissible stress state. This is called the Kuhn–Tucker condition.

Note that, if $\dot{f} > 0$, then $f(t + dt) > 0$, leading to yield violation. So, when $\gamma > 0$, $\dot{f} = 0$ and if $\dot{f} < 0$, $\gamma = 0$, which leads to

$$\gamma \dot{f}(\sigma) = 0,$$

called persistence condition, *ie* for plastic flow ($\dot{\varepsilon}^p \neq 0$, $\gamma > 0$) to occur the stress state must persist on the surface.

We have seen that $\dot{\varepsilon}^p = \gamma \text{sign}(\sigma)$, but

$$\dot{f} = \frac{\partial f}{\partial \sigma} E \left(\dot{\varepsilon} - \dot{\varepsilon}^p \right) = 0,$$

from which it follows, when considering that $\partial f / \partial \sigma = \text{sign}(\sigma)$,

$$\gamma = \dot{\varepsilon} \text{sign}(\sigma),$$

leading to

$$\dot{\varepsilon}^p = \dot{\varepsilon} \quad \text{for} \quad f(\sigma) = 0, \dot{f}(\sigma) = 0.$$

This is to say that, for this simple model, the plastic strain rate equals the applied strain rate, in line with a perfect plastic material model.

An important point from this model is that the relation between the flow rule and the yield condition can also be expressed by

$$\dot{\varepsilon}^p = \gamma \frac{\partial f}{\partial \sigma},$$

characterizing the so called associative flow rule.

We refer to linear isotropic hardening when the yield criterion is of the form

$$f(\sigma, \alpha) = |\sigma| - (\sigma_y + K\alpha) \leq 0,$$

with K being the plastic modulus and α an internal hardening variable which represents the amount of plastic flow (or slip in our simple analogy), whose simplest evolution is taken as

$$\dot{\alpha} = |\dot{\varepsilon}^p|.$$

Typical non–linear and linear strain hardening material stress–strain curves are depicted in the next figure. Note that such a hardening model implies in an expansion of the yield surface (points, for the 1D case). It is also interesting to note that, in the closed loading cycle shown, the final strain is zero but the hardening variable α accumulates the strain.

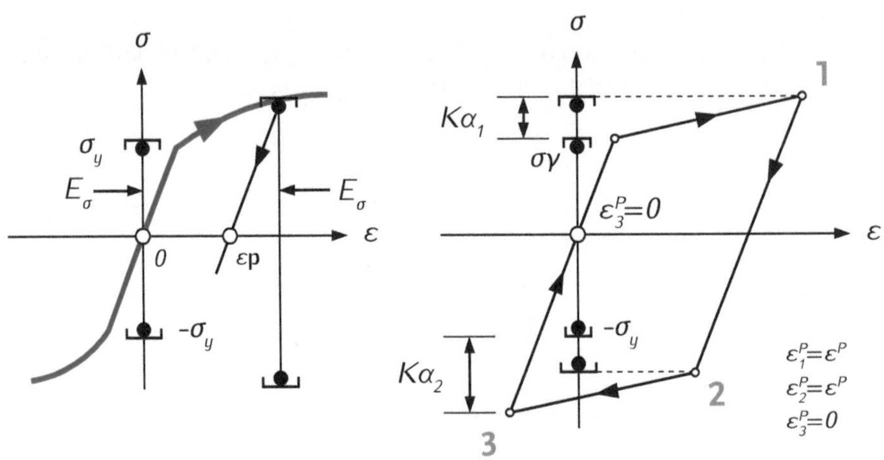

Figure 9.8 A strain hardening material behaviour.

For this material model, the consistency condition allows us once again to explicitly solve for the slip rate γ, *ie*

$$\dot{f} = \frac{\partial f}{\partial \sigma}\dot{\sigma} + \frac{\partial f}{\partial \alpha}\dot{\alpha} = \text{sign}(\sigma) E\dot{\varepsilon} - \gamma(E + K) \leq 0,$$

which gives, for $f = \dot{f} = 0$,

$$\gamma = \frac{\text{sign}(\sigma) E\dot{\varepsilon}}{E + K}.$$

Since $\dot{\sigma} = E(\dot{\varepsilon} - \dot{\varepsilon}^p) = E(\dot{\varepsilon} - \gamma \text{sign}(\sigma)$, it follows that

$$\dot{\sigma} = \frac{EK}{E+K}\dot{\varepsilon}, \quad \text{if} \quad \gamma > 0,$$

$$\dot{\sigma} = E\dot{\varepsilon}, \quad \text{if} \quad \gamma = 0,$$

with $EK/(E+K)$ being called the elastoplastic tangent modulus.

This expression puts in evidence the advantage of using a linear hardening law, as much as it allows an explicit calculation of the stress increment from the overall strain increment. Also, by knowing, or imposing, a strain increment, we obtain the slip rate and with the slip rate we obtain the plastic contribution to the total strain. Let us further develop this idea by studying the so called return mapping algorithm.

9.4.1 Unidimensional return mapping (radial) algorithm

Consider as a result of loading a structure that the strain state in a point was changed by $\Delta\varepsilon_n$, such that

$$\varepsilon_{n+1} = \varepsilon_n + \Delta\varepsilon_n.$$

This leads to a new stress state

$$\sigma_{n+1} = E\varepsilon^e_{n+1} = E(\varepsilon_{n+1} - \varepsilon^p_{n+1}).$$

Let us now consider a discrete counterpart of $\dot{\varepsilon}^p = \gamma\text{sign}(\sigma)$ and $\dot{\alpha} = \gamma$ as

$$\varepsilon^p_{n+1} = \varepsilon^p_n + \Delta\gamma\text{sign}(\sigma_{n+1}) \quad \text{and} \quad \alpha_{n+1} = \alpha_n + \Delta\gamma.$$

σ_{n+1} and α_{n+1} are constrained by the Kuhn–Trucker conditions in their discrete version according to

$$f_{n+1} = |\sigma_{n+1}| - (\sigma_y + K\alpha_{n+1}) \leq 0, \qquad \Delta\gamma \geq 0,$$

$$\Delta\gamma f_{n+1} = 0.$$

The loading causes a change in the total strain of $\Delta\varepsilon_n$ and let us assume that this change is purely elastic. This is a trial configuration given by

$$\sigma_{n+1}^{trial} = E(\varepsilon_{n+1} - \varepsilon_n^p) = \sigma_n + E\Delta\varepsilon_n,$$

$$\varepsilon_{n+1}^{trial} = \varepsilon_n^p \qquad \alpha_{n+1}^{trial} = \alpha_n,$$

$$f_{n+1}^{trial} = |\sigma_{n+1}^{trial}| - (\sigma_y + K\alpha_n)$$

and we proceed by checking to which yield region it leads to.

We check if $f_{n+1}^{trial} <>= 0$. If $f_{n+1}^{trial} \leq 0$, yield did not take place (elastic step with $\Delta\gamma = 0$), which agrees with a purely trial elastic loading. Therefore, in this case, the initial stress state σ_n and the strain $\Delta\varepsilon_n$ resulted in a stress state σ_{n+1} that is the actual trial one, σ_{n+1}^{trial}, as illustrated in the next figure.

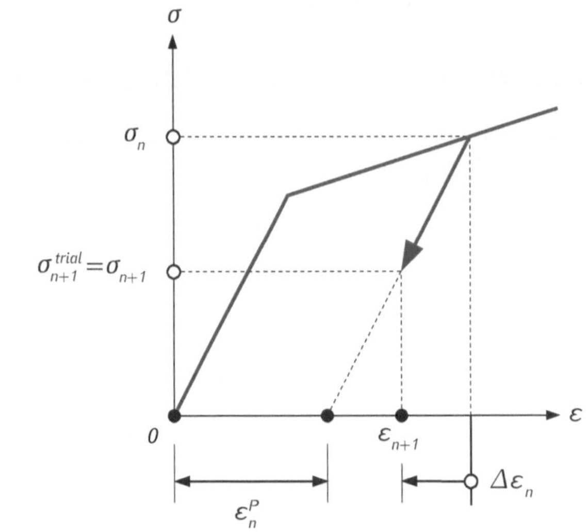

Figure 9.9 Loading strain lead from a plastic stress state to a pure elastic one.

Now, let us consider the case when the loading strain, $\Delta\varepsilon_n$, leads to a stress state from σ_n to σ_{n+1}^{trial} lying outside the yield surface (yield point), see next figure.

In this case, $f_{n+1}^{trial} > 0$, which violates the yield condition, meaning that the trial stress state is not a solution to the incremental problem. Since $f_{n+1} = 0$, the Kuhn–Tucker condition, $\Delta\gamma f_{n+1}$, implies that $\Delta\gamma > 0$, *ie* plastic flow occurred here. This is to say that it is only necessary to observe the magnitude of f_{n+1} to judge whether the step is elastic or plastic.

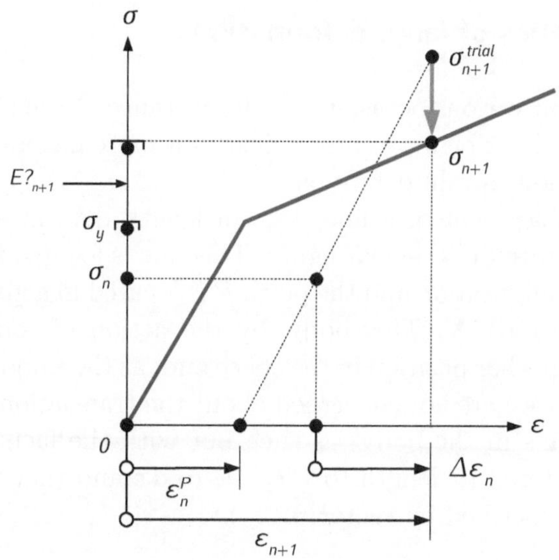

Figure 9.10 Loading strain lead to a plastic stress state.

It follows that

$$\begin{aligned}\sigma_{n+1} &= E(\varepsilon_{n+1} - \varepsilon_n^p)\\ &= E(\varepsilon_{n+1} - \varepsilon_n^p) - E(\varepsilon_{n+1}^p - \varepsilon_n^p)\\ &= \sigma_{n+1}^{trial} - E\Delta\gamma\text{sign}(\sigma_{n+1}^{trial}),\\ \varepsilon_{n+1}^p &= \varepsilon_n^p + \Delta\gamma\text{sign}(\sigma_{n+1}^{trial}),\\ \alpha_{n+1} &= \alpha_n + \Delta\gamma\end{aligned}$$

when noticing that $\text{sign}(\sigma_{n+1}) = \text{sign}(\sigma_{n+1}^{trial})$ and that

$$\Delta\gamma = \frac{f_{n+1}^{trial}}{E+K},$$

which comes from $f_{n+1} = |\sigma_{n+1}| - (\sigma_y + K\alpha_{n+1}) = 0$.

Note that we can write

$$\sigma_{n+1} = \left(1 - \frac{\Delta\gamma E}{|\sigma_{n+1}^{trial}|}\right)\sigma_{n+1}^{trial},$$

which indicates that the actual stress state is scaled, or projected, onto the yield surface.

9.5 Kinematics of large deformations

We have seen on various occasions the importance of the kinematics of a body. We now describe a kinematic field in a more general way, leading the way to various strain definitions.

Consider a segment of a line, PQ, of length $\mathrm{d}\boldsymbol{X}$, in a body of any shape, as illustrated in the next figure. This line is located in an assumed undeformed configuration and the point P is located in a given referential system by a vector $\boldsymbol{X}$. This body, by the action of some loading, is translated to another position in the space and, at the same time, changes its shape. Here, we are not concerned about the translation (since it does not cause strains in the body) as such but with the facts that the line segment has altered its length to $P'Q'$, *ie* to $\mathrm{d}\boldsymbol{x}$ and that the point P is translated as described by vector $\boldsymbol{u}$.

Note that we refer to the original (current) configuration using upper (lower) cases.

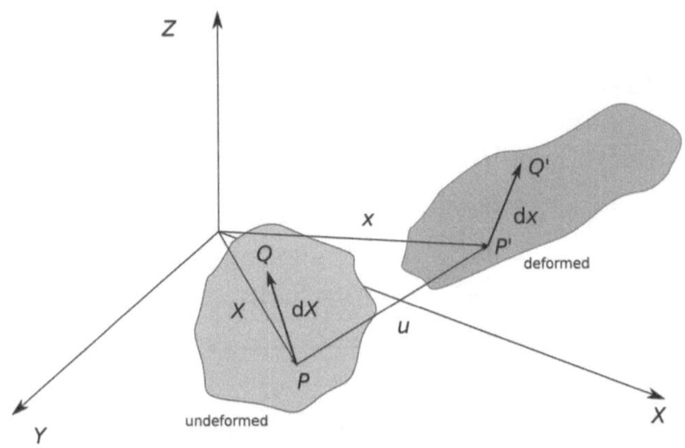

Figure 9.11 A generic body in an underformed and current configurations.

Now, since $\boldsymbol{x} = \boldsymbol{X} + \boldsymbol{u}$, $\mathrm{d}\boldsymbol{X}$ is recovered from $\mathrm{d}\boldsymbol{x}$ via the transformation

$$\mathrm{d}\boldsymbol{x} = \boldsymbol{F}\mathrm{d}\boldsymbol{X},$$

with $\boldsymbol{F}$ being the deformation gradient (not to be confused with the force vector). $\boldsymbol{F}$ describes in full what happened with the line segment PQ as it experienced a change of configuration. Note that such a change is not only due to stretching of the segment but also due to its rotation. This can be seen by decomposing the second order tensor $\boldsymbol{F}$ in its rotation, $\boldsymbol{R}$ and stretching parts, $\boldsymbol{U}$ and/or $\boldsymbol{V}$, as

$$\boldsymbol{F} = \boldsymbol{R}\boldsymbol{U} = \boldsymbol{V}\boldsymbol{R}.$$

This equation shows that we can first rotate our segment and then stretch it or stretch and then rotate, in either case transforming $\mathrm{d}\boldsymbol{X}$ into $\mathrm{d}\boldsymbol{x}$.

■ SHEAR OF A PLANE ELEMENT

Given the original and current configuration of the unit square shown in the figure, obtain the gradient of deformation.

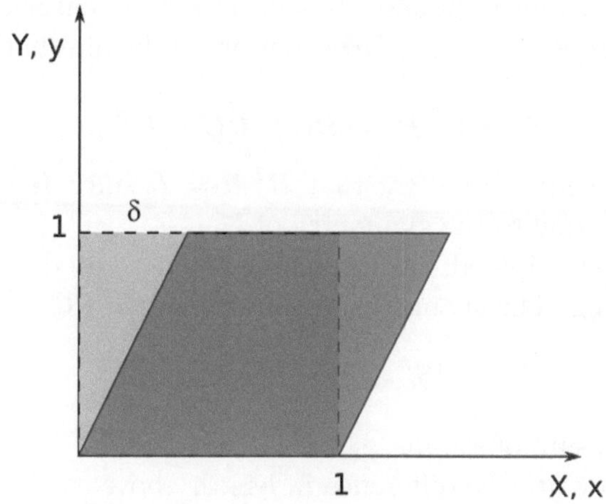

Figure 9.12 Simple shear of a plane element.

The deformed and undeformed configuration are related by

$$x = X + \delta Y, \quad y = Y,$$

leading to the deformation gradient

$$\boldsymbol{F} = \begin{pmatrix} 1 & \delta \\ 0 & 1 \end{pmatrix}$$

since $\frac{\partial x}{\partial X} = 1, \quad \frac{\partial x}{\partial Y} = \delta, \quad \frac{\partial y}{\partial X} = 0, \quad \frac{\partial y}{\partial Y} = 1.$

■

In Chapter 7 we defined stretching and some measures of deformation. We can now lend more generality to these concepts by considering that the undeformed and deformed segments $\mathrm{d}\boldsymbol{X}$ and $\mathrm{d}\boldsymbol{x}$ have lengths $\mathrm{d}S$ and $\mathrm{d}s$.

Let us express $\mathrm{d}S$ as a function of $\mathrm{d}\boldsymbol{x}$ and $\mathrm{d}s$ as a function of $\mathrm{d}\boldsymbol{X}$, *ie*

$$\mathrm{d}S = \sqrt{\mathrm{d}\boldsymbol{X}.\mathrm{d}\boldsymbol{X}} \qquad \rightarrow$$

$$\mathrm{d}S^2 = \mathrm{d}\boldsymbol{X}^T\mathrm{d}\boldsymbol{X} = (\boldsymbol{F}^{-1}\mathrm{d}x)^T\boldsymbol{F}^{-1} = \mathrm{d}\boldsymbol{x}^T\boldsymbol{B}^{-1}\mathrm{d}\boldsymbol{x}$$

and

$$\mathrm{d}s = \sqrt{\mathrm{d}\boldsymbol{x}.\mathrm{d}\boldsymbol{x}} \qquad \rightarrow$$

$$\mathrm{d}s^2 = \mathrm{d}\boldsymbol{x}.\mathrm{d}\boldsymbol{x} = \mathrm{d}\boldsymbol{X}^T\mathrm{d}\boldsymbol{F}^T\mathrm{d}\boldsymbol{F}\mathrm{d}\boldsymbol{X} = \mathrm{d}\boldsymbol{X}^T\boldsymbol{C}\mathrm{d}\boldsymbol{X},$$

with

$$\boldsymbol{B}^{-1} = (\boldsymbol{F}^{-1})^T\boldsymbol{F}^{-1} \qquad \text{and} \qquad \boldsymbol{C} = \boldsymbol{F}^T\boldsymbol{F}$$

We use again $\boldsymbol{A}.\boldsymbol{B} = \boldsymbol{B}^T\boldsymbol{A}$ and $(\boldsymbol{A}\boldsymbol{B})^T = \boldsymbol{B}^T\boldsymbol{A}^T$.

being referred to as the right and left Cauchy–Green tensors, respectively.

Note that the left Cauchy–Green tensor can be alternatively written as

$$\boldsymbol{C} = \boldsymbol{F}^T\boldsymbol{F} = (\boldsymbol{R}\boldsymbol{U})^T\boldsymbol{R}\boldsymbol{U} = \boldsymbol{U}^2,$$

where use was made of the fact that $\boldsymbol{R}^T\boldsymbol{R} = \boldsymbol{I}$, since $\boldsymbol{R}$ is orthogonal, and $\boldsymbol{U}^T\boldsymbol{U} = \boldsymbol{U}^2$ due to the symmetry of $\boldsymbol{U}$.

As we have anticipated, the stretching could be used to quantify the material straining. The change in length of the line PQ,

$$\mathrm{d}s^2 - \mathrm{d}S^2 = \mathrm{d}\boldsymbol{x}.\left(\boldsymbol{I} - \boldsymbol{B}^{-1}\right)\mathrm{d}\boldsymbol{x},$$

is in itself a measure of strain.

We can also write the difference in length above as

$$\mathrm{d}s^2 - \mathrm{d}S^2 = \mathrm{d}\boldsymbol{x}^{\mathrm{T}}\mathrm{d}\boldsymbol{x} - \mathrm{d}\boldsymbol{X}^{\mathrm{T}}\mathrm{d}\boldsymbol{X} = \mathrm{d}\boldsymbol{X}^{\mathrm{T}}(\boldsymbol{C} - \boldsymbol{I})\mathrm{d}\boldsymbol{X} = \mathrm{d}\boldsymbol{X}^{\mathrm{T}}(2\boldsymbol{E})\mathrm{d}\boldsymbol{X},$$

with $\boldsymbol{E} = (1/2)(\boldsymbol{C} - \boldsymbol{I})$ being the Green–Lagrange strain tensor.

As we did for the unidimensional case, we can also have here a strain family according to

$$\mathbf{E}_{(m)} = \frac{1}{2m}\left(\mathbf{U}^{2m} - \mathbf{I}\right) = \frac{1}{2m}\left[\mathbf{C}^m - \mathbf{I}\right].$$

For different values of m, the following table lists the main strain measures used in the literature.

■ STRAIN IN A PLANE ELEMENT

For the previous shear plane, obtain the Green–Lagrange strain, $\boldsymbol{E}$.

Although we could obtain $\boldsymbol{E}$ straight from $\boldsymbol{F}$, we evaluate first the Cauchy–Green tensor, $\boldsymbol{C}$,

$$\boldsymbol{C} = \boldsymbol{F}^{\mathrm{T}}\boldsymbol{F} = \begin{pmatrix} 1 & 0 \\ \delta & 1 \end{pmatrix}\begin{pmatrix} 1 & \delta \\ 0 & 1 \end{pmatrix} = \begin{pmatrix} 1 & \delta \\ \delta & 1+\delta^2 \end{pmatrix},$$

TABLE 9.1 3D deformation family.

m	equation	name
1	$\mathbf{E}_{(1)} = \frac{1}{2}\left(\mathbf{U}^2 - \mathbf{I}\right) = \frac{1}{2}(\mathbf{C} - \mathbf{I})$	Green strain
1/2	$\mathbf{E}_{(1/2)} = (\mathbf{U} - \mathbf{I}) = \mathbf{C}^{1/2} - \mathbf{I}$	Biot strain
0	$\mathbf{E}_{(0)} = \ln \mathbf{U} = \frac{1}{2} \ln \mathbf{C}$	Logarithmic, Natural, True or Hencky strain
-1	$\mathbf{E}_{(-1)} = \frac{1}{2}\left[\mathbf{I} - \mathbf{U}^{-2}\right]$	Almansi strain

and from there we obtain

$$\boldsymbol{E} = \frac{1}{2}(\boldsymbol{C} - \boldsymbol{I}) = \frac{1}{2}\left(\begin{pmatrix} 1 & \delta \\ \delta & 1+\delta^2 \end{pmatrix} - \begin{pmatrix} 1 & 0 \\ 0 & 1 \end{pmatrix}\right) = \frac{1}{2}\begin{pmatrix} 0 & \delta \\ \delta & \delta^2 \end{pmatrix}.$$

Let us take the chance to obtain the rigid body rotation, $\boldsymbol{R}$, and the stretching, $\boldsymbol{U}$, presented in the deformation gradient of this problem. For that, we write

$$\boldsymbol{U} = \begin{pmatrix} U_{xx} & U_{xy} \\ U_{yx} & U_{yy} \end{pmatrix}, \quad \boldsymbol{R} = \begin{pmatrix} \cos\varphi & \sin\varphi \\ -\sin\varphi & \cos\varphi \end{pmatrix}.$$

This gives

$$\begin{pmatrix} 1 & \delta \\ 0 & 1 \end{pmatrix} = \begin{pmatrix} \cos\varphi & \sin\varphi \\ -\sin\varphi & \cos\varphi \end{pmatrix} \begin{pmatrix} U_{xx} & U_{xy} \\ U_{yx} & U_{yy} \end{pmatrix},$$

with $\sin\varphi = \delta/\sqrt{2^2 + \delta^2}$ and $\cos\varphi = 2/\sqrt{2^2 + \delta^2}$, so that

$$\boldsymbol{U} = \frac{1}{\sqrt{2^2 + \delta^2}}\begin{pmatrix} 2 & \delta \\ \delta & 2+\delta^2 \end{pmatrix} \quad \text{and} \quad \boldsymbol{R} = \frac{1}{\sqrt{2^2 + \delta^2}}\begin{pmatrix} 2 & \delta \\ -\delta & 2 \end{pmatrix}.$$

■

We insist a little more on exploring the gradient $\boldsymbol{F}$. Since it can be decomposed in stretching and rotation, it is interesting to know how to obtain $\boldsymbol{U}$ and $\boldsymbol{R}$ from $\boldsymbol{F}$. We have seen that $\boldsymbol{C} = \boldsymbol{F}^T\boldsymbol{F} = \boldsymbol{U}^2$ and from this we can obtain $\boldsymbol{U}$. To this end, it is necessary to perform the operation on what is called the principal values of the tensor. Hence, we need to find the eigenvalues and vectors of the given tensor, perform the operation, in this case the square root. Let us call the matrix containing the eigenvectors of $\boldsymbol{U}$ as $\boldsymbol{H}$, the modal matrix, and the matrix containing in its diagonal the eigenvalues as $\hat{\boldsymbol{U}}$. Then, it can be shown that the following is valid,

$$\boldsymbol{U}^2 = \boldsymbol{H}\hat{\boldsymbol{U}}\boldsymbol{H}^{-1}.$$

Any operation on the left side of this expression equals that operation on the leading diagonal $\hat{\boldsymbol{U}}$ only. Hence,

$$\boldsymbol{U} = \boldsymbol{H}\sqrt{\hat{\boldsymbol{U}}}\boldsymbol{H}^{-1}.$$

Note that this result can be used to compute the true strain by simply extracting the log values of the modal matrix of $\boldsymbol{B}^{-1}$.

9.5.1 Rate of deformation

We have seen on various occasions the importance of the rate of the deformation, which can be quantified from the time variation of the gradient, written as

$$\dot{\boldsymbol{F}} = \frac{\partial}{\partial t}\left(\frac{\partial \boldsymbol{x}}{\partial \boldsymbol{X}}\right) = \frac{\partial \boldsymbol{v}}{\partial \boldsymbol{X}} = \frac{\partial \boldsymbol{v}}{\partial \boldsymbol{x}}\frac{\partial \boldsymbol{x}}{\partial \boldsymbol{X}} = \boldsymbol{L}\boldsymbol{F},$$

with $\boldsymbol{L} = \partial \boldsymbol{v}/\partial \boldsymbol{x}$ being the velocity gradient (not to be confused with the linear momentum symbol, $\boldsymbol{L}$), which can be expressed as

$$\boldsymbol{L} = \dot{\boldsymbol{F}}\boldsymbol{F}^{-1}.$$

It is useful to decompose $\boldsymbol{L}$ in its symmetric and antisymmetric parts as

Any tensor $\boldsymbol{A}$ can be represented as $(1/2)(\boldsymbol{A}+\boldsymbol{A}^{\mathsf{T}}) + (1/2)(\boldsymbol{A}-\boldsymbol{A}^{\mathsf{T}}) = \mathsf{sym}(\boldsymbol{A}) + \mathsf{asym}(\boldsymbol{A})$.

$$\boldsymbol{L} = \boldsymbol{D} + \boldsymbol{W},$$

with $\boldsymbol{D} = \frac{1}{2}\left(\boldsymbol{L} + \boldsymbol{L}^{\mathrm{T}}\right)$ being the (symmetric) rate of deformation and $\boldsymbol{W} = \frac{1}{2}\left(\boldsymbol{L} - \boldsymbol{L}^{\mathrm{T}}\right)$ the (antisymmetric) continuum spin.

Let us express the continuum spin as a function of the velocity gradient,

$$\boldsymbol{W} = \frac{1}{2}\left(\boldsymbol{L} - \boldsymbol{L}^{\mathrm{T}}\right) = \frac{1}{2}\left(\dot{\boldsymbol{F}}\boldsymbol{F}^{-1} - (\dot{\boldsymbol{F}}\boldsymbol{F}^{-1})^{\mathrm{T}}\right),$$

which gives, using $\boldsymbol{F} = \boldsymbol{R}\boldsymbol{U}$ and the time derivative of $\boldsymbol{R}\boldsymbol{R}^{\mathrm{T}} = \boldsymbol{I}$,

$$\boldsymbol{W} = \boldsymbol{\Omega} + \frac{1}{2}\boldsymbol{R}\left(\dot{\boldsymbol{U}}\boldsymbol{U}^{-1} - (\dot{\boldsymbol{U}}\boldsymbol{U}^{-1})^{\mathrm{T}}\right)\boldsymbol{R}^{\mathrm{T}},$$

with

$$\boldsymbol{\Omega} = \dot{\boldsymbol{R}}\boldsymbol{R}^{\mathrm{T}}$$

being called the angular velocity tensor, which is independent of the stretch.

Note here that, if, and only if, the rotation is zero,

$$\boldsymbol{R} = 0 \rightarrow \boldsymbol{W} = 0 \rightarrow \boldsymbol{L} = \boldsymbol{D},$$

ie the velocity gradient equals the rate of deformation.

9.6 Stresses

We have dealt with the concept of stress before but we will now give some generality to it.

We start by stating that there are quite a few different stress definitions, according to the way force and area ratio are calculated. The main stress definitions are

- Cauchy stress, $\boldsymbol{\sigma}$
- Kirchhoff stress, $\boldsymbol{\tau}$
- Nominal stress, $\boldsymbol{N}$
- First Piola–Kirchhoff stress, $\boldsymbol{P}$
- Second Piola–Kirchhoff stress, $\boldsymbol{S}$
- Biot or Jaumann stress, $\boldsymbol{T}$

The Cauchy stress, $\boldsymbol{\sigma}$, was defined in Chapter 7 and it is essentially force divided by area in the current configuration, *ie* it is a true stress. We can weight this definition by multiplying it by the determinant of the gradient $J = \det\boldsymbol{F}$, so to obtain the Kirchhoff stress

$$\boldsymbol{\tau} = J\,\boldsymbol{\sigma}.$$

As for the First Piola—Kirchhoff stress, $\boldsymbol{P}$, it relates forces in the present configuration with the area in the reference configuration. It is our engineering stress seen in previous chapters in a 3D framework. It is connected to the initial area, being the transpose of the nominal stress, $\boldsymbol{N}$ and it can be related to the Cauchy stress via

$$\boldsymbol{P} = J\,\boldsymbol{\sigma}\,\boldsymbol{F}^{-T}$$

The Second Piola—Kirchhoff stress, $\boldsymbol{S}$, relates forces in the reference configuration to areas also in the reference configuration and it is related to the Cauchy stress according to

$$\boldsymbol{S} = J\,\boldsymbol{F}^{-1} \cdot \boldsymbol{\sigma} \cdot \boldsymbol{F}^{-T} \,.$$

As for the Biot or Jaumann stress, $\boldsymbol{T}$, it is defined as the symmetric part of $\boldsymbol{P}^T \cdot \boldsymbol{R}$, *ie*

$$\boldsymbol{T} = \frac{1}{2}\left(\boldsymbol{R}^T \cdot \boldsymbol{P} + \boldsymbol{P}^T \cdot \boldsymbol{R}\right)$$

We list next the relationships among these stress definitions.

TABLE 9.2 Relationship among various stresses definitions.

stress	$\boldsymbol{\sigma}$	$\boldsymbol{P}$	$\boldsymbol{S}$	$\boldsymbol{\tau}$	$\boldsymbol{T}$
$\boldsymbol{\sigma}$	$\boldsymbol{\sigma}$	$J^{-1}\boldsymbol{P}\boldsymbol{F}^T$	$J^{-1}\boldsymbol{F}\boldsymbol{S}\boldsymbol{F}^T$	$J^{-1}\boldsymbol{\tau}$	$J^{-1}\boldsymbol{R}\boldsymbol{T}\boldsymbol{F}^T$
$\boldsymbol{P}$	$J\boldsymbol{\sigma}\boldsymbol{F}^{-T}$	$\boldsymbol{P}$	$\boldsymbol{F}\boldsymbol{S}$	$\boldsymbol{\tau}\boldsymbol{F}^{-T}$	$\boldsymbol{R}\boldsymbol{T}$
$\boldsymbol{S}$	$J\boldsymbol{F}^{-1}\boldsymbol{\sigma}\boldsymbol{F}^{-T}$	$\boldsymbol{F}^{-1}\boldsymbol{P}$	$\boldsymbol{S}$	$\boldsymbol{F}^{-1}\boldsymbol{\tau}\boldsymbol{F}^{-T}$	$\boldsymbol{U}^{-1}\boldsymbol{T}$
$\boldsymbol{\tau}$	$J\boldsymbol{\sigma}$	$\boldsymbol{P}\boldsymbol{F}^T$	$\boldsymbol{F}\boldsymbol{S}\boldsymbol{F}^T$	$\boldsymbol{\tau}$	$\boldsymbol{R}\boldsymbol{T}\boldsymbol{F}^T$
$\boldsymbol{T}$	$J\boldsymbol{R}^T\boldsymbol{\sigma}\boldsymbol{F}^{-T}$	$\boldsymbol{R}^T\boldsymbol{P}$	$\boldsymbol{U}\boldsymbol{S}$	$\boldsymbol{R}^T\boldsymbol{\tau}\boldsymbol{F}^{-T}$	$\boldsymbol{T}$

We state finally that

- The First Piola-–Kirchhoff stress is energy conjugate to the deformation gradient
- The Second Piola-–Kirchhoff stress tensor is energy conjugate to the Green-–Lagrange finite strain tensor
- The Biot stress is energy conjugate to the right stretch tensor $\boldsymbol{U}$
- The Cauchy stress is energy conjugate to rate of deformation tensor, $\boldsymbol{D}$

9.7 Stress rates

We will introduce the stress rate by exploring the stress in a rod that is allowed to rotate, see next figure.

It is clear that the magnitude of the stress depends on the reference system they are related to. It is even possible to pull the rod axially and a shear stress to appear, as indicated in the figure. Such a situation is obviously difficult to deal with and it is far better to use a co–rotational reference system that rotates together with the body, or with a given finite element.

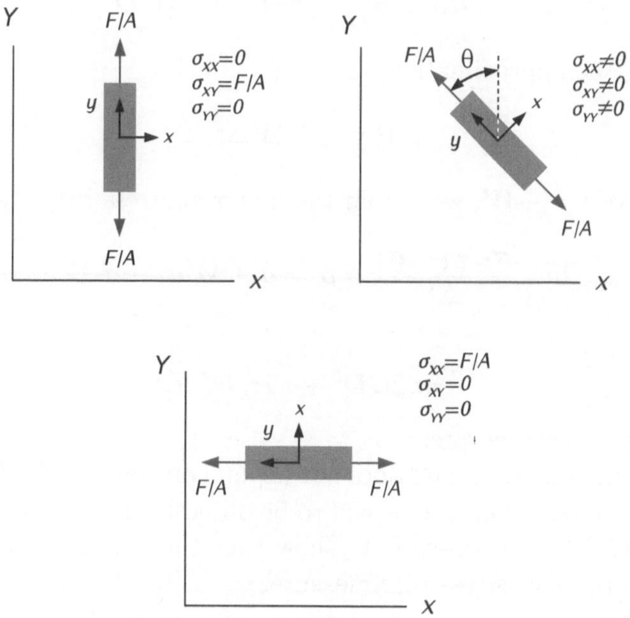

Figure 9.13 A rod undergoing rotation.

The concept of coordinate transformation seen in Chapter 8 help us here to write

$$\boldsymbol{\sigma}' = \boldsymbol{R}\boldsymbol{\sigma}\boldsymbol{R}^{\mathrm{T}},$$

with the dash representing the stress in the co–rotational system.

This section is adapted from F. Dunne and N. Petrinic, Introduction to Computational Plasticity, Oxford Press, 2007.

Now, regardless of the rigid body motion of the rod under consideration, the stress magnitude is the same and we refer to the Cauchy stress, $\boldsymbol{\sigma}'$, as an objective stress. The important aspect is that an objective measure does not depend on the rigid body motion. Hence, we must use objective measures when dealing with large strains, rotations and displacements.

For small time increment, Δt, we can write

$$\sigma'_{t+\Delta t} = \sigma' + \Delta\sigma'$$

and

$$\Delta\sigma' = \frac{\Delta\sigma}{\Delta t}\Delta t = [2G\boldsymbol{D}^e + \lambda \mathrm{Tr}(\boldsymbol{D}^e)\boldsymbol{I}]\Delta t,$$

where we use Hooke's law with $\boldsymbol{D}^e \approx \dot{\boldsymbol{\varepsilon}}^e$, $G = E/[2(1+\nu)]$ and $\lambda = E\nu/[(1-2\nu)(1+\nu)]$.

Now, we can write the co–rotational stress tensor at time $t + \Delta t$ as

$$\boldsymbol{\sigma}'_{t+\Delta t} = \Delta \boldsymbol{R} \boldsymbol{\sigma}_t \Delta \boldsymbol{R}^T + [2G\boldsymbol{D}^e + \lambda \text{Tr}(\boldsymbol{D}^e)\boldsymbol{I}]\Delta t.$$

By making the approximation

$$\Delta \boldsymbol{R} = \boldsymbol{I} + \boldsymbol{W}\Delta t,$$

and using $\boldsymbol{W}^{\mathrm{T}} = -\boldsymbol{W}$, we obtain the material stress rate

$$\lim_{\Delta t \to 0} \frac{\boldsymbol{\sigma}_{t+\Delta t} - \boldsymbol{\sigma}_t}{\Delta t} = \dot{\boldsymbol{\sigma}} = \overset{\triangledown}{\boldsymbol{\sigma}} + \boldsymbol{W}\boldsymbol{\sigma}_t - \boldsymbol{\sigma}_t \boldsymbol{W},$$

where

$$\overset{\triangledown}{\boldsymbol{\sigma}} = 2G\boldsymbol{D}^e + \lambda \text{Tr}(\boldsymbol{D}^e)\boldsymbol{I}.$$

is the Jaumann stress rate.

Note that the Cauchy stress rate $\dot{\boldsymbol{\sigma}}$ depends on the rigid body rotation and it is not objective, although the Cauchy stress is objective.

The Jaumann stress rate does not depend on the rigid body rotation. Considering that a tensor $\boldsymbol{T}$ is said to be objective if it rotates according to $\boldsymbol{T'} = \boldsymbol{R}\boldsymbol{T}\boldsymbol{R}^T$, it is possible to show that the Jaumann stress rate is indeed an objective stress rate measure.

■ STRESS IN A ROTATING ROD

Determine the stress related to the rotating frame x, y for the rod in the previous figure.

Our rotation tensor for this problem is given by

$$\boldsymbol{R} = \begin{pmatrix} \cos\theta & \sin\theta \\ -\sin\theta & \cos\theta \end{pmatrix}$$

and transforms the stress tensor according to

$$\boldsymbol{\sigma}' = \boldsymbol{R}\boldsymbol{\sigma}\boldsymbol{R}^{\mathrm{T}},$$

such that

$$\boldsymbol{\sigma} = \frac{P}{A}\begin{pmatrix} \sin^2\theta & \sin\theta\cos\theta \\ \sin\theta\cos\theta & \cos^2\theta \end{pmatrix}.$$

■

9.8 Plastic deformation gradient

If we load a structure in such a severe way that plastic strains take place, upon removal of the load, we can assume (or imagine) that the structure is left permanently deformed but in a state of free stress. Hence, from the initial to a current given configuration, we can imagine that there is an intermediate state in which a vector $\mathrm{d}\boldsymbol{X}$ has experienced only plastic deformation, to become $\mathrm{d}\boldsymbol{p}$, as illustrated in the next figure. We can relate these two vectors via

$$\mathrm{d}\boldsymbol{p} = \boldsymbol{F}^{\mathrm{p}}\mathrm{d}\boldsymbol{X},$$

with the plastic deformation gradient, $\boldsymbol{F}^{\mathrm{p}}$, being given by

$$\boldsymbol{F}^{\mathrm{p}} = \frac{\partial \boldsymbol{p}}{\partial \boldsymbol{X}}.$$

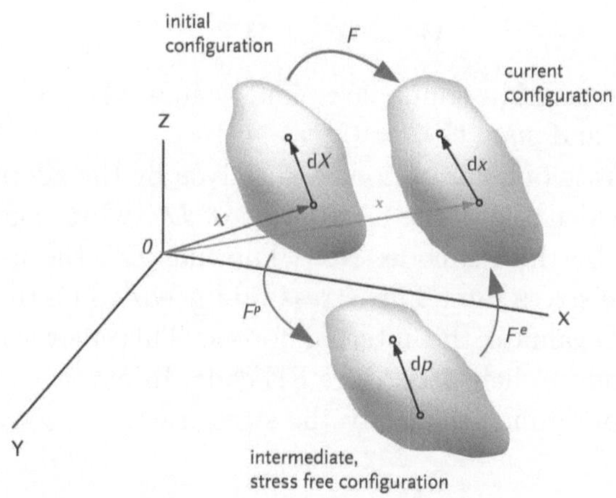

Figure 9.14 An illustration of the multiplicative decomposition.

The elastic deformation gradient, $\boldsymbol{F}^{\mathrm{e}}$, transforms $\mathrm{d}\boldsymbol{p}$ into the deformed line vector $\mathrm{d}\boldsymbol{x}$, via

$$\boldsymbol{F}^{\mathrm{e}} = \frac{\partial \boldsymbol{x}}{\partial \boldsymbol{p}},$$

so that

$$\mathrm{d}\boldsymbol{x} = \boldsymbol{F}^{\mathrm{e}}\boldsymbol{F}^{\mathrm{p}}\mathrm{d}\boldsymbol{X},$$

with

$$\boldsymbol{F} = \boldsymbol{F}^{\mathrm{e}}\boldsymbol{F}^{\mathrm{p}},$$

which is the multiplicative decomposition of the gradient $\boldsymbol{F}$. We see here the importance in operating $\boldsymbol{F}$ and other tensors; some plastic counterpart will now gain importance.

Note that both deformation gradients contain the cooperation of stretching and rigid body rotation, but we assume that the latter goes all into $\boldsymbol{F}^p$, leaving $\boldsymbol{F}^e$ with only stretching, *ie*

$$\boldsymbol{F}^e = \boldsymbol{V}^e, \quad \boldsymbol{F}^p = \boldsymbol{V}^p\boldsymbol{R}.$$

It is possible to show that

$$\boldsymbol{D} = \boldsymbol{D}^e + sym(\boldsymbol{V}^e\boldsymbol{D}^p\boldsymbol{V}^{e^{-1}}) + sym(\boldsymbol{V}^e\boldsymbol{W}^p\boldsymbol{V}^{e^{-1}})$$

$$\boldsymbol{W} = \boldsymbol{W}^e + sym(\boldsymbol{V}^e\boldsymbol{D}^p\boldsymbol{V}^{e^{-1}}) + sym(\boldsymbol{V}^e\boldsymbol{W}^p\boldsymbol{V}^{e^{-1}}),$$

ie the elastic and plastic rates of deformation add as

$$\boldsymbol{D} = \boldsymbol{D}^e + \boldsymbol{D}^p,$$

$$\boldsymbol{W} = \boldsymbol{W}^e + \boldsymbol{W}^p,$$

only when the elastic strains are small since, $\boldsymbol{V}^e = \boldsymbol{V}^{e^{-1}} \approx \boldsymbol{I}$, $sym(\boldsymbol{D}^p) = \boldsymbol{D}^p$ and $sym(\boldsymbol{W}^p) = 0$.

The plastic rate of deformation, $\boldsymbol{D}^p$ is given by the adopted constitutive equation, which allows the calculation of $\boldsymbol{D}^e$, when considering that $\boldsymbol{D}$ is evaluated by the FE code. Now, knowing $\boldsymbol{D}^e$, the use of Hooke's law will give the stress rate. This stress rate needs to be time integrated and is used to evaluate the internal forces. This allows one to check equilibrium in an implicit procedure FE code. In other words, it is very important to determine accurately the stress rate.

9.9 Computational plasticity – three-dimensional case

We have seen in a previous section the plasticity algorithm for 1D cases and of course it is of most importance to expand it for 3D cases. This is what we aim with this section.

We know that, as we progress in loading a structure, the strains change and their magnitude must be captured. Recall that the strains can be split in elastic and plastic. Recall also that Hooke's law can be written as

$$\boldsymbol{\sigma}_{t+\Delta t} = 2G\boldsymbol{\varepsilon}^e_{t+\Delta t} + \lambda\,\mathrm{Tr}\left(\boldsymbol{\varepsilon}^e_{t+\Delta t}\right)\boldsymbol{I},$$

at the end of a time step, *ie* at $t + \Delta t$, and that the elastic strains are

$$\varepsilon^e_{t+\Delta t} = \varepsilon^e_t + \Delta\varepsilon^e = \varepsilon^e_t + \Delta\varepsilon - \Delta\varepsilon^p,$$

leading to

$$\boldsymbol{\sigma}_{t+\Delta t} = \boldsymbol{\sigma}^{tr} - 2G\Delta\boldsymbol{\varepsilon}^p = \boldsymbol{\sigma}^{tr}_{t+\Delta t} - 2G\Delta\varepsilon_{eq}\boldsymbol{n}_{t+\Delta t},$$

where

$$\boldsymbol{\sigma}^{tr}_{t+\Delta t} = 2G\left(\boldsymbol{\varepsilon}^e_t + \Delta\boldsymbol{\varepsilon}\right) + \lambda\,\mathrm{Tr}\left(\boldsymbol{\varepsilon}^e_t + \Delta\boldsymbol{\varepsilon}\right)\boldsymbol{I},$$

is called trial stress, $2G\Delta\boldsymbol{\varepsilon}^p$ is a plastic corrector and use of $\mathrm{Tr}\left(\Delta\boldsymbol{\varepsilon}^p\right) = 0$ was made.

This section is adapted from F. Dunne and N. Petrinic, Introduction to Computational Plasticity, Oxford Press, 2007.

By noting from Chapter 7 that

$$\Delta\boldsymbol{\varepsilon}^p = \frac{3\boldsymbol{s}}{2\sigma_{eq}}\Delta\varepsilon_{eq}$$

we can write

$$\boldsymbol{\sigma}_{t+\Delta t} = \boldsymbol{\sigma}^{tr}_{t+\Delta t} - 2G\Delta\varepsilon_{eq}\frac{3}{2}\frac{\boldsymbol{s}_{t+\Delta t}}{\sigma_{eq_{t+\Delta t}}},$$

or

$$\left(1 + 3G\frac{\Delta\varepsilon_{eq_{t+\Delta t}}}{\sigma_{eq}}\right)\boldsymbol{s}_{t+\Delta t} = \boldsymbol{\sigma}^{tr}_{t+\Delta t} - \frac{1}{3}(\boldsymbol{\sigma}_{t+\Delta t} : \boldsymbol{I})\boldsymbol{I} = \boldsymbol{s}^{tr}_{t+\Delta t},$$

since $\boldsymbol{\sigma} = \boldsymbol{s} + (\boldsymbol{\sigma} : \boldsymbol{I})\boldsymbol{I}/3$.

We can now operate this last expression taking the tensor product, left with $(1 + 3G)\boldsymbol{s}_{t+\Delta t}$ and right with $\boldsymbol{s}^{tr}_{t+\Delta t}$, resulting in

$$\sigma_{eq_{t+\Delta t}} + 3G\Delta\varepsilon_{eq} = \sigma^{tr}_{eq_{t+\Delta t}}.$$

This equation offers an expression for the equivalent stress as a function of the trial equivalent stress, which was generated by an increase of the equivalent strain as a reaction to the load.

We proceed by defining our yield surface boundary as

$$f = \sigma_{eq} - (\sigma_y + r) = 0,$$

where r is a function, usually of the equivalent strain, describing the strain hardening of the material. This leads to the equation

$$f = \sigma^{tr}_{eq} - 3G\Delta\varepsilon_{eq} - r(\varepsilon_{eq}) - \sigma_y = 0,$$

which, being usually nonlinear, requires a numerical method to be solved, here the Newton one.

By making the expansion

$$f + \frac{\partial f}{\partial \Delta\varepsilon_{eq}} \mathrm{d}\Delta f + \frac{\partial f}{\partial \Delta\varepsilon_{eq}} \mathrm{d}\Delta\varepsilon_{eq} + \cdots = 0$$

and adopting a linear strain hardening law, $r = K\varepsilon_{eq}$, K being the hardening modulus, so that $\partial r / \partial \Delta\varepsilon_{eq} = K$, it follows

$$\mathrm{d}\Delta\varepsilon_{eq} = \frac{\sigma_e^{tr} - 3G\Delta\varepsilon_{eq} - r - \sigma_y}{3G + K}.$$

An algorithm would run like: At each iteration i, evaluate

- $r_i = r_t + K\Delta\varepsilon_{eq_i}$
- $\mathrm{d}\Delta\varepsilon_{eq} = \frac{\sigma_{eq}^{tr} - 3G\Delta\varepsilon_{eq_i} - r_i - \sigma_y}{3G+K}$
- $\Delta\varepsilon_{eq_{i+1}} = \Delta\varepsilon_{eq_i} + \mathrm{d}\Delta\varepsilon_{eq}$

With the so calculated $\Delta\varepsilon_{eq}$ we obtain

$$\sigma_{eq_{t+\Delta t}} = \sigma_{eq_{t+\Delta t}}^{tr} - 3G\Delta\varepsilon_{eq}$$

and

$$\boldsymbol{s} = \left(1 + 3G\frac{\Delta\varepsilon_{eq}}{\sigma_e}\right)^{-1} \boldsymbol{s}^{tr},$$

leading to the plastic strain tensor increment

$$\Delta\boldsymbol{\varepsilon}^p = \frac{3}{2}\Delta\varepsilon_{eq}\frac{\boldsymbol{s}_{t+\Delta t}}{\sigma_{eq_{t+\Delta t}}},$$

from which it follows at once the elastic strain tensor

$$\Delta\boldsymbol{\varepsilon}^e = \Delta\boldsymbol{\varepsilon} - \Delta\boldsymbol{\varepsilon}^p$$

and the respective stress increment

$$\Delta\boldsymbol{\sigma} = 2G\Delta\boldsymbol{\varepsilon}^e + \lambda\boldsymbol{I}\Delta\boldsymbol{\varepsilon}^e : \boldsymbol{I}.$$

It can be noted from this algorithm that it is an implicit one. Explicit algorithms are also possible but they are conditionally stable and less used. It is important to understand that this implicit scheme takes place within the material loop of the Finite Element analysis. Outside this loop one can have an explicit procedure updating the structure configuration.

By promoting such calculations we can step out of the material modulus of a Finite Element program. These calculations and of the previous sections fit in together in a single explicit nonlinear finite element code framework.

9.10 Closure

With this chapter we have advanced on the analysis of complex structures under dynamic loads. We have presented many basic aspects of the Finite Element Method, allowing us to work intelligently with professional software as well as developing in house codes for specific problems. We covered from unidimensional to 3D finite elements and various procedures of solutions, exploring the nonlinear realms of plasticity and large deformations.

All the subject here covered is, in its essence, the search for a model response. We inquire then if it is possible to obtain the overall response of a full prototype from the behaviour of a model, usually scaled down in size. This is the subject of the next chapter under the broad name of scaling.

9.11 Problems

1. Proof that
$$\frac{A_n}{A_0} = \left(\frac{L_n}{L}\right)^{-2\nu} = \lambda^{-2\nu}.$$

2. Show that $\boldsymbol{D} = \boldsymbol{D}^e + sym(\boldsymbol{V}^e\boldsymbol{D}^p\boldsymbol{V}^{e^{-1}}) + sym(\boldsymbol{V}^e\boldsymbol{W}^p\boldsymbol{V}^{e^{-1}})$ and $\boldsymbol{W} = \boldsymbol{W}^e + sym(\boldsymbol{V}^e\boldsymbol{D}^p\boldsymbol{V}^{e^{-1}}) + sym(\boldsymbol{V}^e\boldsymbol{W}^p\boldsymbol{V}^{e^{-1}})$. Hint: start from the definition for $\boldsymbol{L}$ and define $\boldsymbol{L}^e = \boldsymbol{D}^e + \boldsymbol{W}^e$ and $\boldsymbol{L}^p = \boldsymbol{D}^p + \boldsymbol{W}^p$.

3. Proof that the rate of deformation, $\boldsymbol{D}$ is objective.

4. Proof that the Jaumann stress rate is objective.

5. Obtain the Green, E_{ij}, and Almansi, e_{ij}, strains for the rectangular block of the figure. (*eg* from Reddy book, $e_{12} = e_o(a+b)/(ab - e_0^2) + e_0^3(a+b)/(ab - e_0^2)^2$.)

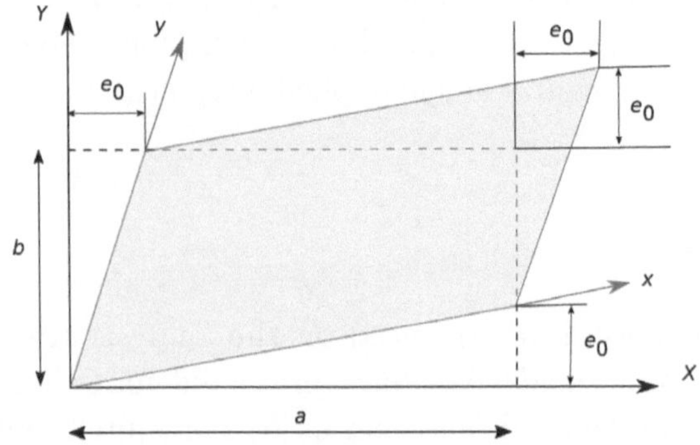

Figure 9.15 Deformation pattern of a rectangle.

6. For a rod of length L_0 deformed into the plastic regime to the length L show that the gradient $\boldsymbol{F} = \lceil \sqrt{L_0/L} \quad L/L_0 \quad \sqrt{L_0/L} \rfloor$. For this problem, calculate the rate of deformation and show that the continuum spin is zero.

7. Doctoral studies require original investigation. At the same time, your supervisor impels you to publish the data of your work ahead of your thesis submission date so to have a possible precedence in a discovery. How compatible do you see these two points of view? How about if your publication were upload to a public archive instead to a journal? Finally, what will be the role of journals in the future when most important publications will be readily available in open access electronic archives?

CHAPTER 10

Scaling

Figure 10.1 Collision of two ships.

Scaling is one of the most important subjects in science, in such that it deals with the fundamental reasons why we have our present size, why we stop growing, why we age, etc. Scaling explains the size of our cities and companies and even the crime rate.

This chapter is about the scaling laws in the context of structural mechanics. It is obvious that collision tests, as of the ship collision above, would immensely benefit from scaling their sizes down and experimenting with ship models rather than with their actual size. Tests on small scales serve as a motivation for our studies here and we start by presenting some interesting aspects of the scaling laws of living creatures. We move

For a broad introduction to scaling see G. West, Scale: the universal laws of growth, innovation, sustainability, and the pace of life in organisms, cities, economies, and companies, Penguin Press, 2017.

DOI: 10.1201/9781003485292-10

on by introducing basic structural engineering quantities and how they are related by the scaling factor β. We point out the reasons for the non scalability of strain rate sensitive structures under impact loads and when models and prototypes are made from different materials. We show how this problem can be overcome, presenting convincing numerical and experimental data that helps one to establish the use of scaling models as a powerful tool to study the behaviour of large structures under impact loads.

A literature review can be found in Reduced scale models based on similitude theory: A review up to 2015, by C.P. Coutinho, A.J. Baptista and J.D. Rodrigues, Engineering Structures 119 (2016) 81—94. The subject is growing at a fast pace.

10.1 Size in nature

The study of scaling goes back to Galileo who, in his book *Dialogue Concerning the Two Chief World Systems*, puts in the mouths of Salviati and Sagredo the following:

SALV. ... we asked the reason why they employed stocks, scaffolding and bracing of larger dimensions for launching a big vessel than they do for a small one; and he answered that they did this in order to avoid the danger of the ship parting under its own heavy weight, a danger to which small boats are not subject ...

SAGR. ... many devices which succeed on a small scale do not work on a large scale. Now, since mechanics has its foundation in geometry, where mere size cuts no figure, I do not see that the properties of circles, triangles, cylinders, cones and other solid figures will change with their size. If, therefore, a large machine be constructed in such a way that its parts bear to one another the same ratio as in a smaller one, and if the smaller is sufficiently strong for the purpose for which it was designed, I do not see why the larger also should not be able to withstand any severe and destructive tests to which it may be subjected.

SALV. ... you would do well to change the opinion which you, and perhaps also many other students of mechanics, have entertained concerning the ability of machines and structures to resist external disturbances, thinking that when they are built of the same material and maintain the same ratio between parts, they are able equally, or rather proportionally, to resist or yield to such external disturbances and blows. For we can demonstrate by geometry that the large machine is not proportionately stronger than the small.

...

SALV. ... if we take a wooden rod of a certain length and size ... may be reduced to such a length that it will just support itself; ... Thus if, for instance, its length be a hundred times its breadth, you will not be able to find another rod whose length is also a hundred times its breadth and which, like the former, is just able to sustain its own weight and no more: all the larger ones will break while all the shorter ones will be strong enough to support something more than their own weight. ... Who does not know that a horse falling from a height of three or four cubits will break his bones, while a dog falling from the same height or a cat from a height of eight or ten cubits will suffer no injury? Equally harmless would be the fall of a grasshopper from a tower or the fall of an ant from the distance of the moon. Do not children fall with impunity from heights which would cost their elders a broken leg or perhaps a fractured skull? And just as smaller animals are proportionately stronger and more robust than the larger, so also smaller plants are able to stand up better than larger. ... I have sketched a bone whose natural length has been increased three times and whose thickness has been multiplied until, for a correspondingly large animal, it would perform the same function which the small bone performs for its small animal. From the figures here shown you can see how out of proportion the enlarged bone appears. Clearly then if one wishes to maintain in a great giant the same proportion of limb as that found in an ordinary man he must either find a harder and stronger material for making the bones, or he must admit a diminution of strength in comparison with men of medium stature; for if his height be increased inordinately he will fall and be crushed under his own weight. Whereas, if the size of a body be diminished, the strength of that body is not diminished in the same proportion; indeed the smaller the body the greater its relative strength. Thus a small dog could probably carry on his back two or three dogs of his own size; but I believe that a horse could not carry even one of his own size.

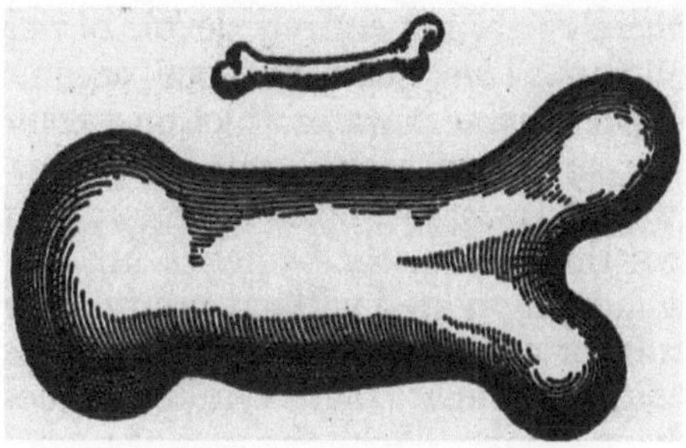

Figure 10.2 Bones sketches by Galileo.

M. Fowler justifies some of the above dialogue as follows. "What happens if we consider a stone of the same material and shape, but one–tenth the radius? It falls much more slowly. Its weight is down by a factor of one–thousand, but the surface area, which gives rise to the frictional retardation, is only down by a factor of one hundred. Thus a fine powder in water may take days to settle, even though a stone of the same material will fall the same distance in a second or two ... gravity becomes less and less important compared with viscosity, or air resistance — this is why an insect is not harmed by falling from a tree"

In galileo.phys.virginia.edu /classes/109/lectures/scaling.html

Also "... Warm blooded creatures (unlike insects) must devote a substantial part of their food energy simply to keeping warm. For an adult human, this is a pound or two of food per day. For a mouse, which has about one-twentieth the dimensions of a human, and hence twenty times the surface area per unit volume, the required food for maintaining the same body temperature is twenty times as much as a fraction of body weight, and a mouse must consume a quarter of its own body weight daily just to stay warm."

Now, suppose a unit density cuboid animal of side 10 cm, supported by one leg of 1 cm $\times$ 1 cm, cross–section area. The stress in this bone leg is $10 \times 10 \times 10/(1 \times 1) = 1000$. If we now increase all dimensions by a factor of 10 we have $100 \times 100 \times 100/(10 \times 10) = 10000$. If the bone of a mouse and of an elephant have similar strength, it can be concluded that the mouse leg is less stressed. It seems that the superior limit of the size of an animal is dictated, but not only, by its bone strength.

There are many other phenomena that are size dependent and follow the so called quarter–power scaling. Hence, a cat, which is 100 times greater in size than a mouse, lives about 100 to the power of one-quarter, or about three times, longer.

Heartbeat scales to mass to the power of minus one-quarter so the cat's heart beats a third as fast as a mouse's. It was Max Kleiber in the early 1930s that found that the metabolic rate scales with body mass to the three-quarter power and this seems to be valid for bacteria, to whales and even to the plant world. It seems that, as organisms grow in size they become more efficient in terms of using energy.

Some interesting examples of allometric (when the objects are dissimilar) scaling laws in the living world, related to the body mass, BM, are:

- heart mass: $HM = 0.006(BM)^{1.0}$
- blood volume: $BV = 0.055(BM)^{0.99}$
- heart beat: $HB = 241(BM)^{0.25}$.

These are complemented by the next figure, which shows the (maximum) number of heart beats per lifetime of a wide spectrum of animals. For mammals this number has been measured to be remarkably constant: an average of 7.3E8 heart beats per lifetime. Note in the figure the wide range of one million in weight.

10.2 Relating model to prototype

We turn now attention to solid mechanics phenomena and we define β as the scaling factor, *ie* a number that relates some representative length

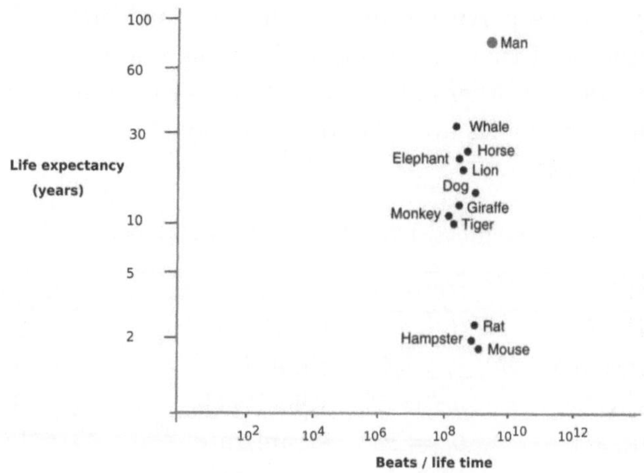

Figure 10.3 Heart beat and life span. Adapted from HJ Levine, Rest Heart Rate and Life Expectancy, J Am Coll Cardiol 1997;30:1104–6.

in the prototype, L_p, to an equivalent one in the model, L_m

$$\beta = \frac{L_m}{L_p},$$

where $_m$ variables are always related to the model and $_p$ to the prototype, with the model being smaller than the prototype, such that $\beta < 1$. All the comments here are equally valid for $\beta > 1$.

Let us see how we could calculate the strains in a prototype. This would be

$$e_p = \frac{\Delta L_p}{L_p},$$

while in the model we have

$$e_m = \frac{\Delta L_m}{L_m} = \frac{\beta \Delta L_p}{\beta L_p} = \frac{\Delta L_p}{L_p} = e_p.$$

Thus, we see that model and prototypes experience the same strain level. If they are made of the same strain rate insensitive material, they will also experience the same stress level.

On the other hand, the time an elastic perturbation takes to travel in a prototype is given by

$$t_p = \frac{L_p}{c_p},$$

whereas in the model it is

$$t_m = \frac{L_m}{c_m} = \frac{\beta L_p}{c_m} = \frac{\beta L_p}{c_p} = \beta t_p,$$

since an elastic wave perturbation has the same velocity, c, in a model and in a prototype if they are made of the same material.

The fact that time scales means that a pressure pulse has to be scaled in terms of duration (but not in amplitude) by β. Time varies with size!

Similar reasons for other parameters allow us to obtain the next table, listing how these variables are related in the model and in the prototype by the scaling factor β.

TABLE 10.1 Factors relating model variables to the prototype.

variable	factor	variable	factor
length, L	β	wave velocity, c	$\beta_c = 1$
displacement, δ	$\beta_\delta = \beta$	time, t	$\beta_t = \beta$
mass, G	$\beta_G = \beta^3$	velocity, V	$\beta_V = 1$
strain, ε	$\beta_\varepsilon = 1$	strain rate, $\dot{\varepsilon}$	$\beta_{\dot{\varepsilon}} = 1/\beta$
stress, σ	$\beta_\sigma = 1$	acceleration, A	$\beta_A = 1/\beta$
force, F	$\beta_F = \beta^2$	density, ρ	$\beta_\rho = 1$

Implicit in the above table is an incongruence between two mathematical statements, one that stress in the model and in the prototype are the same whereas the strain rate has a $1/\beta$ relation. Take for instance the Cowper–Symmonds constitutive equation. The dynamic stress ratio between model and prototype made for the same strain rate sensitive material reads

$$\frac{(\sigma_d)_\mathrm{m}}{(\sigma_d)_p} = \frac{\sigma_0\left[1+\left(\frac{\dot{\varepsilon}_\mathrm{m}}{D}\right)^{1/q}\right]}{\sigma_0\left[1+\left(\frac{\dot{\varepsilon}_p}{D}\right)^{1/q}\right]} = \frac{1+\left(\frac{\dot{\varepsilon}_p}{\beta D}\right)^{1/q}}{1+\left(\frac{\dot{\varepsilon}_p}{D}\right)^{1/q}} \neq 1,$$

which clearly contradicts the equality of stresses in the model and in the prototype. We show in the next figure how such a stress ratio varies with the strain rate and with the scaling factor, being evident that it increases with the decrease in the model size. Hence, we come to our first problem in scaling: how to obtain accurate stress information for a given prototype from the same respective and proportional spatial position values in a tested model.

This idea was first proposed by R.E. Oshiro and M. Alves, Scaling impacted structures, Archive of Applied Mechanics, 74, p. 130-–145, 2004.

To circumvent this shortcoming, let us keep in mind that, traditionally, a given variable is expressed as a function of the basic parameters: M, length, L, and time, T. Hence, for instance, energy units are

$$E = F.\delta = MLT^{-2}L = ML^2T^{-2}.$$

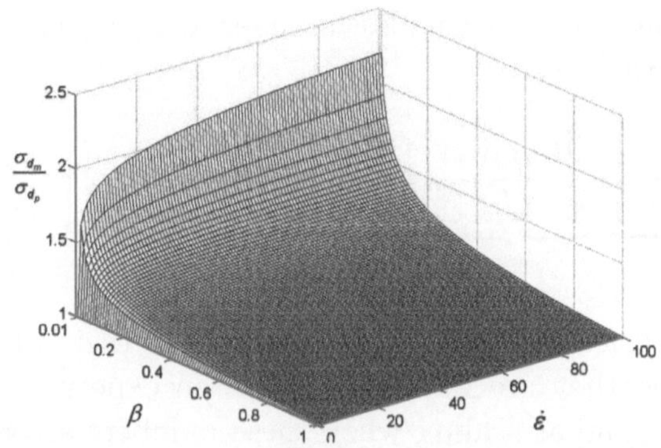

Figure 10.4 Prediction of the dynamic stress ratio for various strain rates and scaling factors.

But we can employ another basis set: impact velocity, V_0, dynamic yield stress, σ_d, and impact mass, G, so that energy units becomes

$$E = F.\delta = ML^2T^{-2} = V_0^d \sigma^e G^f = (L/T)^d (ML/T^2L^2)^e (M)^f,$$

from which it follows that $d = 2$, $e = 0$ and $f = 1$. This and other main variables in a structural analysis are listed in the next table as a function of the new elected basis.

TABLE 10.2 Factors to express variables in the basis V_0, σ_d and G.

	energy	accel.	time	disp.	strain rate	stress
	E	A	T	δ	$\dot{\varepsilon}$	σ
V_0	2	4/3	-1/3	2/3	1/3	0
σ_d	0	1/3	-1/3	-1/3	1/3	1
G	1	-1/3	1/3	1/3	-1/3	0

We can also obtain, from this new basis, dimensionless numbers, Π_i, which are used to handle scaled models. For instance, from the acceleration, A, we can express the product

$$A \times V_0^a \times \sigma_d^b \times G^c$$

in terms of its fundamental dimensions

$$LT^{-2}(LT^{-1})^a(ML^{-1}T^{-2})^b M^c$$

which is equal to 1 because the expression we seek is dimensionless. By comparing the exponents we obtain $a = -4/3$, $b = -1/3$ and $c = 1/3$, allowing us to write the dimensionless number

$$\Pi_1 = \frac{AG^{1/3}}{V_0^4 \sigma_d} = \frac{A^3 G}{V_0^4 \sigma_d}.$$

The following five dimensionless numbers can be generated using this procedure and they read

$$\underbrace{\left[\frac{A^3G}{V_0^4\sigma_d}\right]}_{\Pi_1}, \underbrace{\left[\frac{T^3\sigma_d V_0}{G}\right]}_{\Pi_2}, \underbrace{\left[\frac{\delta^3\sigma_d}{GV_0^2}\right]}_{\Pi_3}, \underbrace{\left[\dot{\varepsilon}\left(\frac{G}{\sigma_d V_0}\right)^{1/3}\right]}_{\Pi_4}, \underbrace{\left[\frac{\sigma}{\sigma_d}\right]}_{\Pi_5}.$$

These dimensionless sets can be used, from their product, to generate others and they are valuable to plan an experiment. We refer to a geometrically similar scaling when these numbers are the same for a model and for a prototype.

Famous in the scaling or similitude science is the Buckingham theorem. It postulates that a problem expressed using k Y_i variables such as

$$\phi\left(Y_1, Y_2, ..., Y_{\mathrm{k}}\right) = 0,$$

can be restated in terms of j independent physical units (usually mass, length, and time for structural analysis problems) as

$$\Phi\left(\Pi_1, \Pi_2, ..., \Pi_{\mathrm{j}}\right) = 0,$$

with $k > j$.

Π_i are dimensionless numbers constructed from Y_i using $k - j$ equations of the form

$$\Pi_{\mathrm{i}} = Y_1{}^{a_1} Y_2{}^{a_2} ... Y_{\mathrm{j}}{}^{a_{\mathrm{j}}}$$

with the exponents a_i being rational numbers.

We can define different scaling factors other than the β we used for size. First, we note that complete similarity or similarity of first kind of a phenomenon is achieved when all governing dimensionless numbers of the model match the corresponding dimensionless numbers of the prototype, *ie* $(\Pi_{\mathrm{i}})_{\mathrm{m}} = (\Pi_{\mathrm{i}})_{\mathrm{p}}$. When applied, for instance, to Π_3 we have

$$\left[\frac{\delta_m^3\sigma_{d_m}}{G_m V_{0_m}^2}\right] = \left[\frac{\delta_p^3\sigma_{d_p}}{G_p V_{0_p}^2}\right],$$

or

$$\left(\frac{\delta_m}{\delta_p}\right)^3\left(\frac{\sigma_{d_m}}{\sigma_{d_p}}\right)\left(\frac{G_p}{G_m}\right)\left(\frac{V_{0_p}}{V_{0_m}}\right)^2 = 1.$$

Defining $\beta_G = G_m/G_p$ as the impact mass ratio between model and prototype, and so on, we write the above equation as

$$\beta_\delta^3 \beta_\sigma = \beta_G \beta_V^2,$$

from which it follows that

$$\beta_V = \sqrt{\frac{\beta^3 \beta_{\sigma_d}}{\beta_G^2}}.$$

This procedure can be developed further by defining another dimensionless number relating the mass of the impacted structure, M, and the impact mass, G,

$$\Pi_6 = \left[\frac{M}{G}\right],$$

from which it follows that

$$\beta_G = \beta_\rho \beta^3,$$

so the impact velocity factor now reads,

$$\beta_V = \sqrt{\frac{\beta_{\sigma_d}}{\beta_\rho}}.$$

This is first suggested in L.M. Mazzariol, R.E. Oshiro and M. Alves, A method to represent impacted structures using scaled models made of different materials, International Journal of Impact Enginnering, 90, p. 81–94, 2016.

This and other expressions listed in the next table and obtained in a similar way are of most importance to scale correctly an impacted model. It is the model that is tested, its variables monitored and up scaled so as to yield the prototype response. Note the important fact that it is possible to obtain information of the prototype behaviour from the model even when they are made of different materials.

Now, consider the case when the scaling factor for the dynamic stress reads

$$\beta_{\sigma_d} = \frac{\sigma_{d_m}}{\sigma_{d_p}}.$$

If we now work with the Π_3 number, for similar scaling we must have that

$$\frac{(\Pi_3)_m}{(\Pi_3)_p} = 1 \rightarrow \frac{\beta^3 \beta_{\sigma_d}}{\beta_G {\beta_V}^2} = 1 \rightarrow \beta_{\sigma_d} = {\beta_V}^2$$

and it is somewhat simple to obtain the relations

$$\beta_{\dot{\varepsilon}} = \frac{\beta_V}{\beta}, \qquad \beta_t = \frac{\beta}{\beta_V}, \qquad \beta_A = \frac{{\beta_V}^2}{\beta}, \qquad \beta_\sigma = \beta_{\sigma_d}.$$

TABLE 10.3 Factors relating the model variables to the prototype in the initial Velocity–Yield Stress–Density (VSG-D) Basis.

Variable	Factor	Variable	Factor
length, L	β	time, t	$\beta_t = \beta/\beta_V$
displacement, δ	$\beta_\delta = \beta$	velocity, V	β_V
impact mass, G	$\beta_G = \beta_\rho \beta^3$	strain-rate, $\dot{\epsilon}$	$\beta_{\dot{\epsilon}} = \beta_V/\beta$
strain, ϵ	$\beta_\epsilon = 1$	acceleration, a	$\beta_a = \beta_V^2/\beta$
stress, σ	$\beta_\sigma = \beta_{\sigma_0}\beta_{\sigma_d}$	energy, E'	$\beta_{E'} = \beta_\rho \beta_V^2 \beta^3$
force, F	$\beta_F = \beta_\rho \beta_V^2 \beta^2$	density, ρ	β_ρ

Now, the relation $\beta_{\sigma_d} = {\beta_V}^2$ can be further manipulated to

$$\beta_{\sigma_d} = \frac{(\sigma_d)_m}{(\sigma_d)_p} = \frac{\sigma_0(\dot{\varepsilon}_m/\dot{\varepsilon}_0)^q}{\sigma_0(\dot{\varepsilon}_p/\dot{\varepsilon}_0)^q} = \left(\frac{\dot{\varepsilon}_m}{\dot{\varepsilon}_p}\right)^q = \left(\frac{\beta_V}{\beta}\right)^q,$$

where use was made of the Norton–Hoff constitutive law,

$$\sigma_d = \sigma_0\left(\frac{\dot{\varepsilon}}{\dot{\varepsilon}_0}\right)^q,$$

already presented in Chapter 7.

See R.E. Oshiro and M. Alves, Scaling of structures subject to impact loads when using a power law constitutive equation, International Journal of Solids and Structures, 46 p. 3412–3421, 2009.

From this, we can finally obtain the important relation

$$\beta_V = \beta^{q/(q-2)}.$$

This relation allows the calculation of the velocity at which a mass should hit a model such that the respective prototype behaves the same.

10.3 Analytical example

We now apply this general framework of scaling to the case of a beam under a transverse velocity loading, as shown in the next figure. We explore how the overall final displacement of the beam, as measured in a small model, can be used to infer the prototype behaviour. From the theory seen in Chapter 5 it is possible to obtain the final transverse central displacement of a beam under transverse velocity loading as

$$w_f = \frac{1}{2}\left\{(1 + 3\lambda/4)^{1/2} - 1\right\},$$

with $w_f = W_f/h$ and

$$\lambda = \frac{4\rho V_0^2 L^2}{\sigma_d h^2},$$

being ρ the beam density. An average strain rate for this impact event is

$$\dot{\varepsilon} = \frac{V_0 W_f}{3\sqrt{2}L^2}.$$

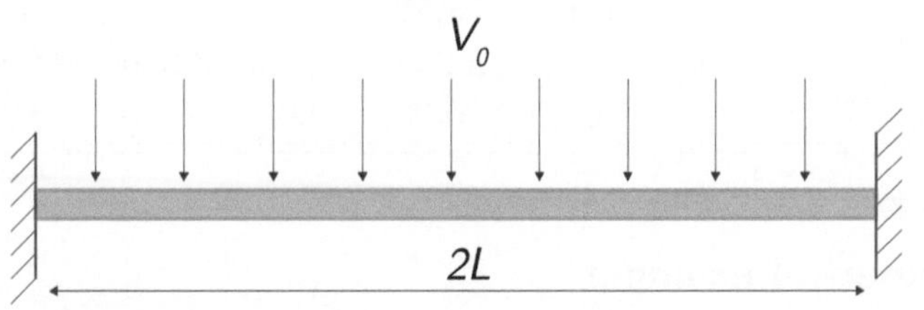

Figure 10.5 A beam under transverse velocity loading.

We can evaluate the relevant variables for a beam of $2L = 130$ mm, $H = 2.5$ mm, $\sigma_0 = 235$ MPa, $q = 0.077$, $\dot{\varepsilon}_0 = 0.001$ s^{-1}, $\rho = 7800$ kg.m^{-3} and subjected to an initial velocity of 50 m/s. The results are listed in the next table, where it can be seen that, for the uncorrected model, *ie* the model to which the traditional scaling laws are adopted, the transverse dimensionless beam displacement becomes smaller and smaller as the scaling factor increases. This can be explained because the small beam exhibits a larger strain rate and therefore larger strength, leading to smaller transverse displacements. A comparison with Galileo's statements in the opening of this chapter would be interesting.

To overcome this limitation of the traditional scaling laws, we are impelled to adopt the approach explained here. Indeed, it is possible to directly correct the initial velocity that should be applied to the model such that it will displace the same β fraction of the beam height as in the prototype. This requires an increase in magnitude of the initial velocity, *ie* β_V, so as to compensate the larger strain rate (and hence material strength) as the model decreases in size. By doing so, we can really infer the final displacement of the beam prototype by probing the behaviour of the beam model, whatever scaling factor used, as indicated in the table. This is the key aspect of the use of scaling models and laws.

TABLE 10.4 Comparison between the results for the prototype and the model for a clamped beam subject to a uniformly distributed velocity pulse.

Scale factor	1	1/2	1/4	1/10	1/20	1/2000
Uncorrected	model					
w_h	3.9060	3.7962	3.6891	3.5519	3.4511	2.8447
Error (%)		2.81	5.55	9.07	11.65	27.17
Corrected	model					
β_V	1	1.0281	1.0571	1.0966	1.1274	1.3557
w_h	3.9060	3.9060	3.9060	3.9060	3.9060	3.9060
Error (%)		0.00	0.00	0.00	0.00	0.00

10.4 Numerical example

We investigate now the use of the above scaling procedure to the axial impact of a cylindrical tube. The prototype tube has external diameter of 20 mm, thickness of 1 mm and height of 300 mm. The elastic, perfectly plastic material adopted has an elastic modulus of $E = 210$ GPa, a static yield stress of $\sigma_0 = 235$ MPa, a density of $\rho = 7800$ kg.m^{-3} and a Poisson ratio of $\nu = 0.33$. These are properties representative of a mild steel. The material strain rate sensitive response was modelled by the Cowper–Symonds equation, with the parameters $C = 40\text{s}^{-1}$ and $q = 5$. The rigid impact mass hits the free top end of the shell with an initial velocity V_0.

The tube rests freely on a semi infinite rigid base and was modelled with shell finite elements in Abaqus/Explicit. A friction coefficient of 0.25 at both ends was adopted. Self contact of the inner and the outer surfaces of the shell was also taken into account with a friction coefficient of 0.1. In order to trigger asymmetric buckling patterns, initial imperfections corresponding to the first eight elastic buckling modes of a tube were introduced. The two first modes have a maximum magnitude of $0.0005L$ and the last ones have magnitude of $0.000025L$. Initial impact velocities were 30, 40, 50 and 60 m/s.

Under the normal scaling procedure, we would not alter these initial impact velocities for the models and prototype: as per the above table, $\beta_V = 1$. By keeping the impact velocities the same, however, we would not scale this impact event correctly due to the material strain rate response, see next figure. This is when the correction procedure outlined before should be applied. The corrected impacted velocity can be obtained by solving the equation

$$\beta_{V_0} = \sqrt{\frac{1 + [\beta_{V_0} V_0/(4CR)]^{1/q}}{1 + [\beta V_0/(4CR)]^{1/q}}},$$

obtained when adopting an average strain rate, as presented in Chapter 6, of

$$\bar{\dot{\varepsilon}} = \frac{V_0}{4R},$$

with R being the radius of the shell.

Note the transcendental nature of this equation. The use of the Norton–Hoffman one leads to an explicit expression for β_{V_0}.

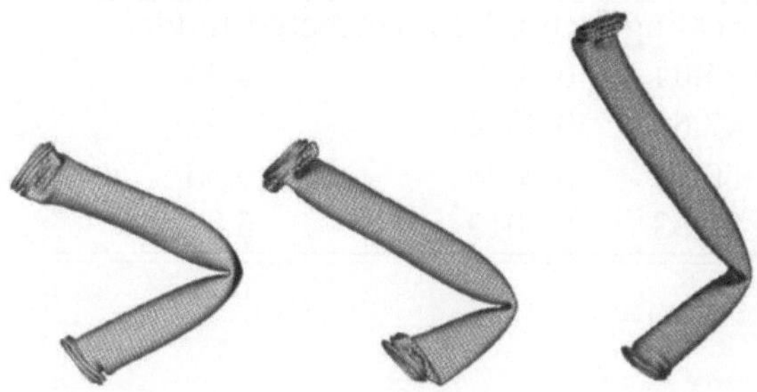

Figure 10.6 From left to right: prototype impacted at 30 m/s; corrected model enlarged 10 times; non corrected model enlarged 10 times.

The next table lists the main results of this study. First, an initial kinetic energy of 3500 J was maintained for the prototype tube, resulting in different combinations of impact mass and velocity. Now, if we scale by a factor of 10 and keep the same impact velocity, as it should be the case if the standard scaling laws were to be obeyed, the impact masses scale by β^3, as also listed below. Keeping these impact masses and applying the correction procedure we obtain the so listed impacted velocities, which were then used in the simulations.

It is clear in the sequence of the next figure that the scaling procedure estimates well the final dimensionless displacement of the tubes such that, by testing the model one could infer the prototype behaviour. The same could be said about the important structure response force *versus* displacement, which is also well represented by the model, including its typical strong variations.

TABLE 10.5 Main parameters used in the simulation of the axial impact of a tube.

V_0 (m/s)	Impact mass (kg)	Initial kin. en. (J)
Scaling factor 1, prototype		
30.00	7.78	3500.00
40.00	4.38	3500.00
50.00	2.80	3500.00
60.00	1.94	3500.00
Scaling factor 1:10, non corrected model		
30.00	0.0078	3.50
40.00	0.0044	3.50
50.00	0.0028	3.50
60.00	0.0019	3.50
Scaling factor 1:10, corrected model		
35.64	0.0078	4.94
47.68	0.0044	4.97
59.74	0.0028	5.00
71.83	0.0019	5.01

10.5 Model and prototype made of different materials

Let us now work on the important case when the model and the prototype are made from different materials. This is rather challenging because we know well that the material stress–strain curve, density, Poisson ratio, are all important in the structure response. How to relate model and prototype in this case is the subject of this section.

This section is based on M. Alves and R.E. Oshiro, Scaling impacted structures when the prototype and the model are made of different materials, International Journal of Solids and Structures 43, p 2744–2760, 2006.

We start by dealing with the flow stress, σ_s, assumed different in the model, σ_{s_m} and in the prototype, σ_{s_p}. As we learned, the scaling velocity factor can be transformed to

$$\beta_{V_0} = \sqrt{\beta_{\sigma_d}} = \sqrt{\frac{\sigma_{s_m}\left[1 + \left(\frac{\dot{\varepsilon}_m}{D}\right)^{1/q}\right]}{\sigma_{s_p}\left[1 + \left(\frac{\dot{\varepsilon}_p}{D}\right)^{1/q}\right]}}.$$

This equation can be applied to various theoretical and numerical problems, as the next example demonstrate.

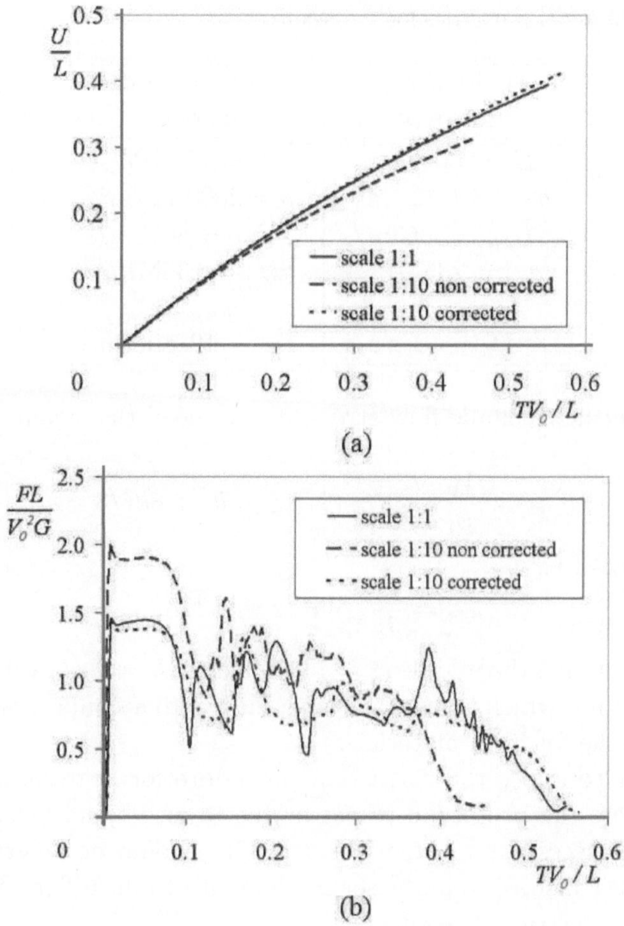

Figure 10.7 Results for tubes impacted at 30 m/s: (a) dimensionless displacement at impacted end. (b) Dimensionless load *versus* dimensionless time.

■ SCALING AN IMPACTED BEAM

See M. Alves and N. Jones, Impact failure of beams using damage mechanics: Part I — Analytical model, International Journal of Impact Engineering, 27, p. 837–861, 2002.

Consider a clamped aluminium and mild steel beams hit at the centre by a mass G travelling with a velocity V_0. Obtain the mid–displacement for these beams and various scaling factors.

The material properties of the two beam types are listed in the next table and we the steel beam as the reference one, *ie* the prototype.

The final beam mid–span displacement, W_f, is given by

$$w_f = \frac{W_f}{H} = \frac{H}{2L}\left(\sqrt{1+\frac{2GV_0^2L}{BH^3\sigma_d}}-1\right),$$

TABLE 10.6 Material properties for the beams.

mild steel prototype	aluminium models
$E = 200$ GPa	$E = 72$ GPa
$\rho = 7800$ kg/m^3	$\rho = 2800$ kg/m^3
$\sqrt{E/\rho} = 5063$m/s	$\sqrt{E/\rho} = 5070$m/s
$\sigma_{s_p} = 235$MPa	$\sigma_{s_m} = 135$MPa
$q_p = 5$	$q_m = 4$
$D_p = 40$/s	$D_m = 1288000$/s

To calculate the dynamic flow stress, σ_d, we need the strain rate in the beam,

$$\dot{\varepsilon}_{eq} = \frac{V_0}{L_1}\sqrt{(9/8)\left(1+\xi^2\right)^2 h^2 + 8k^2/3},$$

which reduces to

$$\dot{\varepsilon}_{eq} = \frac{V_0}{L}\sqrt{9h^2/2 + 8k^2/3},$$

when the mass hits the middle ($\xi = 1$) and where $h = H/L$, with $k = 0.26$ being a constant which takes transverse shear into account, assumed here to be the same for both materials.

In order to apply the correction procedure for a given test configuration where the prototype response due to an impact velocity V_{0_p} is desired, it is necessary to calculate β_{V_0}. This should be different from 1 since there is a distortion in the scaling laws due to the different materials used and/or to strain rate effects.

We use

$$\beta_{V_0} = \sqrt{\frac{\sigma_{s_m}\left[1+\left(\frac{\dot{\varepsilon}_m}{D}\right)^{1/q}\right]}{\sigma_{s_p}\left[1+\left(\frac{\dot{\varepsilon}_p}{D}\right)^{1/q}\right]}},$$

with the strain rate given by the previous equation. The corrected new impact velocity, to which the aluminium beam model is subjected to, V_{0_m}, is then obtained, which leads to new corrected scaled displacement.

The next figure shows the dimensionless maximum beam displacement for prototype and models when no correction procedure whatsoever is adopted. The prototype beam is $2L$=100mm long, with B =7.94mm, H =8.84mm, subjected to an impact velocity of $V_0 = 50$ m/s by a mass of $G = 6.5$ kg. Prototype and model have mechanical strengths typical of mild steel and aluminium, respectively. It is evident that the models, scaled by $\beta = 1/2$, $1/4$, $1/10$, $1/20$, are rather distorted in relation to the prototype in the sense that the higher the impact velocity, V_0, and the smaller the scaling factor β, the larger the difference between the dimensionless mid-displacement of models and prototype. This is all expected since the usual scaling laws cannot cope with models and prototype made of different materials.

The figure also shows the same dimensionless mid–span beam displacement for the mild steel prototype and for the aluminium models but now subjected to the correction outlined before. It is rather evident and convincing that models and prototypes behave the same; indeed the error in this case is zero.

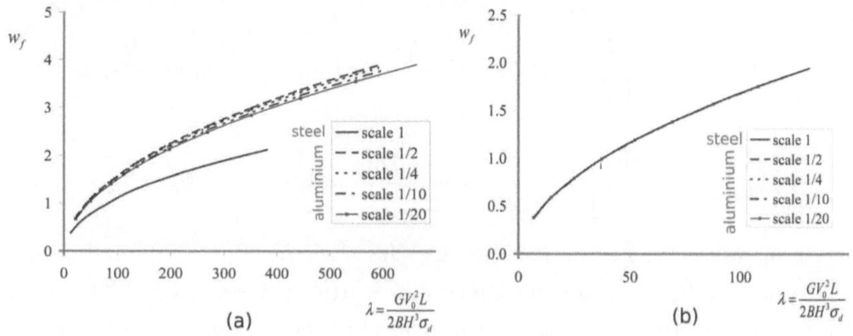

Figure 10.8 Mid–displacement evolution of an impacted clamped beam versus impact energy. (a) no correction, (b) corrected solution.

■

Consider now the temperature independent Johnson–Cook constitutive equation

$$\sigma_d = (A + B\epsilon^n)\,(1 + C \ln \dot{\epsilon}^*)\,,$$

with ϵ being the equivalent plastic strain, $\dot{\epsilon}^* = \dot{\epsilon}/\dot{\epsilon}_0$ is the dimensionless plastic strain-rate for $\dot{\epsilon}_0 = 1/s$, a reference strain-rate, A, B, C and n material constants. We can write

$$\beta_\sigma = \frac{(A_m + B_m \epsilon_m^{n_m})\,(1 + C_m \ln \dot{\epsilon}_m^*)}{(A_p + B_p \epsilon_p^{n_p})\,(1 + C_p \ln \dot{\epsilon}_p^*)}$$

and, from $\dot{\epsilon}_p = (\beta/\beta_V)\dot{\epsilon}_m$, it follows that

$$\beta_\sigma = \frac{(A_m + B_m \epsilon_m^{n_m})\,(1 + C_m \ln \dot{\epsilon}_m^*)}{(A_p + B_p \epsilon_p^{n_p})\left(1 + C_p \ln\left(\frac{\beta \dot{\epsilon}_m^*}{\beta_V}\right)\right)} = \beta_{\sigma_0}\beta_{\sigma_d},$$

with

$$\beta_{\sigma_0} = \frac{A_m + B_m\, \epsilon_m^{n_m}}{A_p + B_p\, \epsilon_p^{n_p}},$$

and

$$\beta_{\sigma_d} = \frac{1 + C_m\, \ln \dot{\epsilon}_m^*}{1 + C_p\, \ln\left(\frac{\beta \dot{\epsilon}_m^*}{\beta_V}\right)},$$

representing quasi-static and viscoplastic material characteristics, respectively.

It follows that

$$\beta_V = \sqrt{\frac{\beta^3 \beta_{\sigma_0} \beta_{\sigma_d}}{\beta_G}}$$

For more details see L.M. Mazzariol, R.E. Oshiro and M. Alves, A method to represent impacted structures using scaled models made of different materials, International Journal of Impact Engineering, 90, p. 81–94, 2016.

with the other Π numbers being

$$\Pi_1 = \left[\frac{a^3 G}{\sigma_d V_0^4}\right] \rightarrow \qquad \frac{\beta_a^3 \beta_G}{\beta_{\sigma_d} \beta_V^4} = 1 \rightarrow \qquad \beta_a = \frac{\beta_V^2}{\beta},$$

$$\Pi_2 = \left[\frac{t^3 \sigma_d V_0}{G}\right] \rightarrow \qquad \frac{\beta_t^3 \beta_{\sigma_d} \beta_V}{\beta_G} = 1 \rightarrow \qquad \beta_t = \frac{\beta}{\beta_V}.$$

$\beta_F = \beta_G \beta_a$ and the factor for energy $\beta_E = \beta_F \beta$.

We can go further along these lines and investigate if the density could be scaled in these laws and for that consider the dimensionless number which relates the mass of the impacted structure, M, and the impact mass, G,

$$\Pi_6 = \left[\frac{M}{G}\right].$$

Since $(\Pi_6)_m = (\Pi_6)_p$, it follows that

$$\left[\frac{M_m}{G_m}\right] = \left[\frac{M_p}{G_p}\right] \rightarrow \beta_G = \frac{M_m}{M_p}$$

$$\beta_G = \beta_\rho\, \beta^3$$

and

$$\beta_V = \sqrt{\frac{\beta_{\sigma_d}}{\beta_\rho}} = \sqrt{\frac{\beta_{\sigma_0}\, \beta_{\sigma_d}}{\beta_\rho}}$$

■ Scaling for different densities

For a clamped beam subjected to a uniformly distributed velocity pulse, prove that perfect similarity can be achieved achieved when model and prototype have different densities and strengths.

The final displacement at the mid–span, W_f, reads

$$w_H = \frac{1}{2}\left[\left(1 + \frac{3\rho V_0^2 L^2}{\sigma_d H^2}\right)^{1/2} - 1\right]$$

being $w_H = W_f/H$, H the beam height, ρ the material density, and L standing for the half length of the beam. Since displacement and geometry should be scaled by the same factor, β, a model geometrically scaled in relation to the prototype implies that $(w_H)_m = (w_H)_p$ or

$$\frac{(w_H)_m}{(w_H)_p} = \frac{\frac{1}{2}\left\{\left[1 + \left(\frac{3\rho_m V_{0m}^2 L_m^2}{\sigma_{dm} H_m^2}\right)\right]^{1/2} - 1\right\}}{\frac{1}{2}\left\{\left[1 + \left(\frac{3\rho_p V_{0p}^2 L_p^2}{\sigma_{dp} H_p^2}\right)\right]^{1/2} - 1\right\}}.$$

Inserting the corresponding scaling factors into the model variables, one has

$$\frac{(w_H)_m}{(w_H)_p} = \frac{\left[1 + \frac{3(\beta_\rho \rho_p)(\beta_V V_{0p})^2 (\beta L_p)^2}{(\beta_{\sigma_d} \sigma_{dp})(\beta H_p)^2}\right]^{1/2} - 1}{\left[1 + \frac{3\rho_p V_{0p}^2 L_p^2}{\sigma_{dp} H_p^2}\right]^{1/2} - 1}$$

$$= \frac{\left[1 + \left(\frac{\beta_\rho \beta_V^2 \beta^2}{\beta_{\sigma_d} \beta^2}\right)\left(\frac{3\rho_p V_{0p}^2 L_p^2}{\sigma_{d_p} H_p^2}\right)\right]^{1/2} - 1}{\left[1 + \left(\frac{3\rho_p V_{0p}^2 L_p^2}{\sigma_{d_p} H_p^2}\right)\right]^{1/2} - 1}.$$

Using the relation between the velocity factor and impact mass factor we write the above equation as

$$\frac{(w_H)_m}{(w_H)_p} = \frac{\left[1 + \left(\frac{\beta_\rho \beta^3 \beta_{\sigma_d}}{\beta_{\sigma_d} \beta_G}\right)\left(\frac{3\rho_p V_{0p}^2 L_p^2}{\sigma_{d_p} H_p^2}\right)\right]^{1/2} - 1}{\left[1 + \left(\frac{3\rho_p V_{0p}^2 L_p^2}{\sigma_{d_p} H_p^2}\right)\right]^{1/2} - 1}$$

and introducing the mass distortion expression, it follows that

$$\frac{(w_H)_m}{(w_H)_p} = \frac{\left[1+\left(\frac{\beta_\rho\beta^3}{\beta_\rho\beta^3}\right)\left(\frac{3\rho_p V_{0p}^2 L_p^2}{\sigma_{d_p} H_p^2}\right)\right]^{1/2} - 1}{\left[1+\left(\frac{3\rho_p V_{0p}^2 L_p^2}{\sigma_{d_p} H_p^2}\right)\right]^{1/2} - 1} \to \frac{(w_H)_m}{(w_H)_p} = 1.$$

Therefore, the condition for perfect similarity between model and prototype is fulfilled, even if different materials are considered. Note that the material model adopted here does not consider neither elasticity nor strain hardening.

■

10.6 Geometrically distorted scaled models

Let us now ponder on the problem depicted in the next figure. Here we have two beams to be used in an experimental program to study scaling. We have a prototype beam and a model scaled by $\beta = 1/4$. The holes are just to help clamping the beams and will not be important to the analysis to follow. We can immediately see that it is easy to scale the length and the breadth of the beam by this 1/4 factor. But to find two metal sheets scaled by 1/4 may not be possible. In the present case, our model was scaled in the thickness by a little less than 1/3. By so proceeding, we will not be able to apply the scaling laws seen above and we are facing what we call a geometrically distorted scaled model. How to handle these models is the subject of this section.

This section is based on the article by R.E. Oshiro and M. Alves, Predicting the behaviour of structures under impact loads using geometrically distorted scaled models, Journal of the Mechanics and Physics of Solids, 60, p. 1330–1349, 2012.

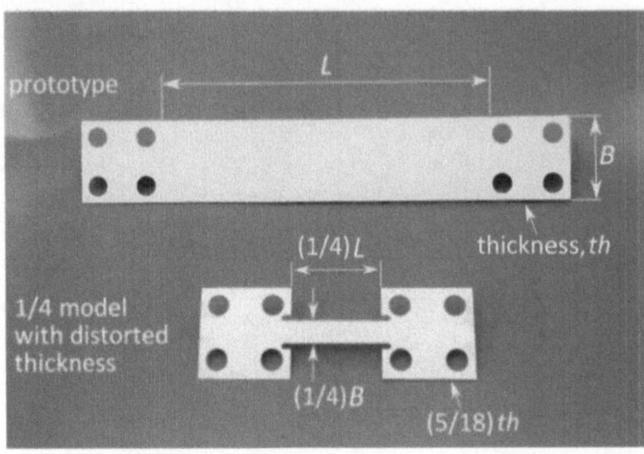

Figure 10.9 Two beams scaled in length and in breadth by 1/4 but the thickness by 5/18.

Let us first expand a little the dimensionless numbers we obtained before, *ie*

$$\underbrace{\left[\frac{\delta^3\sigma_{\rm d}}{GV_0^2}\right]}_{\Pi_1}, \underbrace{\left[\frac{\dot{\varepsilon}^3 G}{\sigma_{\rm d}V_0}\right]}_{\Pi_2}, \underbrace{\left[\frac{A^3 G}{V_0^4\sigma_{\rm d}}\right]}_{\Pi_3}, \underbrace{\left[\frac{t^3\sigma_{\rm d}V_0}{G}\right]}_{\Pi_4}, \underbrace{\left[\frac{\sigma}{\sigma_{\rm d}}\right]}_{\Pi_5}, \underbrace{\left[\frac{F^3}{{V_0}^4\sigma_{\rm d}G^2}\right]}_{\Pi_6}, \underbrace{\left[\frac{E'}{G{V_0}^2}\right]}_{\Pi_7},$$

with E' being the energy, t the time and F the force.

By equating the same dimensionless numbers for the prototype and for the model we obtain expressions like

$$(\Pi_2)_{\rm m} = (\Pi_2)_{\rm p} \rightarrow \frac{{\beta_{\dot{\varepsilon}}}^3\beta_G}{\beta_{\sigma_{\rm d}}\beta_V} = 1 \rightarrow \beta_{\dot{\varepsilon}} = \beta^{2/(q-2)},$$

$$(\Pi_3)_{\rm m} = (\Pi_3)_{\rm p} \rightarrow \frac{{\beta_A}^3\beta_G}{{\beta_V}^4\beta_{\sigma_{\rm d}}} = 1 \rightarrow \beta_A = \beta^{(q+2)/(q-2)},$$

$$(\Pi_4)_{\rm m} = (\Pi_4)_{\rm p} \rightarrow \frac{{\beta_t}^3\beta_{\sigma_{\rm d}}\beta_V}{\beta_G} = 1 \rightarrow \beta_t = \beta^{2/(2-q)},$$

$$(\Pi_5)_{\rm m} = (\Pi_5)_{\rm p} \rightarrow \beta_\sigma = \beta^{2q/(q-2)},$$

$$(\Pi_6)_{\rm m} = (\Pi_6)_{\rm p} \rightarrow \frac{{\beta_F}^3}{{\beta_V}^4\beta_{\sigma_{\rm d}}\beta^6} = 1 \rightarrow \beta_F = \beta^{4(q-1)/(q-2)}$$

and

$$(\Pi_7)_{\rm m} = (\Pi_7)_{\rm p} \rightarrow \frac{\beta_{E'}}{\beta_G{\beta_V}^2} = 1 \rightarrow \beta_{E'} = \beta^{3+(2q)/(q-2)},$$

where use was made of the Norton–Hoff constitutive equation, $\sigma_{\rm d} = \sigma_0(\dot{\varepsilon}/\dot{\varepsilon}_0)^q$.

Now, in order to handle a distorted geometry of a model, we need to create a new factor,

$$\beta_X = (X)_{\rm m}\Big/(X)_{\rm p},$$

so the new impact velocity factor, β_V', becomes a function of the type

$$\beta_V{}' = f\left(\beta, \beta_X/\beta\right) = f_1 f_2.$$

Here,

$$f_1 = \beta_V = \beta^{q/(q-2)}$$

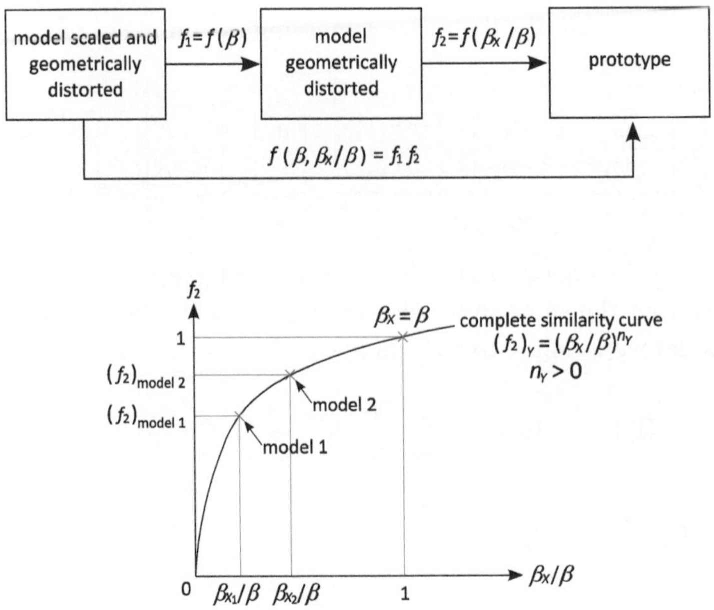

Figure 10.10 An illustration of the basic procedure for scaling distorted models.

is the previously obtained impact velocity factor and f_2 denotes the effect of the geometric distortion, β_X/β, assumed to be calculated by

$$f_2 = f\,(\beta_X/\beta) = (\beta_X/\beta)^{n_V},$$

with n_V being an exponent to be evaluated.

The figure above helps to understand the calculation of n_V. We use two geometrically distorted models

$$\beta_{X_1} = (X)_{\text{m1}}\Big/(X)_{\text{p}} \quad \text{and} \quad \beta_{X_2} = (X)_{\text{m2}}\Big/(X)_{\text{p}},$$

with their respective velocity factors

$$\beta_{V_1}{}' = (V)_{\text{m1}}\Big/(V)_{\text{p}} \quad \text{and} \quad \beta_{V_2}{}' = (V)_{\text{m2}}\Big/(V)_{\text{p}}.$$

The condition for complete similarity between the two models implies a common factor n_V, such that

$$\beta_{V_1}{}' = \beta_V(\beta_{X_1}/\beta)^{n_V}, \qquad \beta_{V_2}{}' = \beta_V(\beta_{X_2}/\beta)^{n_V},$$

which leads to

$$\beta_{V_2}{}'/\beta_{V_1}{}' = (\beta_{X_2}/\beta_{X_1})^{n_V} \rightarrow (V)_{\text{m2}}/(V)_{\text{m1}} = (\beta_{X_2}/\beta_{X_1})^{n_V},$$

with the remark that β_{X_2}/β_{X_1} is known.

Let us now take a given variable Y measured in the model and which we want to transfer to the prototype. The scaling factor here is

$$\beta_Y{}' = (Y)_{\mathrm{m}} \Big/ (Y)_{\mathrm{p}} = f_{1Y} f_{2Y}.$$

f_{1_Y} comes from one of the various equations from the table seen before. f_{2_Y} is similar to f_2 with the exponent evaluated, for a given variable, according to

$$n_Y = \frac{\log\left[(Y)_{\mathrm{m}2}/(Y)_{\mathrm{m}1}\right]}{\log\left(\beta_{Y_2}/\beta_{Y_1}\right)},$$

which finally results in

$$\beta_Y{}' = (f_1)_Y (f_2)_Y = \beta_Y (\beta_X/\beta)^{n_Y}.$$

We can then obtain the relations in the next table for the variables there listed, with n_σ, n_F, n_t, $n_{\dot{\varepsilon}}$ and n_a being the exponents of f_2 for stress, force, time, strain rate and acceleration, respectively. Note in the next table that, if the material is not strain rate sensitive, $q = 0$ and the equations are now only functions of β and β_X. If we further assume that there is no geometrical distortion then $\beta_X = 1$ and the equations are reduced to the ones in the table seen before.

TABLE 10.7 Scaling factors for the VSGd model with dimension X initially distorted by β_X/β.

Variable	Scaling factor	Variable	Scaling factor
length, L	β	Force, F	$\beta^{4(q-1)/(q-2)}(\beta_X/\beta)^{n_F}$
distortion, X	β_X	time, t	$\beta^{2/(2-q)}(\beta_X/\beta)^{n_t}$
displac., δ	β	veloc., V	$\beta^{q/(q-2)}(\beta_X/\beta)^{n_V}$
mass, G	β^3	str. rt., $\dot{\varepsilon}$	$\beta^{2/(q-2)}(\beta_X/\beta)^{n_{\dot{\varepsilon}}}$
strain, ε	β/β_X	accel., A	$\beta^{(q+2)/(q-2)}(\beta_X/\beta)^{n_A}$
stress, σ	$\beta^{\frac{2q}{q-2}}(\beta_X/\beta)^{n_\sigma}$	energy, E'	$\beta^{3+2q/(q-2)}(\beta_X/\beta)^{2n_V}$

As an example of the above procedure, let us examine again the case of a beam subjected to a blast load, characterized by a transverse velocity applied along all its span. We deal here with two beams, the reference one, the prototype, and the model, with a a scaling factor of 1/20 for its length and breadth. The model beam height is however not scaled by $\beta = 1/20$ but rather by the scaling factors of $\beta_H = 1/2$, case I, and $\beta_H = 2$, case II. These are extremely large geometrical distortions and we explore whether it is possible to obtain accurate information of the prototype behaviour by testing these models.

We have seen that the final maximum dimensionless displacement, $w_H = W_f/H$, of this beam, reads

$$w_H = \frac{1}{2}\left[\left(1 + \frac{3\rho V_0^2 L^2}{\sigma_{\mathrm{d}} H^2}\right)^{1/2} - 1\right]$$

and using the Norton–Hoff equation to correct the material strength based on a strain rate of

$$\dot{\varepsilon} = \frac{V_0 W_{\mathrm{f}}}{3\sqrt{2}L^2},$$

it follows that

$$w_H = \frac{1}{2}\left\{\left[1 + \frac{3\rho {V_0}^2 L^2}{\sigma_0 H^2}\left(\frac{3\sqrt{2}L^2\dot{\varepsilon}_0}{V_0 W_{\mathrm{f}}}\right)^q\right]^{1/2} - 1\right\}.$$

Let us take, for example, a beam of 400 mm span and a square cross–section of side 20 mm, made from the same material as in the previous beam case and subjected to an initial velocity of 100 m/s.

After three interactions, we found $n_V = -1.1$ and $n_V = 0.1$ for the cases of β_X/β equal 0.5, case I, and 2.0, case II, with X standing for height or length. Using these n_V values, we obtain the solutions summarized in the next table. There, it can be appreciated that the larger distortions for the beam model in height and in length do not refrain us of obtaining quite good quantitative information on the final prototype beam transverse displacement.

10.7 Scaling of distorted models using finite elements

The approach detailed before was applied so far to analytical models. The question raised is how to obtain information on a prototype behaviour from a scaled model whose behaviour cannot be predicted by analytical methods. We next show that it is indeed possible to obtain the behaviour of a prototype from the model response obtained analytically.

Take for example the frontal collision of a ship. We see in the next figure the complexity of the problem, the advantage of the use of a ship model in the frontal collision test being obvious.

TABLE 10.8 Results for the problem of a beam subjected to a uniformly distributed velocity pulse.

distorted	β	β_V/β	n_V	f_1	f_2	β_V'	w_H	error (%)
				prototype				
	1			1	1	1	2.8763	
				MLT model				
length	0.05	$\frac{\beta_L}{\beta} = 2.0$	0	1	1	1	5.6369	95.98
height	0.05	$\frac{\beta_H}{\beta} = 2.0$	0	1	1	1	2.1697	24.57
				VSG model				
length	0.05	$\frac{\beta_L}{\beta} = 2.0$	-1.1	1.1274	0.4665	0.5260	2.9190	1.48
height	0.05	$\frac{\beta_H}{\beta} = 2.0$	0.1	1.1274	1.0718	1.2084	2.7144	5.63

We set the task to obtain variables like the frontal displacement of the bow as it collides against a rigid wall. The difficulty lies in obtaining these variables for the actual ship, since it is not feasible to perform such an impact test. With the procedure developed here, it is possible to impact a model and to extrapolate the results with the scaling laws to the actual ship. This is one route, whose accuracy could be verified only at the expense of the real test. Instead, we perform finite element simulations of both, model and prototype, comparing the results.

We first adopt a model 20 times smaller than an actual ship, with the geometry being perfectly scaled. This is to say that any nonscalability is only due to strain rate effects.

Figure 10.12 plots the time evolution of the permanent bow displacement at its bottom and top, based on the MLT and VSG formulations. Two important aspects arise from the figure. First, as we have already seen in other cases, the VSG methodology assures much better results than the traditional MLT approach. The other aspect is that there is not limitation whether the problem is solvable analytically or numerically, provided the initial conditions are known. Here, they are set according to the VSG or MLT approaches and any strain rate effects in the material response can be rightly assimilated. Note that the use of the Norton–Hoff equation is a key factor that allows the calculation of the initial ship model velocity regardless of one knowing a priory the strain rate.

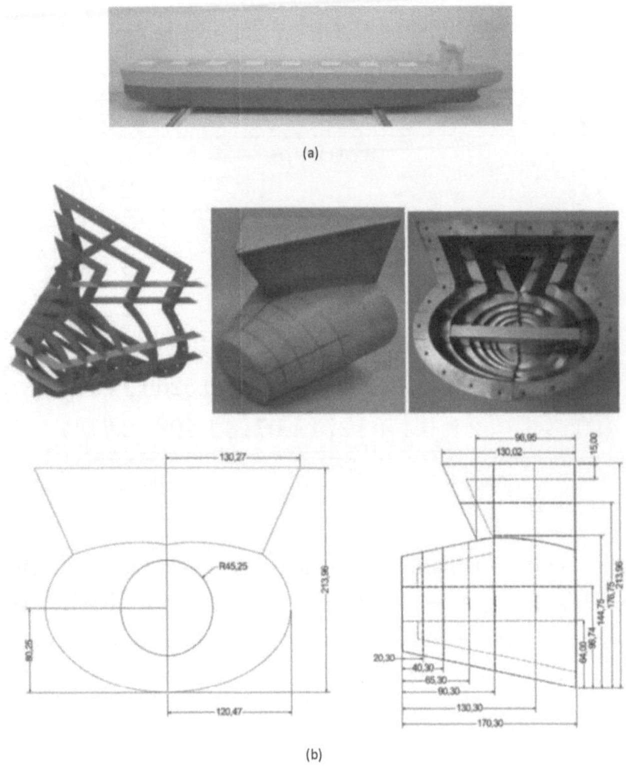

Figure 10.11 (a) A ship model in a frontal collision and (b) details of its bow.

As we have seen, it is not always possible to build a model that scales in all its variables to the prototype by the same single factor. Let us suppose that the thickness of the bow sheet cannot be built to 1/20 of the prototype. Indeed, for $\beta = 1/20$, a ship 100 m long and with a bow made of, say, a 17 mm thick sheet, would be represented by a model whose bow thickness should be $17/20 = 0.85$ mm. For lack of availability, the sheet for building the model is 1 mm thick, which gives $\beta_{th}/\beta = 0.0588$.

It is very important to realize that, in this specific case, the moving mass is the ship so, if the thickness is distorted, it is also the "impact mass" itself, and this should be taken into account.

10.8 Closure

The scaling laws were presented in this chapter to convey the important fact that it is possible to impact test large structures with the aid of small

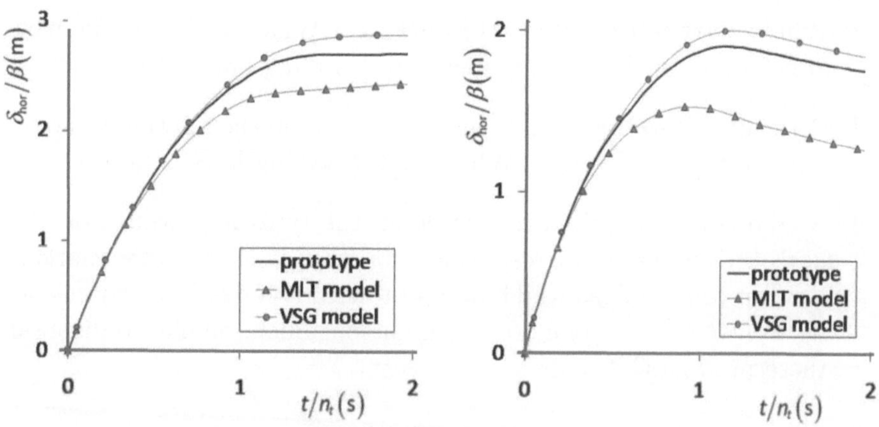

Figure 10.12 (a) Displacement at the (a) top and (b) bottom of the proximal end of the bow with no geometrical distortion.

models. The results obtained from these models can be projected on the actual structure size if the right corrected scaling laws are observed.

The subject of scaling went so far that it is possible to work with a model whose not all dimensions are scaled by a single scaling factor β. This has important practical consequences since, as seen here, it is not possible to find a sheet thickness that scales as desired in many cases.

With the advent of 3D printing, it is also important to take note that new mathematical procedures dealing with the scaling laws allows one to test a structure model whose material is different from the prototype model. Although in its infancy, distorted scaling may well be an important aid to experimental mechanics, for inferring the behaviour of both very large and very small structures.

Having now a more comprehensive knowledge on impact engineering, we can move on to our last chapter, where various application of this engineering field will be explored.

10.9 Problems

1. For a clamped beam under a pulse velocity, obtain an expression for the scaled velocity such that the scaling is exact. Adopt the Norton law to describe the material strain rate sensitivity.

2. Obtain the variable force in the VSGd basis and show that it reduces to β^2 when there is no geometrical distortion or strain rate effects.

3. What are type I and type II structures? What would be the most appropriate of these structures to test scaling laws? Why?
4. Perform a literature search and investigate on the relation between crime rate and city size? What are the scaling laws behind?
5. A company is required to recycle all the tyres it produces on the basis that it should be responsible for what it offers to the market. How do you see this? And how about when it comes to electrical car batteries? Is a company or the final consumer the one to disposal these products?

CHAPTER 11

Impact engineering

with Pouria B. Ataabadi

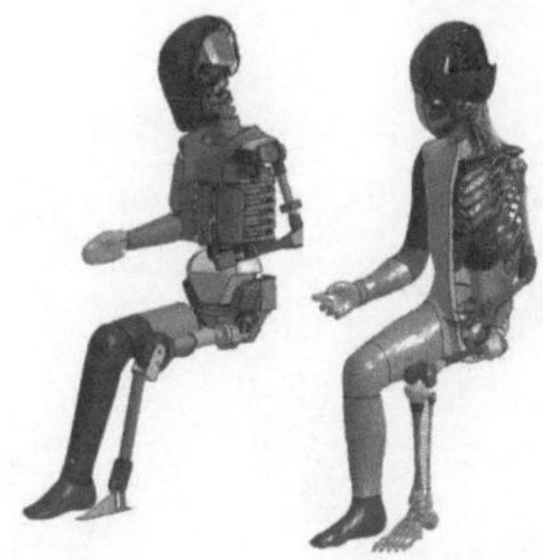

Figure 11.1 Different attempts to model a human being.

This last chapter aims to present some cases in the field of impact engineering. The topics discussed here rely on the various concepts presented in the previous chapters, highlighting their importance. We also take the opportunity to present some new areas, albeit superficially, not seen before, like ship collision and impact on carbon fibre tubes.

We begin with a very practical case of the impact on a tyre. The complex numerical modelling of a tyre is presented together with experiments that correlate well with the model. We move on to an optimization procedure applied to a compressor in order to prevent it from damage

DOI: 10.1201/9781003485292-11

in case of a fall. Plate penetration with projectiles at very high speed are treated next, with emphasis on the numerical model. Unlike Chapter 6, where we studied the impact of metallic tubes, we present here a section on the axial impact of carbon fiber composite tubes, comparing numerical models with experiments. We also provide an application of the scaling laws by presenting a ship collision study and the impact of a car on a guardrail. We further explore the behaviour of tubes under axial impact, but now the tubes ae made of carbon fibres. Finally, we discuss the design of a car, with the implications of meeting crash regulations, use of advanced optimization procedures and the behaviour of dummies and their tolerance to impact loading.

11.1 Tyre impact

Tyres are very complex structures. They are formed from different kinds of rubber and by layers of polyester, nylon or steel reinforcements, all having different spatial orientations, as depicted in the next figure. Tyres are loaded internally by high pressure and externally can be subjected to a variety of loads, including impact.

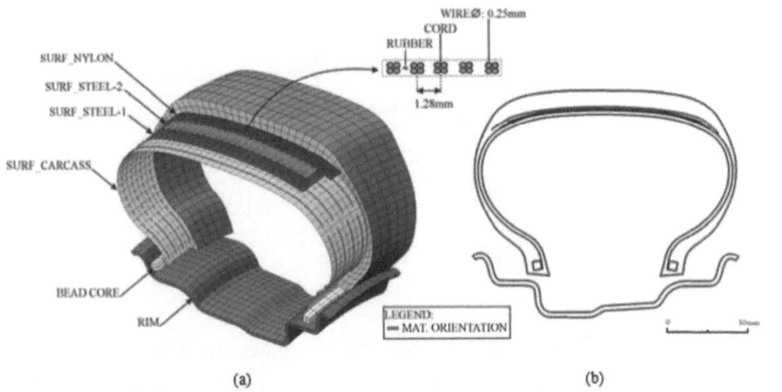

Figure 11.2 A schematic cross–section of an automotive tyre meshed for finite elements analysis.

See R.R.V. Neves, G.B. Micheli and M. Alves, An experimental and numerical investigation on tyre impact, International Journal of Impact Engineering, 37, 685–693, 2010.

Of course one needs to design a tyre so it can resist the high loads generated in the common situation of the tyre impacting a curb or the edge of a hole. A study on this subject reveals that the internal pressure is also an important variable on safety. The analysis of a tyre can be best performed using the finite element method and some results are here described when using the FE software ABAQUS.

The study summarized here had a double approach, with experiments going side by side with the numerical analysis. By so proceeding, it was possible to validate the numerical results, meaning that, for instance, the measured response of the tyre to an impact load was similar to the numerical results. This is a necessary approach for such a complex structure and also for some other problems. Needless to say that an investigation gains confidence if analyses and experiments go side by side.

11.1.1 Experimental and numerical set–up

The experimental set–up in the next figure consists of a heavy support to which different tyres and wheels were assembled. The angles indicated in the figure were chosen to best represent the impact of a tyre against a curb. Impact masses of between 47.24 kg and 74.64 kg were used at impact velocities of 2.42m/s and 1.93m/s, respectively, delivering an energy of 138 J. The tyres internal pressure was the one recommended by the manufacturer, *ie* 206.8 kPa, which was also reduced to 165.5 kPa to explore the tyre response.

Figure 11.3 Impact on a tyre set up and detail of the damage.

Apart from the use of a drop hammer with a large capacity, the experimental set–up also used a laser Doppler to track the motion of the impact mass, see next figure. We have seen that this is a useful procedure because the laser can measure with high accuracy and precision the velocity of the mass during the actual impact event. Once this information is known, the derivative of this signal, performed by software, gives the (des)acceleration which, times the impact mass, gives the impact force along time. On the other hand, the integral of the laser signal yields the mass displacement, allowing us to plot force times displacement of the impact event, whose integral is the actual impact energy the tyre experienced. Finally, a high speed camera was used to film the impact event. This gave important qualitative information of the behaviour of both the tyre and the rig.

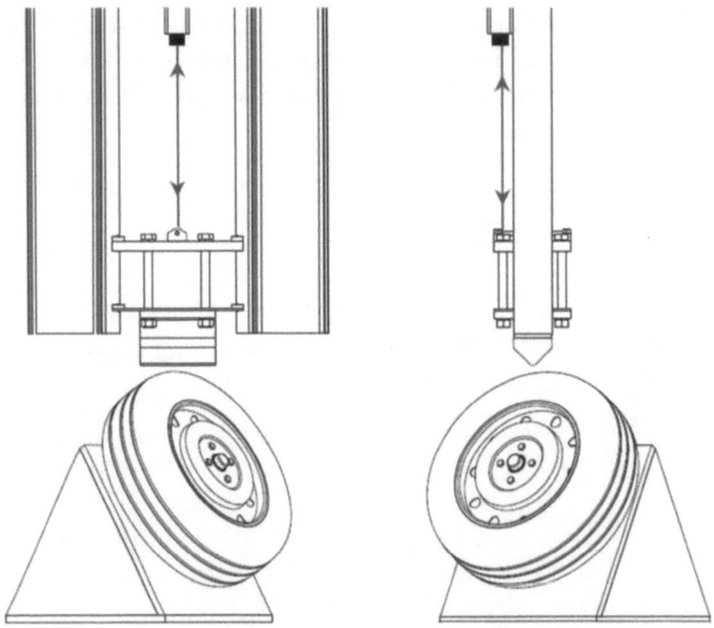

Figure 11.4 A laser Doppler velocimeter used to measure the impact phenomenon.

The material properties of the various components are listed next and were obtained from the open literature. These properties were used by ABAQUS, which run the tyre model with some 86,000 finite solid elements. The rubber part used some 30,000 solid finite elements and had its non– linear behaviour modelled by the Mooney–Rivlin constitutive equation. Possible strain rate effects were not considered in the analysis and extra care had to be taken to correctly define the various directions of the reinforcements. This makes this particular analysis rather complex.

As for the loading, the tyre was inflated to the desired pressure while the striker freely fell. The striker was represented by rigid elements. Of course, there are various contact regions on this event and the mastering of the pre–processor, *ie* the software responsible for meshing, loads, initial and boundary conditions, is a must. Only then can the analysis be run by the solver, with the results being studied in the post–processor phase. This sequence of pre–solver–post is common in the engineering practice.

TABLE 11.1 Material properties used in the tyre finite elements analysis.

Material	E N/m^2	ν	σ_0 N/m^2	σ_u N/m^2	ε_{rup}	E_h N/m^2	ρ kg/m^3
Steel belts	2E11	0.3	3.5E8	4.2E8	0.10	7.0E8	7850
Steel rim	2E11	0.3	2.2E8	3.4E8	0.30	4.0E8	7850
Nylon	3E9	0.3	6.6E7		0.21	0.0	1140
Polyester	9.5E9	0.3	1.3E8		0.16	0.0	1250

11.1.2 Results

The overall behaviour of the tyre under the impact of a mass is shown in the next figure. The results were obtained both from the images of the filming as well as from the finite element results. It can be seen that there was good visual agreement between these two set of results.

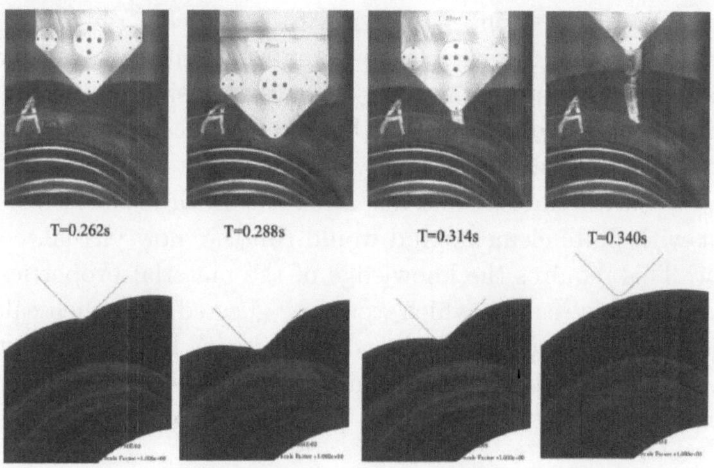

Figure 11.5 Experimental versus numerical results for the overall behaviour of a tyre under impact.

Visual agreement is only part of a validation, which is reassured by comparing the laser information with the FE results. Next figure shows velocity and force information along time of the tyre impact event. It can be seen, first of all, that the two experimental results are nearly identical, which is something desirable and gives confidence on the set–up. As for the numerical results, the velocity signal is well represented. However, there is a measured peak load which was not detected by the simulation.

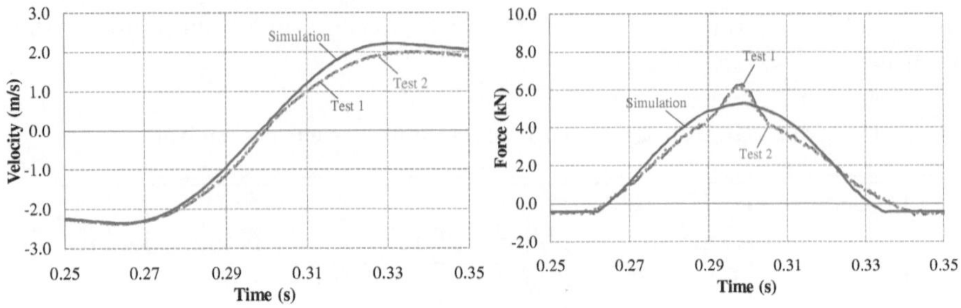

Figure 11.6 Experimental versus numerical results for drop mass velocity and load on a tyre along time.

11.2 Optimizing a compressor to resist a fall

In recent years, several different optimization schemes combined with finite element modelling have been used to improve the energy absorption of structures. The shape optimization method consists of converting the boundaries of a given geometry into splines and polynomial functions, with their parameters being the design variables. Hence, the optimization methods can indicate regions of a structure that are too stressed, so requiring more (or a stronger) material to be added.

As we have seen, the knowledge of material properties is fundamental for a correct structural analysis. However, considering the small pad shown in the next figure, it is not that easy to machine a tensile specimen from it. The easiest way here would be just to compress this component, measuring the load and displacement. In the sequence, we would model it with finite elements and would run the, now virtual, compression test. This requires the knowledge of the material properties in the elastic and plastic regime, which would be guessed. The then calculated load–displacement curve would be compared with the experimental one, with the discrepancies indicating that the guessed material stress–strain curve was not the actual one. The processes is repeated until a good

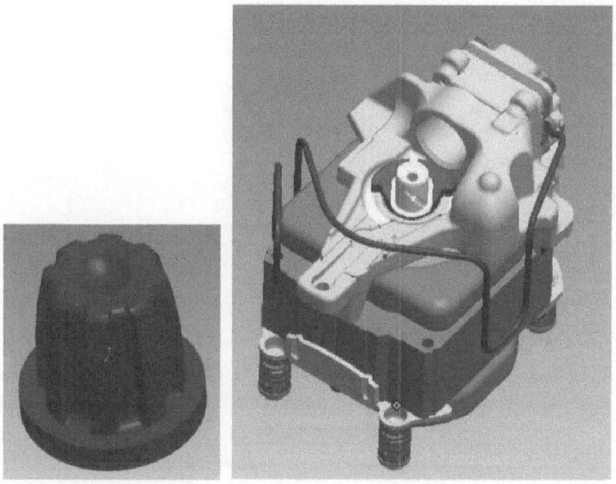

(a) (b)

Figure 11.7 (a) A polymeric pad that fits under the support of a (b) refrigerator compressor.

mismatch is obtained. This procedure is referred to as an inverse modelling technique and it is performed automatically by modern finite element software.

Let us now turn to an application of the compression test, together with the optimization technique introduced here. The problem is to evaluate the performance of the refrigerator compressor shown in the previous figure, when it is dropped from a certain height. The machine is supported by four polymeric pads involved in metal springs. In order to predict the impact performance of the whole assembly, it is necessary also to know the material properties of the pad polymer.

A finite element simulation follows and the discrepancies between both so obtained load–displacement curves are minimized using the optimization technique. It acts on the material parameters input in the finite element, which in this case used the Ramberg–Osgood already seen. Once the good agreement between the experimental and numerical load–displacement curves is obtained, as shown in the next figure, the material parameters are considered to be correct.

Now that we have the curve of this material, and of others obtained from traditional tensile tests, it is possible to examine the free fall of a compressor. We found that the pad studied above could also be shape optimized so as to dissipate the kinetic energy.

See Inverse modelling and optimization of an impact absorber (in Portuguese), D.L. Martins, Master Thesis, University of Campinas, Brazil, 2007.

The pad supports the compressor unity. By allowing the lengths L_1 and L_2 in Figure 11.9 to vary, the optimization procedure indicates what

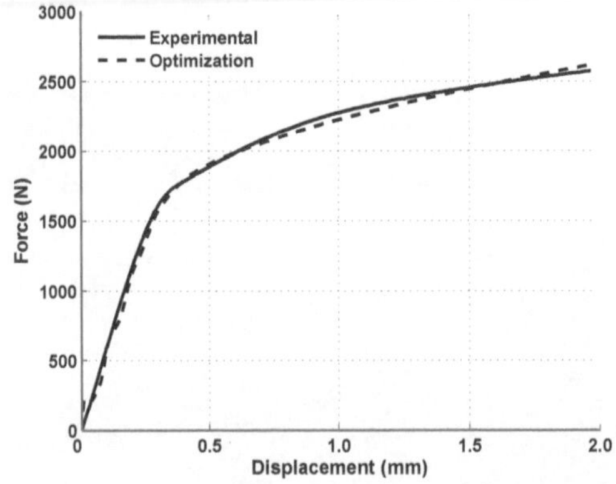

Figure 11.8 Experimental and numerical load–displacement curve of polymeric pad.

are the best values so as to maximize the dissipate kinetic energy by squashing the pad. It turns out that $L_1 = 10.00$ mm and $L_2 = 1.15$ mm are the best values, with the other dimensions being kept as the original. Hence, it was only necessary to cut a bit the steel rod, so allowing space for the pad squashing. When implemented in the compressor, the absorbed kinetic energy evolves as indicated in the next figure and a fall of the unity would be partially absorbed by the pads, so saving the compressor.

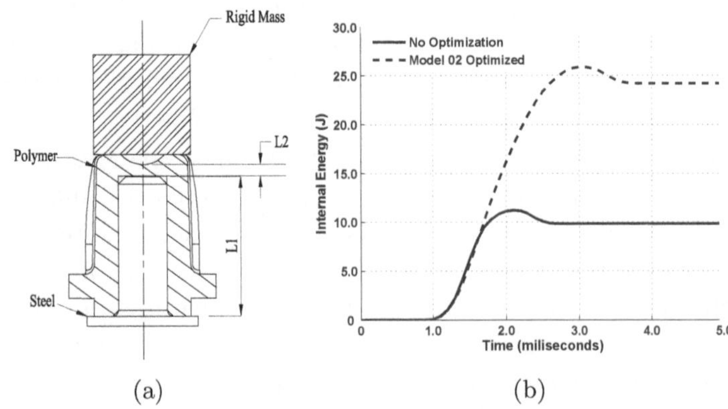

Figure 11.9 (a) The geometry of a pad for optimization. (b) Absorbed energy evolution of a compressor supported by four pads.

11.3 High-velocity impact in metal plates

In this book we have not explored much the collision of bodies at very high velocities, say over 200 m/s. This is an important topic in the area of bullet perforation and protection, impact of birds on air planes, of meteorites on space crafts, etc...

If we take the case of plates, a major concern is the determination of the ballistic limit, here loosely defined as the minimum projectile velocity that allows it to fully cross the plate thickness. The accurate determination of the ballistic limit has important technological and scientific consequences. For example, the determination of the ballistic velocity requires a good failure criterion. This is so because projectile penetration in a structure goes side by side with material shearing and tearing, local increase of temperature, shear bands, strain rate effects, etc ...

We report here some experiments and numerical simulation of the impact at a high velocity, say over 300 km/h, of hard spheres against circular aluminiun plates.

See A finite element–experimental analysis of the impact of spheres on aluminium thin plates, G.B. Micheli, L. Driemeier and M. Alves, Structural Engineering and Mechanics, 2015, 55 (2), 263-280.

11.3.1 Material properties and ballistic tests

The quasi–static mechanical behaviour of the aluminiun alloys was investigated at low and high strain rates using a standard tensile machine and an elastic wave machine. The aluminium alloy is strain rate sensitive, as shown in the next figure.

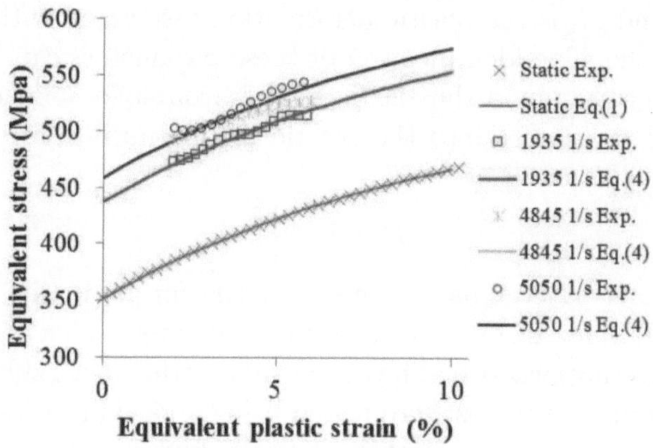

Figure 11.10 Static and dynamic compressive behaviour of an aluminium alloy.

The material plastic behaviour is assumed to be represented by

$$\sigma_y = \sigma_0 + Q_1[1 - \exp(-C_1\varepsilon_p)],$$

where Q_1 and C_1 are material parameters and ε_p is the plastic strain. The optimization simplex method was used to determine the model parameters via the objective function

$$f_{obj} = \sum_i \left(\sigma_0 + x_1[1 - \exp(-x_2\varepsilon_{p(i)})] - \sigma_{y(i)}\right)^2.$$

The Cowper–Symonds (CS) model was adopted to represent the strain rate influence on the material behaviour via the relation

$$\sigma = \sigma_y \left[1 + \left(\frac{\dot{\varepsilon}_p}{Q}\right)^{1/p}\right],$$

where Q and p are material parameters. It is possible to rewrite this relation by defining the parameters $V_k = \sigma_y (1/Q)^{1/p}$ and $V_m = 1/p$, giving

$$\sigma = \sigma_y + V_k \dot{\varepsilon}_p^{V_m}.$$

To predict the material damage, whose accumulation leads to local material rupture, a well known model is the Lemaitre damage model, which postulates that the damage, D, evolves according to

$$\dot{D} = \begin{cases} 0 & \text{if } \varepsilon^p \leq \varepsilon^p_D \\ \dfrac{Y}{S(1-D)}\dot{\varepsilon}^p & \text{if } \varepsilon^p > \varepsilon^p_D \end{cases},$$

with Y being the deformation energy release rate density, S is a material parameter and ε^p_D is the plastic deformation associated with the beginning of damage. The identification of these parameters can be made by measuring the change in the elastic moduli from load–unload curves, as the material progresses into the plastic regime until failure, such that damage is

$$D = 1 - \frac{\tilde{E}}{E}.$$

The various material parameters necessary for predicting the ballistic limit were obtained from somewhat standard tests, which are all desired in the engineering practice. They are listed in the next table.

Thin aluminium plates were tested using a gas gun, which impacted 20 mm diameter steel spheres at velocities as high as 140 m/s. Square 1.6 mm thick plates, were sandwiched by two thick steel rings with internal diameter of 250 mm. At low impact velocities no fracture was detected and the maximum deformation was monitored. At higher velocities, the limit between the partial and total perforations was searched for using a high speed camera. The velocity associated with the macroscopic fracture formation and the ballistic velocity was found to be 120.23 m/s.

TABLE 11.2 Material parameters for an aluminium alloy.

E MPa	σ_0 MPa	$\dot{\varepsilon}_0^p$ s^{-1}	Q_1 MPa	C_1	S MPa	ε_D^p	D_c	V_k MPa·s	V_m
73394.5	351.70	$3.22 \cdot 10^{-3}$	156.48	24.59	3.3	0.05	0.19	15.019	0.231

11.3.2 Finite element model

The finite element method, as implemented in the software LS–DYNA, was used to numerically analyse the plate impact perforation, using the refined mesh shown next.

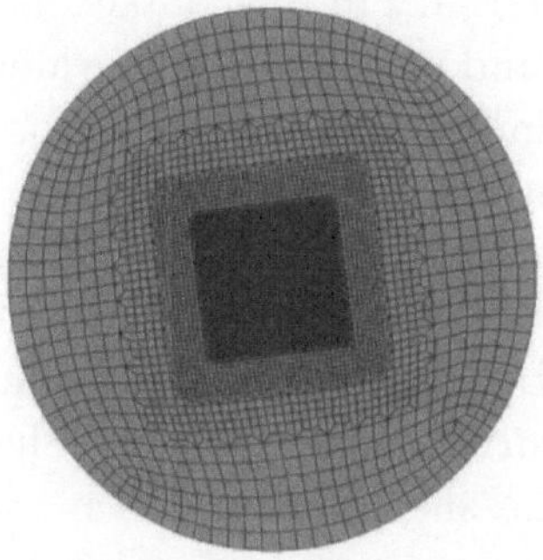

Figure 11.11 The finite element mesh for the plate analysis.

We have not discussed in detail the important issue of mesh sensitivity. We mean that different results can be obtained depending on the size of the finite element. One simple reason is that large finite elements, specially when not p enriched, do not have the capability to describe accurately the displacement field in regions of large gradients. This is why numerical studies always perform various analyses using different mesh sizes.

For the case of the plate perforation discussed here, the next figure shows that, as the finite shell element density increases, the ballistic limit reaches a more or less stable value. Clearly, for low mesh density, the ballistic limit would be wrong, which only can be inferred because experimental data are available. Solid finite elements are preferable in simulation involving failure since they can describe well the strain field through the plate thickness. The drawback is that they are very computer demanding.

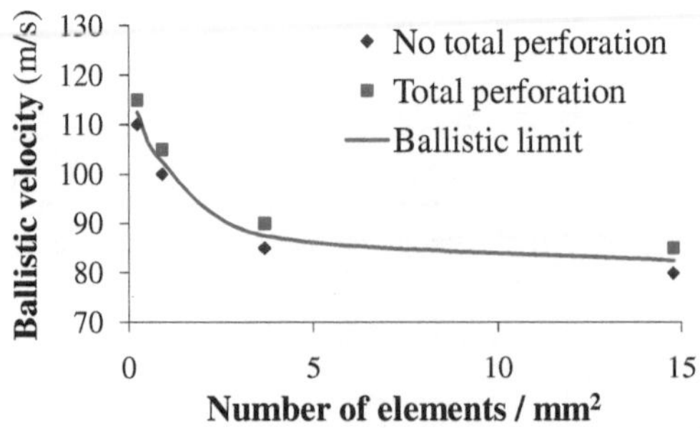

Figure 11.12 Influence of the mesh density on the ballistic velocity.

The results in the figure are not yet definitive. We would like to discuss the influence of some parameters on the results. In a simulation, there are many parameters whose values are not definitely known. For instance, the friction between the projectile and the plate is quite hard to know, and yet it is necessary in the contact model, and it is at least a function of temperature and velocity. Here, a value of 0.02 for the friction coefficient was adopted and further simulation fortunately showed a weak dependency of the ballistic velocity on friction.

Six possible values between 0 and 5 for the integration order in a 3D contact algorithm represent the number of points in the slave surface, in addition to the number of nodes. These values were used to verify whether penetration occurred. The next figure shows how this parameter affects the results for the shell model, which was finally chosen to be 3. Similar studies are also necessary for other parameters, for instance the hours glass control. This just points out the various difficulties one encounters in a simulation. At the same time, it does point out the necessity of the FE user to have strong background on the method so to judge the correctness of an analysis.

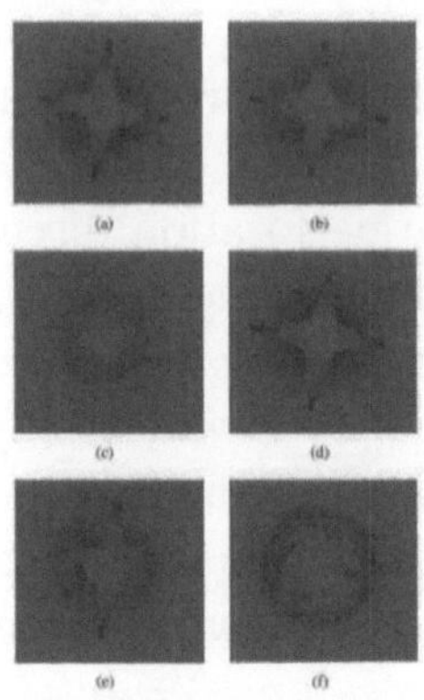

Figure 11.13 Shell mesh configuration (in red) during the sphere (in blue) penetration for different contact integration parameters.

The experimental value for the ballistic limit of this plate was established as 120.23 m/s. After a detailed quasi–static and dynamic material characterization plus a sophisticated numerical analysis, we obtained numerically the values indicated in the next figure as a function of the mesh refinement and of the finite element type. It is clear that the errors are rather large. A similar study in the same reference above but for another aluminium alloy, gave the prediction within an error of 3% for the ballistic velocity now at 133.26 m/s.

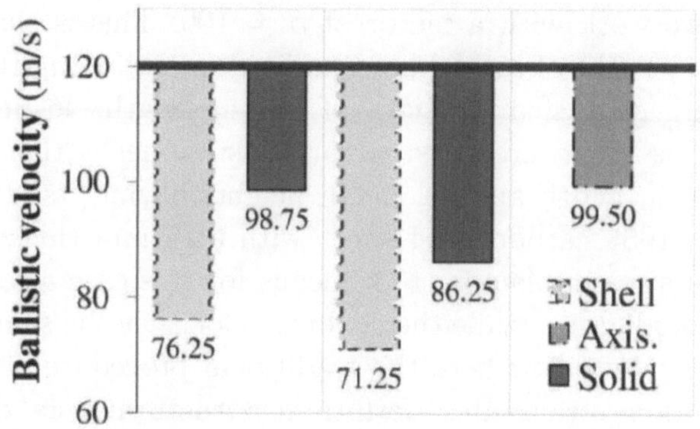

Figure 11.14 Numerical ballistic velocities as a function of the mesh refinement. The experimental value is the continuous line.

One of the reasons for showing here a case of poor performance of a numerical model is to highlight that, without experimental and/or theoretical support, one would tend to accept that the so determined ballistic velocity is the actual one. After all, FE programs, with all their sophistication and post processing would "not go in error in this simple problem". The reason for the poor performance could be attributed to the failure criterion, but have in mind that the same criterion, with the specific parameters, did give a much better prediction for a thick plate, as reported elsewhere. This then highlights the care an analyst should always exercise in dealing with extreme nonlinear phenomena.

11.4 Ship collision

Ship collisions are not so infrequent as it may be imagined. The risk of a ship collision has increased together with the growth of the global fleet of ships. Oil tankers represent nearly half of the entire world fleet and they are the maritime segment in which the largest ships are built. Therefore, it is important to access the behaviour of ships colliding with other ships

See M.A.G. Calle, R.E. Oshiro and M. Alves, Ship collision and grounding: Scaled experiments and numerical analysis, International Journal of Impact Engineering, 103 (2017) 195–210.

or against rigid walls. However, full scale tests are extremely difficult to be performed for obvious reasons and the use of scaled models proves to be a good alternative. Small scale tests can then be used to validate finite element models, which in turn can be scaled up to model actual size ships.

Having prohibitive impact tests of these large structures, we meet here with the opportunity to apply the knowledge we have gained in the previous chapter on scaling. Given the dimensions of a ship, the largest ones being over 350 m, our scaling factor should be very high and in this section we work with a factor of $\beta = 100$. This is similar to the ratio between lengths of an elephant and a mouse. Taking the Suezmax oil tanker with general dimensions of 254 m length, 46 m width and 24 m height, as the reference structure, our model is then 254 cm in length, 46 cm in width and 24 cm in height, having its striking bowl made of SAE 1008 carbon steel sheet with 0.25 mm thickness. As we know, the mass scaling factor of β^3 means for this case a reduced scale model 10^6 times lighter, while the striking velocity is the same for model and prototype. We follow here the traditional procedure of testing the material at various strain rates, performing structural tests, the collision, and using the experimental data as a reference for the finite element simulations.

As we have already encountered, measuring plastic strain hardening of a material using standard uniaxial tensile tests is difficult after necking since there the strain field is no longer uniform and an overall strain measurement does not represent the real local plastic behaviour. We adopted an optimization procedure to calibrate the parameters of the Voce plastic strain hardening model

$$\sigma = \sigma_0 + R_0\varepsilon_p + R_\infty\left(1 - e^{-b\varepsilon_p}\right).$$

We see in the next figure that the numerical and experimental matching is quite good, meaning that the parameters chosen for the numerical simulation are correct. These parameters are listed in the next table.

TABLE 11.3 Material parameters for steel SAE1008.

E (MPa)	ν	ρ (kg/m^3)	k	R_0 (MPa)	R_∞ (MPa)	b	C (1/s)	p
205	0.3	7800	213.59	425.00	125.18	18.757	15673.5	6.787

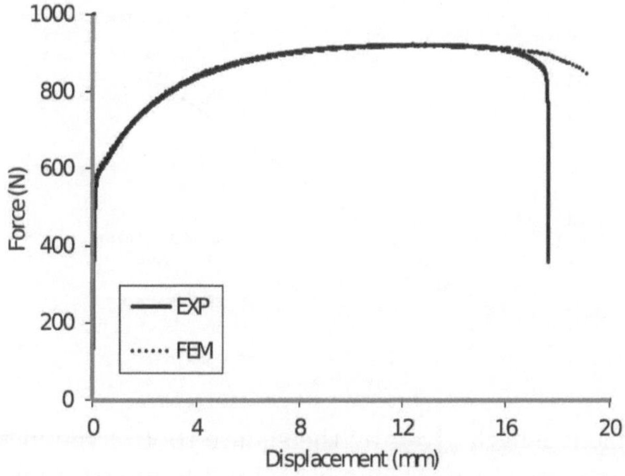

Figure 11.15 Numerical and experimental force–displacement curves for the ship model material.

We also know that carbon steel is a very strain rate sensitive material so it is necessary to perform tensile tests at medium and high strain rate. The medium strain rate tests were executed in a standard, but fast, tensile test machine, while Hopkinson bar was used for the high strain rate tests in the range of 500 to 1200 s^{-1}. These varied and time consuming tests lead to the strain rate hardening parameter, according to the Cowper–Symonds equation, of

$$DIF = \left(1 + \frac{\dot{\varepsilon}}{C}\right)^{1/p}.$$

The next figure shows how the strain rate hardening parameter increases substantially with the strain rate and in a linear–log fashion. It confirms that, for this material, the Cowper–Symonds equation describes the strain rate sensitivity well.

As for the material failure, we know that failure initiates at a microscopic level by void nucleation. But to model such a phenomenon we need not only more sophisticated failure criteria but also a very refined mesh, which is rather prohibitive for large naval structures. Here we adopt a failure criterion based on the maximum plastic strain and dependent on the mesh element size. It reads

$$\bar{\varepsilon}_f = \bar{\varepsilon}_u + \bar{\varepsilon}_n \left(\frac{t}{L_e}\right),$$

with $\bar{\varepsilon}_n$ and $\bar{\varepsilon}_u$ denoting the necking and the uniform strains material parameters, respectively. t is the finite element thickness and L_e is the

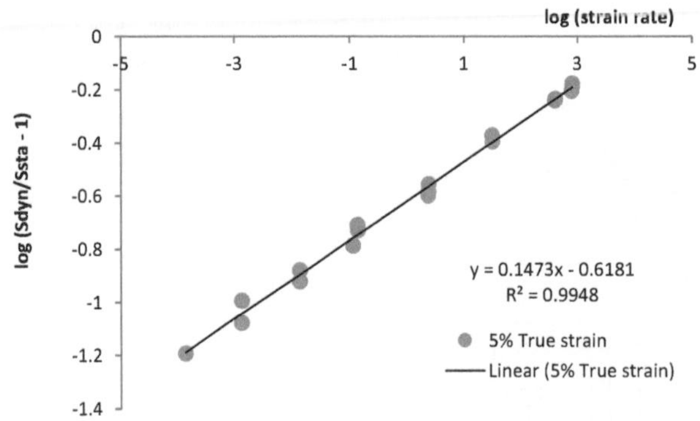

Figure 11.16 Strain rate sensitivity for a SAE 1008 steel.

equivalent element length given by the square root of the mesh element area. These parameters are evaluated having the uniaxial tensile test as reference and adjusting the numerical force–displacement curve to the experimental one. It is called UTT criterion.

Another failure criterion, the TT criterion, uses a tearing test mode III fracture as reference, with the parameters being evaluated by adjusting the numerical plastic energy per displacement (demanded to tear the plate) to that measured in the experiment.

As it can be seen in the next figure, these criteria show a large discrepancy for the plastic strain to failure and so they were averaged, AVE criterion, in what follows. The AVE criteria balances both initiation and propagation of the crack phenomena observed in collision tests.

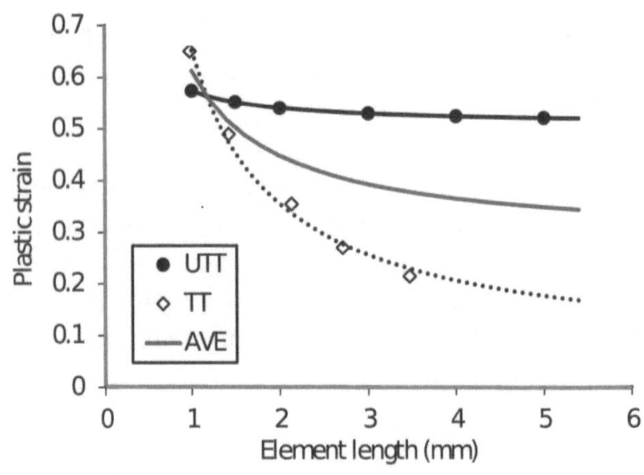

Figure 11.17 Failure plastic strain as a function of the element size for UTT, TT and AVE criteria.

In the simulation results that will follow, care was taken to model the laser welded parts. Experimental tensile tests were performed in welded strips extracted directly from reduced scale naval structures to evaluate the mechanical resistance of weld joints. Experiments on strips were simulated and compared to the experimental ones, being possible to adjust the strain to failure for a match of the curves.

11.4.1 T cross–section beams

A first set of tests deals with a naval substructure scaled to 1:100. It comprises a T cross–section beam clamped at both ends and indented at its mid span by a round headed indenter under quasi-static and dynamic conditions, see next figure.

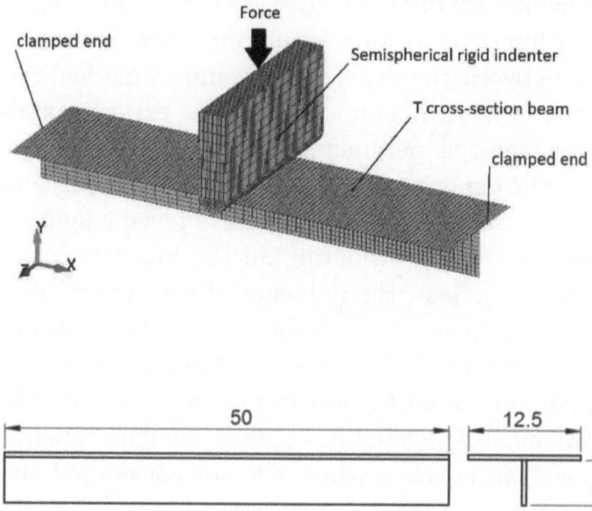

Figure 11.18 T cross–section beam.

With this test we can access the performance of our material models in describing strain hardening and strain rate hardening. No failure is expected for this tests. The 50 mm length T beams were manufactured using 0.25 mm thickness mild steel plate and their parts were cut by laser and assembled using laser welding. A hardened steel indenter with a 3.0 mm diameter hemispherical head was used in both tests.

The quasi-static test was performed using a universal testing machine Instron 3369 at a constant low velocity of 0.0025 mm/s using an acquisition data rate of 0.625 Hz. The impact test of the beam was conducted in a low-energy drop weight machine with a mass of 0.234 kg travelling at a velocity of 3.453 m/s. The mass velocity was measured by a Polytec laser Doppler vibrometer, model OFV-323, at an acquisition data rate of

500 kHz. The impact test took about 2.7 ms. The displacement and the acceleration of the indenter were obtained by integrating and deriving the velocity data, respectively. The force was obtained by multiplying the acceleration signal by the mass. In all tests, the beam ends were clamped using a robust metallic support.

Following the similarity laws, the reduced scale impact test intends to represent a stiffened plate section between two transverse frames or bulkheads, with dimensions of 5 m length and 1.25 m width, when centrally collided by a sharp and rigid structure of 234 tons at 3.453 m/s (6.71 knots). The tests were modelled numerically by LS-DYNA 971 using regular square 0.5 mm × 0.5 mm thin shell elements.

The plastic behaviour of the T cross-section beam structure is presented in the next figure by its force-displacement response. There we see the good match between the experimental and numerical models. This gives credit to the numerical simulation, to the experimental technique of measuring the material parameters together with the optimization procedure and to the experimental technique of obtaining force and displacement from the laser velocity reading. It is indeed a figure that strikes our sense of wonder, when considering all the knowledge behind it. In the case of the dynamic test, the influence of the strain rate sensitivity on the plastic behaviour of the material is also clearly depicted in the force–displacement response. As expected, the numerical model of the dynamic test that considered the sensitivity to strain rate on the material model (FEM w/SRS) agreed more with the experimental response (in force and penetration) than when it is not considered on the model (FEM w/o SRS).

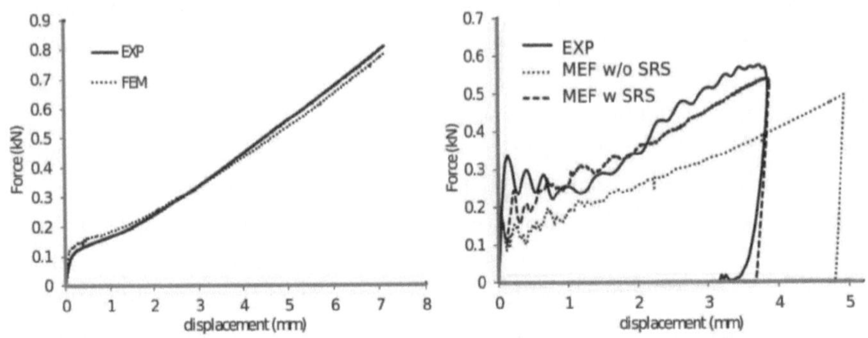

Figure 11.19 Quasi–static and dynamic force–displacement responses of a T cross–section beam structure.

11.4.2 Frontal ship collision against rigid wall

A frontal collision of an oil tanker against a rigid plane structure was reproduced experimentally at a reduced scale of 1:100. This experiment basically involves the crushing collapse of a bulbous bow structure when colliding against a rigid flat obstacle. Given that the bow structure absorbs practically all kinetic energy of the ship by transforming it into plastic deformation in real collision events, only the ship bow is being considered as a deformable structure, with the tanker body being rigid.

The bulbous bow structure was made of 0.25 mm thick mild steel plates. All parts were laser cut and welded. As it can be seen in the next figure, a rail system was used to restrict the tanker movement only to surge direction (horizontal). The velocity propulsion was induced by the vertical movement of a falling mass in a drop weight rig machine. The mass, which freely falls by the gravity force, was connected by a cable to the ships body to accelerate it horizontally along the rail guides for 1.46 m so achieving a final velocity of 1.91 m/s. About 40 cm before ship collision, the applied force by the cable is ceased and the so obtained velocity is the only momentum acting in the test. Following the similarity laws, this reduced scale test corresponds to a head-on collision of a Suezmax tanker against a flat rigid structure, such as a pier or bridge abutment, at a velocity of 1.91 m/s (3.71 knots) and considering a total ship mass of 63,810 tons.

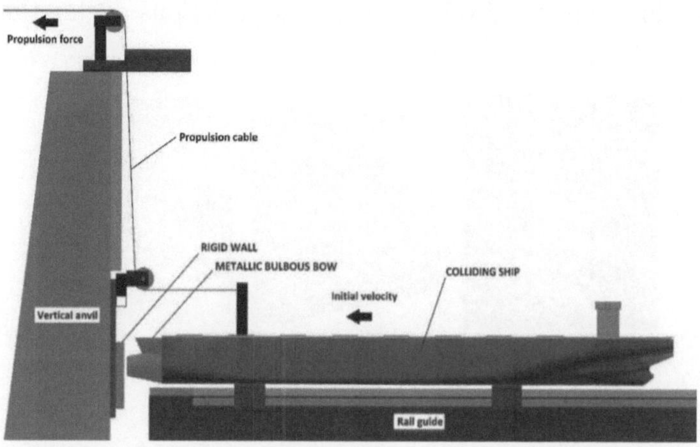

Figure 11.20 Frontal ship collision test scheme in 1:100 reduced scale.

The measurement of the bulbous bow motion during the collision test was performed using a non-contact photogrammetric technique at a frame rate of 20,000 per second. The Pro-analyst software was used for the post-processing of the images. The time for complete plastic deformation of the bulbous bow was about 25 ms. The FE simulations of the frontal collision experiment were performed using the LS-Dyna code. One model considers a straight frontal collision of the ship against the rigid wall and the second model considers a slightly inclined rigid wall, so reproducing the inclination angle measured in the experiment. Shell elements were used in the entire mesh of the tanker model. The next figure presents the force–displacement curve of this test.

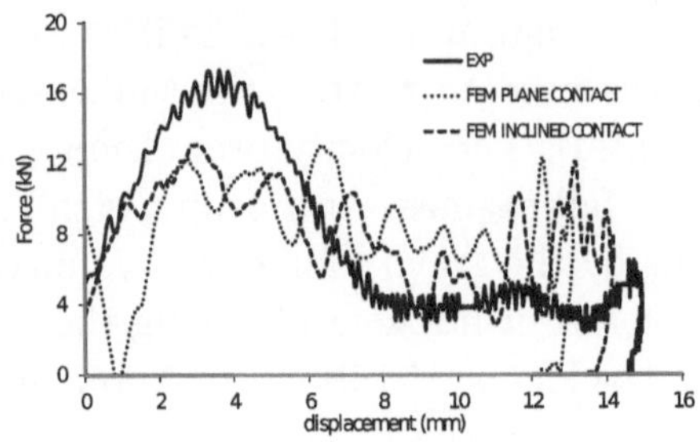

Figure 11.21 Force–displacement responses of frontal ship collision test.

The collapse modes of the bulbous bow during the collision obtained experimentally and numerically are compared in the next figure. The formed folding is composed by small folds around and between the outer plate and the first ring reinforcement. Localized, small joint failures were also detected in the bow specimen (experiment) caused by stress concentration in the internal reinforcement intersections during the folding process.

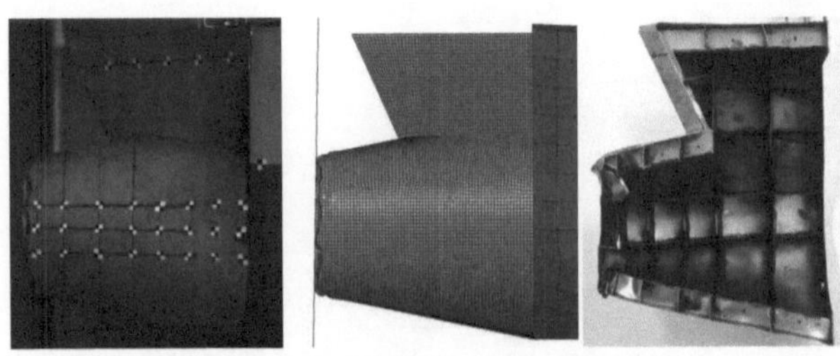

Figure 11.22 Lateral view of bulbous bow crushing at test time of 15 ms.

11.4.3 Lateral ship–to–ship collision

A ship–to–ship collision was investigated experimentally in order to explore the resulting structural damage and the capability of the numerical modelling of predicting different ship impact scenarios. The experimental arrangement is depicted in the next figure. Note that the lateral struck part of the ship comprises a double panel.

Figure 11.23 a) Experimental arrangement and b) numerical model of 1:100 reduced scale ship–to–ship lateral collision experiment.

The impact velocity of the colliding ship was set at 4.72 m/s while the collided ship stood still in a perpendicular position. The colliding and collided ships have a total mass of 123.3 kg and 178.3 kg respectively, achieved using small sandbags distributed uniformly between the bulkheads along the body of both ships. The collided ship was about 40% heavier than the colliding ship in order to recreate the added mass due to the hydrodynamic resistance of the surrounding water caused by the lateral ship movement.

Considering the similarity laws and a scale reduction factor of 0.01, the reduced scale experiment of ship–to–ship collision represents a real-scale perpendicular ship collision test at 4.72 m/s (9.17 knots) of two 123,300 ton Suezmax. Experimental and finite element results are qualitatively compared in the next figure, where a positive agreement can be seen.

11.5 Collision of a car against a guardrail

A road barrier, or guardrail, has as the main function, to redirect back to the road an out of control vehicle, avoiding a frontal collision or a dangerous veering off the road trajectory. It may also serve to absorb part of the kinetic energy of the vehicle, so potentially decreasing the accelerations experienced by the occupants.

See R.R. Neves, H. Fransplass, M. Langseth, L. Driemeier and M. Alves, Performance of some basic types of road barriers subjected to the collision of a light vehicle, Journal of the Brazilian Society of Mechanical Sciences and Engineering, 2018, 40:274.

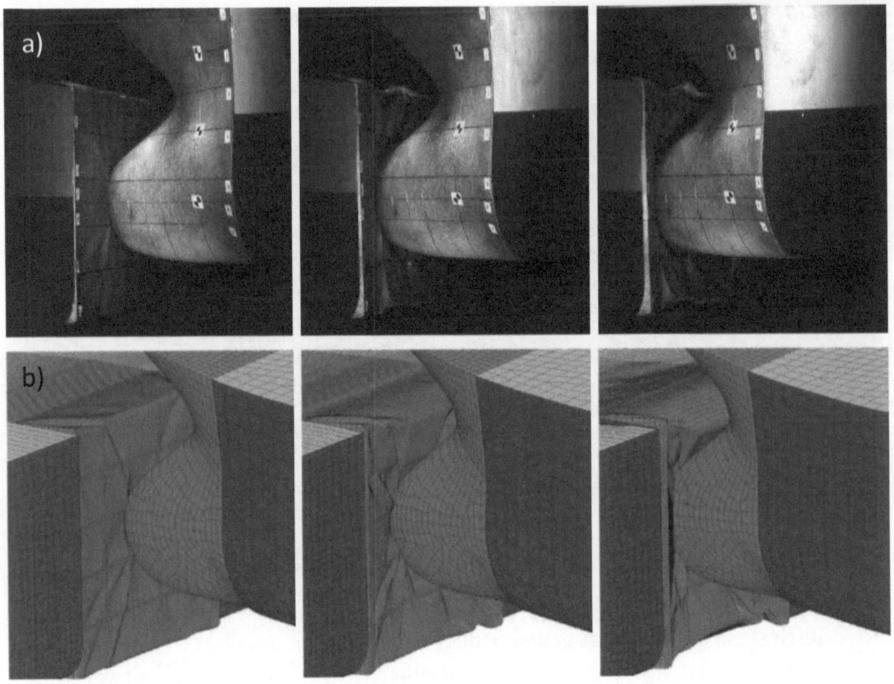

Figure 11.24 Lateral view of bow penetration in a) experiment and b) numerical model of the ship–to–ship collision test obtained at test times: 4, 12 and 20 ms (from left to right).

In this section, we follow a crash scenario as defined by the European Road Federation via the EN 1317 standard. It considers a car of approximately 900 kg with an impact speed of 100 km/h and 20° angle to the road restraint system. This scenario is modelled here by a finite element simulation to evaluate two different road restraint systems and their severity index to a car occupant. The systems analysed are highlighted in the next figure and comprise the collision of a car against a metallic guardrail and against a concrete barrier.

Three severity index parameters were taken into account to define which road restraint system is the safer for a car occupant:

- ASI – Acceleration Severity Index, grade A, ASI < 1.0, grade B, $1.0 <$ ASI ≤ 1.4, grade C, $1.4 <$ ASI ≤ 1.9
- THIV – Theoretical Head Impact Velocity (THIV), should be less then 33 km/h
- PHD – Post–impact Head Deceleration, should be less than 20g

(a) (b)

Figure 11.25 a) Flexible road barrier made from steel and b) concrete barrier.

In a crash test, the combination of these results gives the severity grade A, B or C. Grade A means low severity, while B and C are associated with a serious injury or even fatal to the car occupants. ASI is calculated by placing an (virtual) accelerometer in the car center of gravity. It is computed along the impact event and its maximum value and the maximum acceleration is used to evaluate it. If maximum ASI exceeds 1.0 or 1.4, then the impact event is considered very dangerous or lethal for the passengers. For THIV, the occupants head is considered to be a freely moving object that, as the vehicle changes its speed during contact with the road barrier it continues moving until it strikes a surface within the interior of the vehicle at a certain speed limit. The PHD describes the head deceleration after this impact.

Metallic guardrails composed by w–beam are installed with different cross section posts, like C shape, wood, etc. but here we explore only the so called sigma post placed every 2 m along the guardrail. The inclusion of the soil and their property adds flexibility to the metallic guardrail and makes the behaviour of the system closer to the real installations on highways. The deformable vehicle model with 25037 finite elements is based on a commercial vehicle obtained in the NCAC database. It is a light vehicle of 894 kg with no passengers, as shown in the next figure.

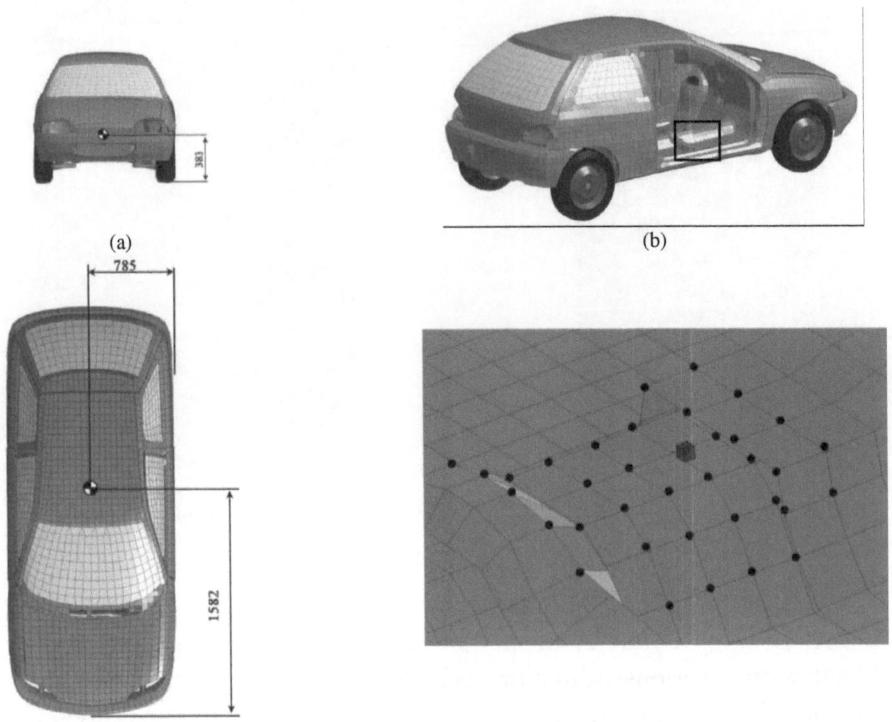

Figure 11.26 a) Front view and c) top view of the car mesh. In b) isometric view in detail the accelerometer position and d) virtual accelerometer.

In order to calculate the severity index for the car occupant, a virtual accelerometer was set to capture the acceleration history and angular displacement of the vehicle. A good quality of severity index post–processing was achieved by setting the data acquisition frequency to 100 kHz, with a SAE 180 Hz filter applied before the index calculation. The ASI index measures the severity of impact to the occupant. It is a scalar calculated from

$$ASI(t) = \sqrt{\left(\frac{\bar{a}_x(t)}{\hat{a}_x}\right)^2 + \left(\frac{\bar{a}_y(t)}{\hat{a}_y}\right)^2 + \left(\frac{\bar{a}_z(t)}{\hat{a}_z}\right)^2},$$

with $\hat{a}_x = 12g$, $\hat{a}_y = 9g$ and $\hat{a}_z = 10g$ being the limits of acceleration for an occupant using seatbelts.

In the calculation of the THIV, it is considered that the vehicle moves only in the horizontal plane and it is assumed that at the beginning of the impact the head and the vehicle have a purely translational motion of the same horizontal speed. During the collision, therefore movements of roll, pitch or vertical translation are neglected. THIV is calculated from

$$THIV = \sqrt{\dot{x}_b^2(T_v) + \dot{y}_b^2(T_v)},$$

the flight time, T_v, being the instant the head reaches any inner surface inside the vehicle.

The deceleration of the head after impact gives the PHD index. It is calculated from the maximum value of the acceleration in the vehicle CG at an interval of 10 ms after instant T_v, as follows,

$$PHD = max\sqrt{\ddot{x}_c^2 + \ddot{y}_c^2}.$$

The W–beam material stress–strain curves for low and high strain rates were measured using a standard tensile machine and a Hopkinson bar. The results were input in the finite element model as implemented in LS–DYNA. A guardrail model of over 300,000 finite elements comprising a W-beam, sigma posts and anchorages was modelled on LS–DYNA similar to the real assembling situation shown in the next figure, which also shows details of the W-beam and post.

An initial speed of 27.78 m/s (100 km/h) was applied to the longitudinal direction of the vehicle and angular speeds of 88.97 rad/s were set to the wheels. The gravitational acceleration of 9.81 m/s^2 was adopted in the model. One bolted connection was used for fixing the posts to the W-beam, while the overlapping ends of the w beams were fixed by eight bolt connections. A massless rigid beam between nodes of the involved components was used to represent the bolts which fail at an average plastic deformation PS. As we can see in Figure 11.28, failure of the bolts connecting the W-beam and posts has an important influence on vehicle trajectory. Considering the results in the figure, $PS = 0.9$ was adopted.

Having all this analysis setup established, we compare in Figure 11.29 the simulations with the experimental results. The various injury criteria are shown there and we see that a quite good estimate was obtained. The results are even more meaningful when we bear in mind the complexity of the car and barrier modelling, including failure. We highlight that THIV is a velocity criterion and therefore comes from integration of the acceleration, which always smooths a signal.

The experimental results are from LIER, Barrier test Robust work package 4 - H2, Lyon: Laboratoire d'essais INRETS Equipements de la Route, 2004.

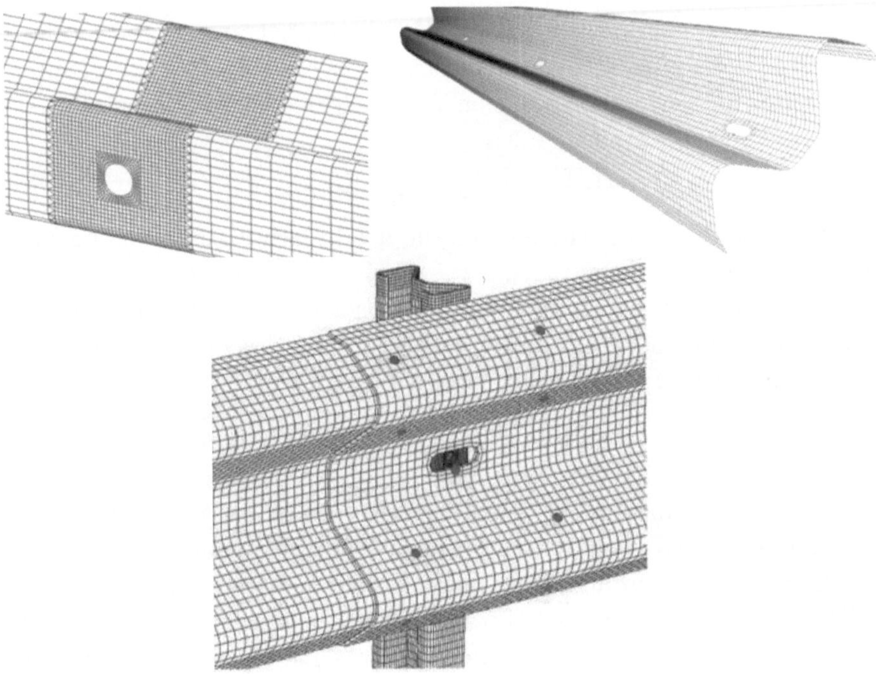

Figure 11.27 W-beam mesh and sigma post with bolts.

11.5.1 Collision against a concrete barrier

The experimental results are from LIER, Test report — Round Robin Phase 2 TB11, Lyon: Laboratoire d'essais INRETS Equipements de la Route, 2002.

We present next some results of a collision of a car against a concrete barrier. The vehicle is a Peugeot 205 Junior of approximately 900 kg, travelling at 100.4 km/h and hitting the barrier at 20°. The barrier is a temporary type, made of precast concrete blocks of 3.15 m in length and having 2,400kg mass each. A total of 10 blocks were assembled together by M20 bolts set at the junction of each module with an angle of 45° with the vertical plane of the barrier. On the impact position and behind the barrier, 3 concrete blocks were added with approximately 1,600 kg each, in order to avoid any displacement of the barrier. The concrete barrier ends were not anchored. Based on these set of boundary conditions from the real test, the concrete barrier numerical model was considered totally fixed to the ground. The concrete barrier was modelled as incompressible material. Thus, a single solid block was meshed with rigid elements, in a total of 848 elements along the 25 m barrier.

The numerical and experimental results for the overall car behaviour and the ASI, THIV and PHD parameters are presented in Figures 11.30 and 11.31, respectively. Here we can notice reasonable

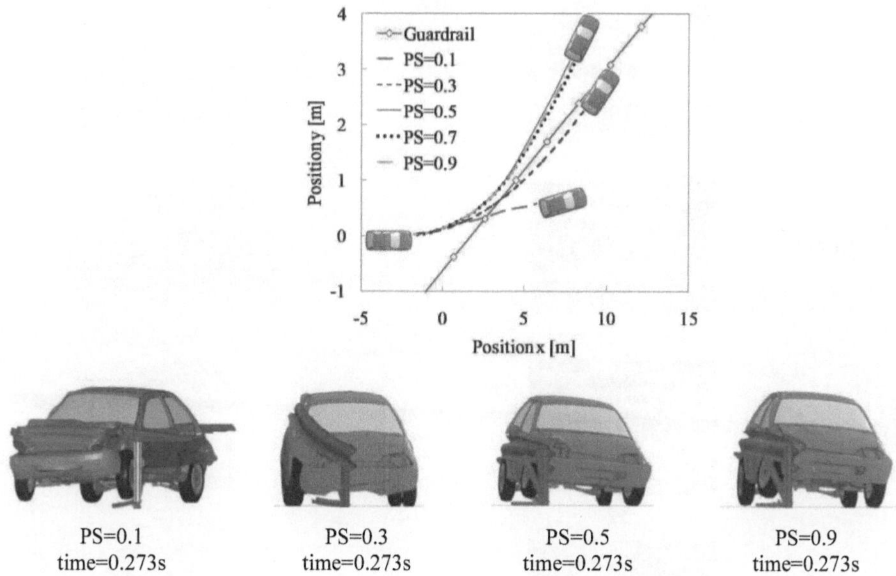

Figure 11.28 CG position of GeoMetro vehicle model in different simulations. PS is the plastic strain failure of the *SPOTWELD connection.

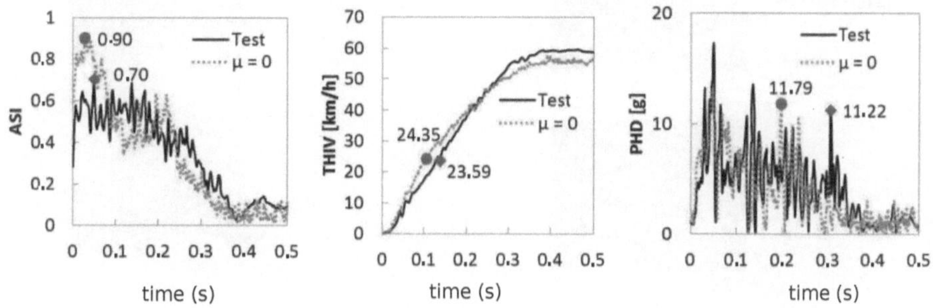

Figure 11.29 Severity indexes history of the car impact against the steel guardrail system.

agreement, specially when friction is rightly considered. The simulation using $\mu = 0.1$ to 0.05 shows an average deviation in all severity index of approximately 4.3%

Now, a broad picture of this study is that the ASI index of 0.7 for a car collision against the metallic guardrail indicates that the acceleration of the occupant head did not exceed human limits. The THIV and PHD indexes were also below the limits of 33 km/h and 20g. Thus, the metal guardrail exhibits a good restraint system for the vehicle and its occupants. The New Jersey concrete barriers exhibit higher ASI indexes, above the unit, while the THIV and PHD indexes are below the limits. In any case, for the concrete barriers, occupant severity level is C. Considering only the ASI index, the rigid barrier offers twice as much risk

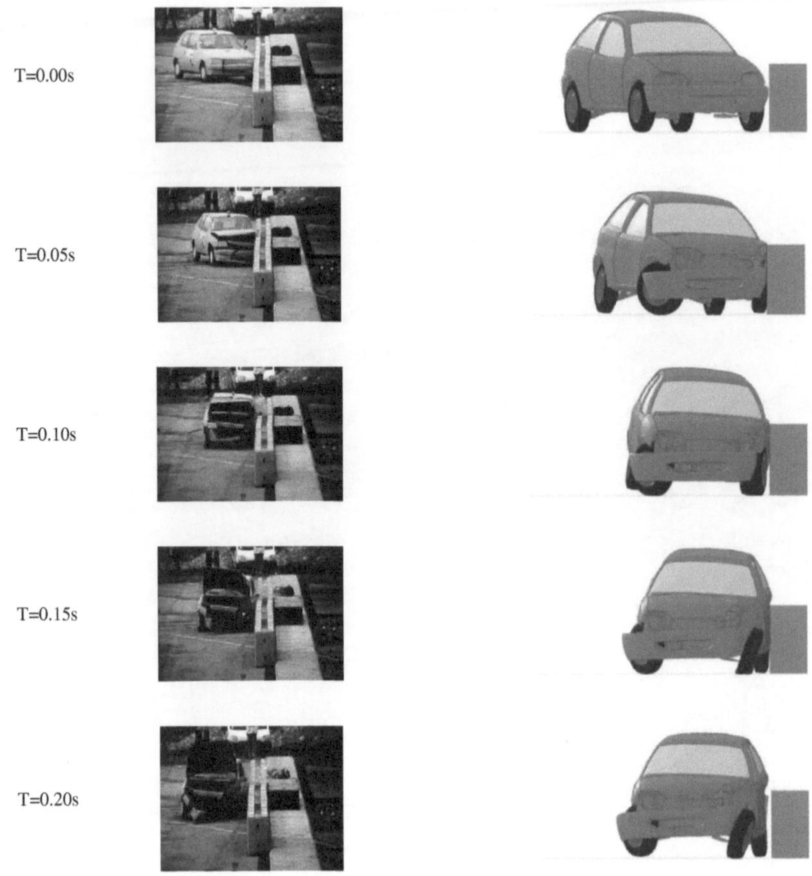

Figure 11.30 Comparison between experimental tests and simulation of a car collision against a barrier.

for the car occupants of a light vehicle, when compared to the metallic guardrail.

11.6 Carbon fibre tubes under axial impact

In this section we present some experimental and numerical studies on fiber-reinforced plastic tubular structures made of carbon fibers, CFRP. The subject has been explored at large in the literature and it is a difficult one, with many more studies being required.

CFRP excel in energy absorption under axial compression, making them valuable for lightweight substitutes in the automotive and aerospace industries. Their lower density compared to metallic materials,

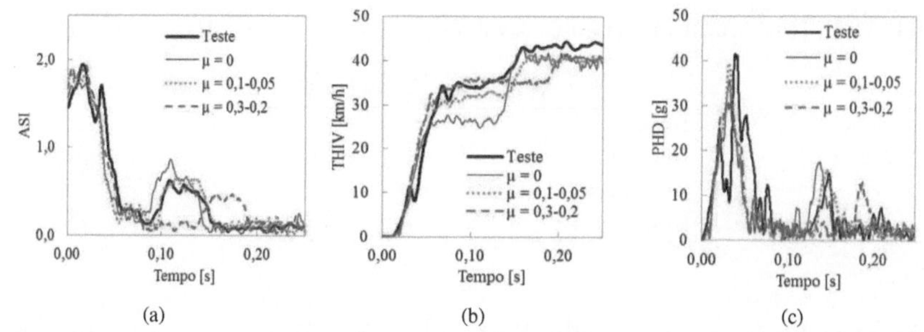

Figure 11.31 History of severity indexes for the car impact against the vertical concrete barrier: (a) ASI, (b) THIV and (c) PHD.

along with tailored lamination designs, allows for innovative energy absorbers with good crashworthy performance. In this section we focus on the axial crushing of CFRP tubes, as revealed by experiments and finite element analyses.

11.6.1 Experimental axial crushing

In the present experimental programme, circular cylindrical tubes were fabricated from unidirectional (UD) CFRP, featuring two distinct stacking sequences while maintaining nearly identical dimensions. The specifications of the specimens are listed in the following table.

Details of the results can be found in P.B. Ataabadi, D. Karagiozova and M. Alves, Crushing and energy absorption mechanisms of carbon fiber-epoxy tubes under axial impact, International Journal of Impact Engineering, 131 (2019) 174–189

TABLE 11.4 Geometry of the tubes.

Item	Angle-ply specimen	Cross-ply specimen
Stacking sequence	$\pm45/0/\pm45/0/\pm45/0/\pm45$	$\mathbf{0}/(\mathbf{90})/\mathbf{0}$
Inner diameter (mm)	50.00	50.00
Outer diameter (mm)	54.70	54.38
Mass per length (g/mm)	0.59	0.54

The specimens underwent axial compression under various loading conditions. The influence of loading rate and initial impact energy on both qualitative aspects (crush morphologies and failure modes) and quantitative crashworthy parameters (Specific Energy Absorption – SEA) were evaluated. The testing conditions are outlined in the following table.

The fiber orientation (stacking sequence) has a significant effect on the crushing behaviour (including bending of each layer and its failure mechanism) of a composite ply in the laminated tubes. The bending/crushing behaviour of composite plies of angle-ply and cross-ply tubes under quasi-static and dynamic compression loads are depicted

TABLE 11.5 Test conditions (nominal values).

Test	Mass (kg)	Velocity (m/s)
Impact#1	∼ 100	∼ 7
Impact#2	∼ 100	∼ 5
Impact#3	∼ 200	∼ 5
Impact#4	∼ 48	∼ 9.5

in the next figure, respectively. Distinctive bending / crushing responses due to different stacking sequences are remarkable.

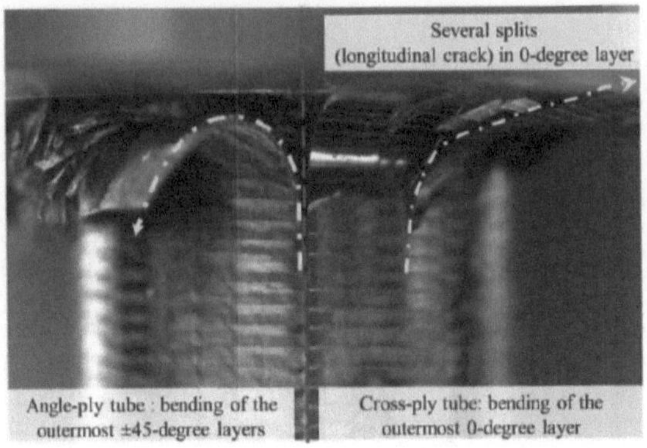

Figure 11.32 Effect of fiber orientations (stacking sequence) on the bending behaviour of composite plies under quasi-static axial compression load.

The crushing behavior of CFRP tubes is characterized by three well-known progressive failure modes: (i) Fragmentation mode, (ii) Splaying mode, and (iii) Brittle fracturing mode. When subjected to compressive loads, each composite layer experiences instability and separates from its adjacent layers, followed by bending deformation. However, the dominant failure mode of one layer can significantly differ from another layer due to variations in boundary conditions and fiber orientations. Figures 11.33 and 11.34 present some details of the tube damage.

Figure 11.35 illustrates the influence of different testing conditions on the axial crushing stresses of angle-ply and cross-ply tubes. The loading rate has an insignificant effect on the quantitative crashworthy parameters such as axial crushing stress and SEA value. Similar comments are valid for the stacking sequence.

11.6.2 Finite Element Analysis

Based on experimental findings, it is evident that both intra-laminar (within composite plies) and inter-laminar (delamination) damages

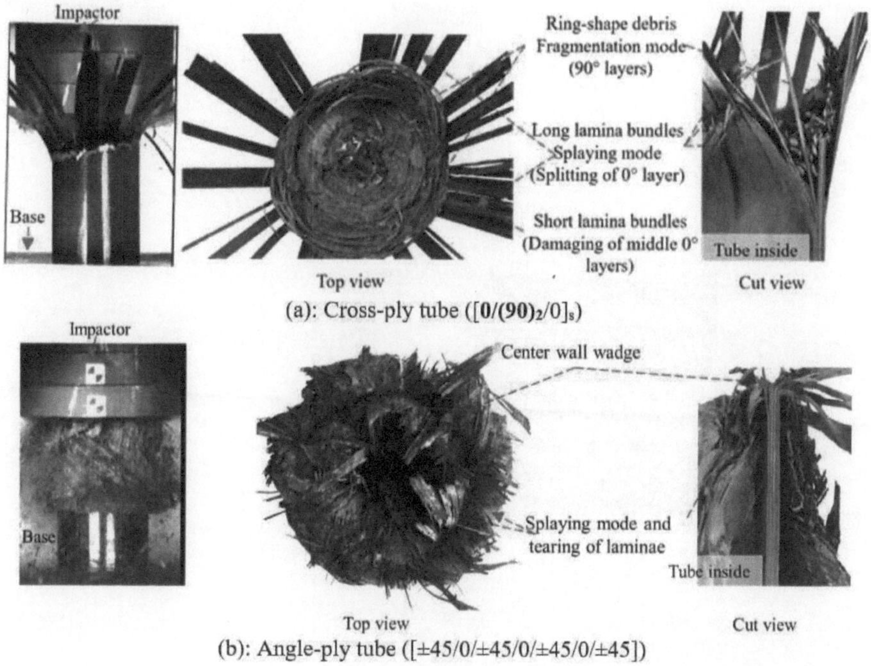

Figure 11.33 Effect of fiber orientations (stacking sequence) on the crushing modes of laminated CFRP tubes under dynamic axial compression.

significantly influence the crushing behavior of laminated tubes under axial impact. Consequently, a multi-layer finite element (FE) model, capable of capturing both intra-laminar and inter-laminar behaviour, is necessary and considered here to model axial crushing force on the CFRP tubes. The concentric layers of shell elements were stacked to simulate the intra-laminar behaviour of laminated tubes under axial impact loading. Additionally, the ABAQUS in-built cohesive element model was utilized to bond the adjacent composite laminae and simulate the delamination failure mode upon damage to the cohesive layers.

For intralaminar damage modelling, selecting the most suitable finite element type, whether 2D or 3D continuum elements, to discretize thin composite plies is crucial. Given that composite lamina in a laminate tube subjected to axial impact loading is likely to undergo significant bending deformation, the finite elements employed to model these layers (intra-laminar model) should accurately capture bending behaviour. In theory, it is feasible to utilize 3D solid elements, with one layer of bricks representing a ply of composite material. However, to correctly capture the lamina bending response, a minimum of three solid elements

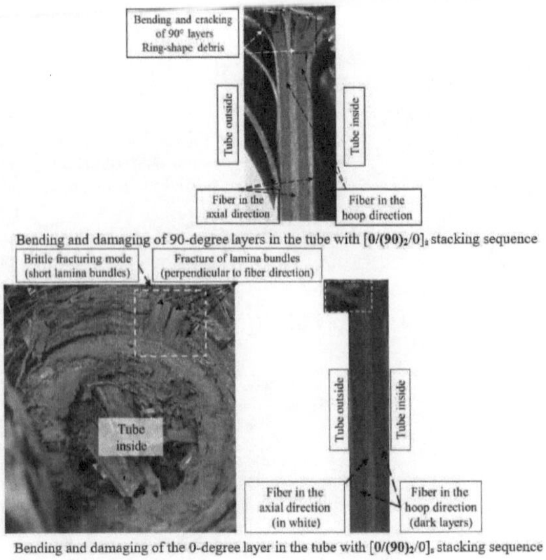

Figure 11.34 Details of bending and damaging behaviour of UD plies with different fiber orientations under axial crushing force.

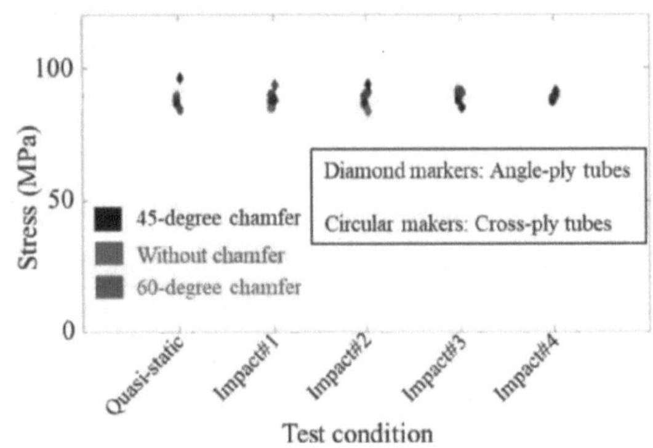

Figure 11.35 Effect of different crush triggers (chamfer angle) and stacking sequences on the average crushing stress.

through the thickness should be employed. Such models are computationally inefficient for relatively thin-walled and large-size structures, as the continuum element thickness affects the stable time increment of the explicit solver. It is noteworthy that a detailed stress analysis may not always be necessary, as the global response in terms of force-displacement characteristics typically provides sufficient information about structural performance, as reported in most relevant literature on axial impact on composite components. Hence, mid-surface shell elements (with 6 degrees of freedom, including three rotational and three translational DOFs at each node) are selected here due to their reasonable bending behaviour and computational efficiency.

In general, the mechanical responses of a composite ply are assumed to be linear elastic until the onset of damage (undamaged behaviour). The elastic response thresholds are typically defined by suitable failure criteria such as Hashin's criterion, maximum stress (maximum strain) criterion, Puck's criterion, and others. Once the damage criterion is satisfied, the elastic properties of the damaged region undergo degradation according to a degradation law (post-damage behaviour). Below three possible post-damage degradation laws are illustrated.

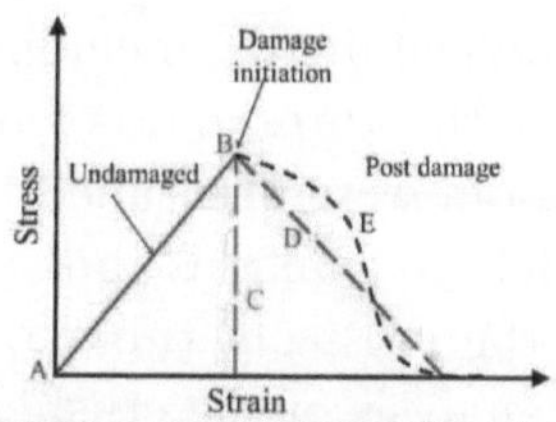

Response / path	Undamaged	Post damage	Type
ABC		Ply discount	Instantaneous degradation
ABD (bilinear)	Linear elastic	Linear degradation	Gradual degradation
ABE (linear-cube)		Cubic degradation	Gradual degradation

Figure 11.36 Examples of the mechanical response of composite materials.

In this study, a bilinear material response is employed to characterize the elastic (AB path) and post-damage (BD path) behaviours of CFRP plies under in-plane tension/compression loading. A nonlinear in-plane shear behaviour is considered for the plies. To capture the post-damage response and material degradation, a continuum damage mechanics (CDM) based model is utilized and had to be implemented as a user material model subroutine (VUMAT) within the ABAQUS/-Explicit solver.

For details see P.B. Ataabadi, D. Karagiozova and M. Alves, Finite element modelling of crushing of CFRP cylindrical tubes under low-velocity axial impact, Composite Structures, 280 (2022) 114902.

As for interlaminar damage, while the dissipation of energy due to the failure of a weak resin-rich interface is relatively insignificant compared to the dissipated energy within composite laminae, inaccuracies in the finite element (FE) model of this interface can significantly compromise the accuracy of simulations. In essence, underestimating or overestimating the mechanical properties of the interface may result in the prediction of excessively low or high crushing forces, as well as incorrect crushing modes (morphologies).

11.6.3 Numerical results

To enhance the efficiency of simulation procedures, a quarter model configuration was employed for both cross-ply and angle-ply tubes.

Evaluation of the drop test outcomes indicated that the chamfer angle, acting as the trigger mechanism, did not yield significant effects on pivotal crashworthiness metrics such as average crushing force and specific energy absorption (SEA) of the tubes upon impact. As a result, only a 45-degree external chamfer with a stepped-wall geometry was chosen for numerical modelling in this investigation.

The crushing modes and force-displacement curves depicted in the subsequent illustrations demonstrate a good correlation between experimental findings and numerical simulations for the cross-ply tube. Various parameters of the numerical model, including the penalty stiffnesses of the cohesive model and the thresholds of deformation gradient for the element deletion criterion, were fine-tuned through a trial-and-error process to effectively predict crush modes and achieve relatively accurate crushing force values for the cross-ply tube compared to experimental results. Consequently, an analysis of the crushing force on the angle-ply tube was conducted to ascertain the capability of the finite element (FE) model to predict the crushing response of another tube with a distinct stacking sequence.

The next two figures depict the main results of the simulation and experiments, for both cross and angle ply. It is clear the good agreement between model and reality.

This study presents a concise overview of the experimental and numerical responses to the axial compressive loads applied to CFRP laminated tubes. The numerical aspect of the study focuses on the development of a 2D finite element model based on Continuum Damage Mechanics (CDM). This model demonstrates the capability to accurately predict the intricate crushing modes exhibited by CFRP tubes with different stacking sequences.

11.7 Final conclusions

This book ends here!

We travelled far to get to this point. We visited the fields of vibration, wave propagation, beams, plates and shells behaviour. We had to explore material behaviour side by side with experimental techniques applied both to test materials and components. We visited Fourier transform and scaling. All this knowledge had to be complemented by what is most modern in terms of analysis of a structure, namely the Finite

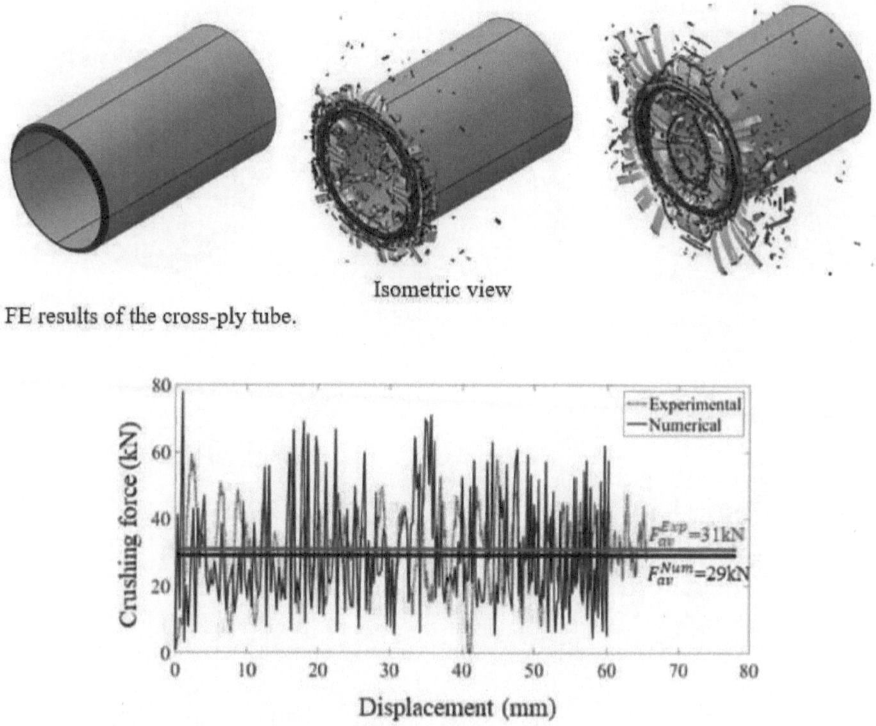

Figure 11.37 Some illustrations of the crushing behaviour of the CFRP tubes and a comparison between force–displacement experimental and numerical data of the axial impact of a mass on a cross-ply tube.

Element Method. We learned on the explicit and implicit time integration methods, the assembly of elements, matrix rotation, stress calculation, etc...

We also explored plasticity and material failure in a computational mechanics context. Nonlinear finite element procedures were also presented together with the classical Newton–Raphson method. The fact that we proposed some ethical queries only stress the complexity of this field of knowledge.

It is believed that this book could put in a single volume knowledge gained along decades of hard work of many researchers. The field of structural impact, by its own characteristics, is not so easy or quick to be grasped in its various aspects. Even so, it is hoped that this book will serve well those who strive to offer to the society the best of impact engineering.

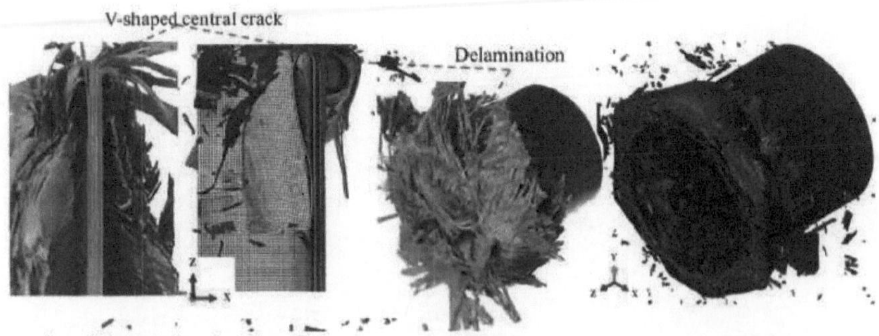

FE results of the angle-ply tube.

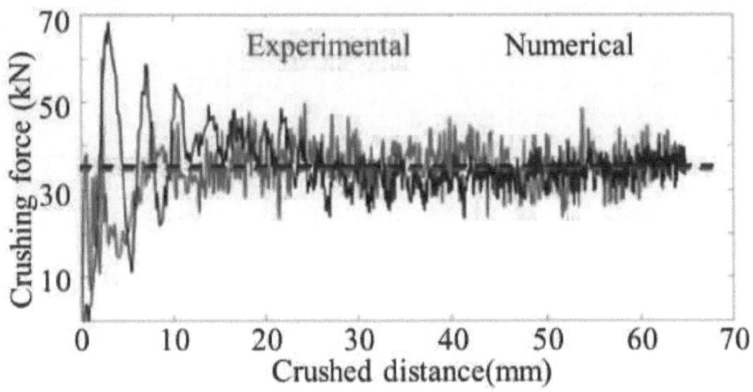

Figure 11.38 Some illustrations of the crushing behaviour of the CFRP tubes and a comparison between force–displacement experimental and numerical data of the axial impact of a mass on a angle-ply tube.

11.8 Problems

1. A team of lecturers invite you to design a car structure. Budget limitations constrain you to select only low performance materials so your car will not be crashworthy. Even being so aware, you go ahead and build the car. Sadly, the test driver is badly hurt in an accident during tests of your design. How would you react? Would you blame your colleagues for not providing you the necessary resources? Would you take the responsibility for the bad design, given full assistance to the pilot? Imagine an ethical test, sometimes called New York Times Test, where the news of this accident is in the headlines of a popular newspaper blaming you. Would this test help you to take another attitude towards this project?

2. In many sophisticate finite element analysis of impacted structures, it is common to adopt a single value for the friction coefficient among different parts of the structure. Perform a literature review and obtain information of how the friction coefficient changes with the pressure and slip rate. Is there some equipment in the market to measure friction between skin and airbags?

3. In the opening of this book, a comment was made on the forces involved in boxing and diving. How would you analyse these phenomena?

4. Additive manufacturing can be used to manufacture sophisticate and efficient structures. Prepare a list of software that would be required to design and make a motorcycle helmet based on AM. How could one generate a human head finite element model? What material properties would be used? Are helmets efficient to avoid skull fractures? And how about brain damage?

Index